Current Aspects of Neutrino Physics

Springer

Berlin
Heidelberg
New York
Barcelona
Hong Kong
London
Milan
Paris
Singapore
Tokyo

Physics and Astronomy ONLINE LIBRARY

http://www.springer.de/phys/

David O. Caldwell (Ed.)

Current Aspects
of Neutrino Physics

With 94 Figures

 Springer

Professor David O. Caldwell

University of California
Department of Physics
Santa Barbara, CA 93106, USA
E-mail: caldwell@slac.stanford.edu

Library of Congress Cataloging-in-Publication Data.
Current aspects of neutrino physics / David O. Caldwell, ed.
p.cm. Includes bibliographical references and index.
ISBN 3540410023 (alk. paper)
1. Neutrinos. I. Caldwell, D. O. (David O.)
OC793.5.N42 C87 2001   539.7'215–dc21   00-069846

ISBN 3-540-41002-3 Springer-Verlag Berlin Heidelberg New York

Springer-Verlag Berlin Heidelberg New York
a member of BertelsmannSpringer Science+Business Media GmbH

© Springer-Verlag Berlin Heidelberg 2001
Printed in Germany

Typesetting: Data conversion by Danny Lewis, Heidelberg
Cover design: design&production, Heidelberg

Printed on acid-free paper   SPIN 10765898   56/3141/tr   5 4 3 2 1 0

Preface

This is a particularly exciting time for neutrino physics. Now providing the first experimental evidence for new physics beyond the Standard Model of particle physics, neutrino studies are leading that larger field in new directions. As a probe for discovery, neutrinos are unique among particles. Being leptons, neutrinos are as far as one knows true elementary particles, but in addition they are unencumbered by charge, and of course they do not have complicating strong interactions. Their unusually small mass also points to their novelty and possibly to their providing a window onto very high energy scales.

The emphasis for neutrino physics at this time is on their masses and mixing with each other, as well as on their basic nature and their role in the Universe. While they have been important tools for studying particle properties, such as structure functions and the nature of the weak interaction, at present this is not the thrust of most research and hence is not covered in this book. Rather, the topics are those currently of most interest at the frontier of particle physics.

Authors have been chosen as internationally recognized authorities on the topic of each chapter so as to provide not only a definitive, up-to-date review, but also a wise oversight. Even some subjects which have been extensively discussed in the literature are presented from fresh standpoints, and there is material which can be found nowhere else. The reader of the whole book will have a comprehensive view of this fast blossoming field, but one who simply wants to acquire current knowledge of a particular topic can do so without having to cover the rest of the material.

The book starts with a history of neutrino physics, followed by a discussion of the theoretical nature and properties of neutrinos (Chap. 2). The next six chapters deal with what and how we have learned about neutrinos via experiment and the guiding theoretical framework: attempts to determine neutrino masses (Chap. 3), observation of neutrinos from the sun (Chap. 4) and from the atmosphere (Chap. 5), experiments on neutrino oscillations using reactors (Chap. 6) and accelerators (Chap. 7), and double beta decay searches (Chap. 8). This information on neutrino masses and mixings is collated in the next two chapters via various schemes, rather phenomenologically in the first, with more theoretical bases being given in the second. The final three chapters go beyond neutrinos' role in particle physics to their effects on element production in the Big Bang and by supernovae, their effects on the formation of structure in the Universe and their role in the new field of neutrino astronomy. In each case what is learned in these large domains also provides new information about the neutrinos themselves.

The interplay between the domains of the largest and smallest distance scales is particularly striking in the case of neutrinos, and not only are both areas advancing rapidly, but also so is the interaction between them. Thus this book can provide a snapshot of where we are now, but only some indication of where we shall be going, since significant changes could occur rapidly. As a speculative example, consider a presently popular four-neutrino mass-mixing scheme. Originally suggested to provide an explanation of the solar electron neutrino deficit (via $\nu_e \to \nu_s$, a sterile neutrino), the anomalous lack of atmospheric muon neutrinos relative to ν_e (via $\nu_\mu \to \nu_\tau$, the tau neutrino) and a neutrino (hot) contribution to dark matter (from the ν_μ and ν_τ, which are much heavier than ν_e and ν_s), this scheme was later supported by the LSND experiment's observation of $\nu_\mu \to \nu_e$ and by the need to rescue heavy-element production in supernovae. The present problem for the scheme, and indeed any scheme, is to understand quantitatively the observations of solar ν_e by several experiments. There is evidence for and against matter-enhanced and vacuum oscillations. If there are large extra dimensions, a currently popular theoretical speculation, the sterile neutrino in $\nu_e \to \nu_s$ could allow vacuum oscillations to its ground state and matter-enhanced oscillations to its tower of higher-mass (Kaluza–Klein) states, enabling all the data to be fitted. While neutrino physics has already achieved much coherence recently, if this recently reported[1] idea presently being worked on were to prove correct, it would both unify the field and provide evidence for a quite different picture of our Universe. That would be motivation for the next edition!

Santa Barbara, CA, USA   *David O. Caldwell*
May 2001

---

[1] D.O. Caldwell, R.N. Mohapatra, and S.I. Yellin, hep-ph/0010353, submitted to Phys. Rev. Lett. and more completely in hep-ph/0102279, submitted to Phys. Rev. D.

Contents

11   Neutrino Flavor Transformation in Supernovae and the Early Universe

12   Hot Dark Matter in Cosmology

13   High Energy Neutrino Astronomy: Towards Kilometer-Scale Detectors

List of Contributors

Vernon Barger
Phenomenology Institute
Department of Physics
University of Wisconsin
1150 University Avenue
Madison, WI 53706, USA
barger@pheno.physics.wisc.edu

Felix Boehm
Department of Physics 161-33
California Institute of Technology
1201 E, California Blvd
Pasadena, CA 91125, USA
boehm@caltech.edu

David O. Caldwell
Department of Physics
University of California
Santa Barbara, CA 93106-9530,
USA
caldwell@slac.stanford.edu

George M. Fuller
Physics Department 3019
University of California
San Diego, LaJolla, CA 92043, USA
gfuller@ucsd.edu

Michael A. K. Gross
Department of Computer Science
University of California
Santa Cruz, CA 95064,USA
gross@ucsc.edu

Francis Halzen
Department of Physics
University of Wisconsin
Madison, WI 53706, USA
halzen@pheno.physics.wisc.edu

Wick C. Haxton
Institute for Nuclear Theory
Box 351550
and
Department of Physics
Box 351560
University of Washington
Seattle, WA 98195, USA
haxton@phys.washington.edu

Boris Kayser
National Science Foundation
4201, Wilson Blvd.
Arlington, VA 22230, USA
bkayser@nsf.gov

John G. Learned
Department of Physics and
Astronomy
2505 Correa Road
Honolulu, HI 96822, USA
jgl@phys.hawaii.edu

Rabindra N. Mohapatra
Department of Physics
University of Maryland
College Park, MD 20742, USA
rmohapat@physics.umd.edu

Joel R. Primack
Physics Department
University of California
Santa Cruz, CA 95064, USA
joel@physics.ucsc.edu

Michael Riordan
Institute for Particle Physics
University of California
Santa Cruz, CA 95064, USA
michael@slac.stanford.edu

R.G. Hamish Robertson
Department of Physics
University of Washington
P.O. Box 351560
Seattle, WA 98195-1560, USA
rghr@u.washington.edu

Petr Vogel
Department of Physics 161-33
California Institute of Technology
1201 E, California Blvd
Pasadena, CA 91125, USA
vogel@citnp12.caltech.edu

K. Whisnant
Department of Physics and
Astronomy
Iowa State University
Ames, IA 50010, USA
whisnant@iastate.ed

John F. Wilkerson
Nuclear Physics Laboratory
University of Washington
P.O. Box 354290
Seattle, WA 98195-1560, USA
jfw@u.washington.edu

1   Pauli's Ghost:
The Conception and Discovery of Neutrinos

Michael Riordan

When Wolfgang Pauli conceived his idea of the neutrino in 1930, it was substantially different from the ghostly particles recognized today. That December he proposed a light, neutral, spin-1/2 particle he at first called the "neutron" as a "desperate remedy" for the energy crisis of that time – the continuous energy spectrum of electrons emitted in nuclear beta decay [1.1].[1] The crisis had grown so severe by the late 1920s and early 1930s, after experiments by Charles Ellis, Lise Meitner and their colleagues [1.3], that Niels Bohr had even begun to contemplate abandoning the sacrosanct law of energy conservation in nuclear processes ([1.4], esp. pp. 382–383). Pauli could not countenance such a radical departure from orthodoxy and suggested instead that such poltergeists might inhabit the nucleus along with protons and electrons. Their mass had to be "of the same order of magnitude as the electron mass and . . . not larger than 0.01 proton mass." They would be electromagnetically bound within nuclei by virtue of an anomalous magnetic moment, and they might have "about 10 times the penetrating capacity of a gamma ray." He could account for the continuous beta decay spectrum by assuming that "in beta decay a neutron is emitted together with the electron, in such a way that the sum of the energies of neutron and electron is constant" [1.1].

Pauli obviously thought of his ghosts as *constituents* of atomic nuclei, with a small mass and substantial interaction strength. He was trying not only to preserve energy conservation in nuclear processes but also to avoid severe problems with spin and statistics that cropped up in nuclei, then imagined to consist only of protons and electrons. The nitrogen nucleus, for example, was widely thought to contain 14 protons and 7 electrons, but it curiously did not obey Fermi statistics, as expected of an object with an odd number of fermions inside. By adding seven such "neutrons" to the heap, Pauli could explain the observation that it behaved like a boson. But nobody could figure out how to cloister such light, speedy particles (including electrons) within the narrow confines of a nucleus.

James Chadwick's 1932 discovery of the heavy fermion that he also dubbed the neutron resolved most of the problems [1.5]. Composed of seven protons and seven of these far more massive neutrons, the nitrogen nucleus now had an even number of fermions inside and could easily behave like a boson. Enrico Fermi's famous theory of beta decay put the capstone on the growing

---

[1] See also [1.2], pp. 7–31, for an accessible account of the origins of the neutrino idea.

Fig. 1.1. Wolfgang Pauli, Niels Bohr, Erwin Schrödinger and Lise Meitner at the 1933 Solvay Conference (courtesy Niels Bohr Archive)

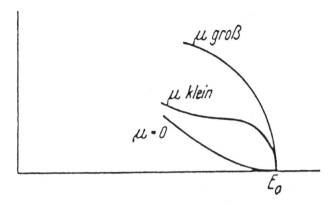

Fig. 1.2. Graph from Fermi's famous paper on the theory of beta decay, showing how the shape of the emitted electron's energy spectrum varies with neutrino mass

edifice [1.6]. Instead of inhabiting the nucleus as constituents, the electron and "neutrino" (a name coined by Fermi in 1931, before Chadwick's discovery, to mean "little neutral object" ([1.2], p. 28), were to be *created* the moment a neutron transformed itself into a proton – just as photons are created in Paul Dirac's theory of atomic radiation.[2] Fermi even went so far as to indicate how the energy spectrum of beta decay electrons depends critically on the neutrino's mass. By comparing his theoretical curves (Fig. 1.2) with the measured spectra near their high-energy end point, he was able to conclude that "the rest mass of the neutrino is either zero, or, in any case, very small in comparison to the mass of the electron" [1.6].

Shortly thereafter, Hans Bethe and Rudolf Peierls used Fermi's theory to show that the interaction of neutrinos with matter had to be essentially negligible [1.7]. In the few MeV range characteristic of beta decay neutrinos, they estimated that the interaction cross section was less than 10^{-44} cm^2, equivalent to a mean free path in water of more than 1000 light years! Bethe and Peierls concluded "there is no practically possible way of observing the neutrino." Pauli (Fig. 1.1) was dismayed. "I have done a terrible thing," he remarked. "I have postulated a particle that cannot be detected." [3]

Thus was the idea of the neutrino born, and it remained mostly an intriguing idea for years. Even after the appearance of Fermi's paper, Bohr was still unconvinced of its physical reality. "In an ordinary way I might say that I do not believe in neutrinos," Sir Arthur Eddington remarked in 1939. "Dare I say that experimental physicists will not have sufficient ingenuity to *make* neutrinos?" [1.9] (quoted in [1.2], p. 35).

---

[2] Other physicists, most notably F. Perrin and D. Iwanenko, had similar ideas as early as 1932, but Fermi was the first to work out a detailed theory of beta decay. See [1.2], p. 28.

[3] Quoted in [1.8], p. 318.

1.1   Detecting Neutrinos

When George Gamow wrote "The reality of neutrinos" in 1948, however, he could discourse about them with confidence that they indeed existed [1.10]. Although nobody had yet detected a neutrino directly, there were several indirect experimental proofs of their reality. Sensitive measurements of the energy and momentum of beta decay electrons and of their recoiling nuclei in Wilson cloud chambers indicated that substantial quantities of energy and momentum were missing. "This means some other particle must have been ejected at the same time as the electron," he wrote. "These single-process experiments leave little doubt that a third particle must be involved."

Even stronger evidence had been obtained during World War II from experiments in which K shell electrons were absorbed by nuclei without the emission of any charged particle at all, just a recoiling nucleus. James S. Allen made one of the most sensitive of these measurements by studying nuclear recoils following such K captures in beryllium [1.11]. The observed recoil energy came in close to the expected energy that would have been produced by the emission of a neutrino during the process.

Gamow even speculated that the neutrino (and its antiparticle) might be involved in the slow disintegrations of the recently discovered pi mesons (or pions) and their lighter counterparts, the mu mesons (muons) [1.10]. After all, some kind of invisible entity was spiriting energy away from these two- and three-body decay processes. Why not the same elusive particle involved in slow nuclear decays?

But little could be said conclusively about a particle that had thus far evaded direct detection. Whether its antiparticle was a completely distinct entity or merely a different spin state of the very same poltergeist could not then be determined. And the best attempts at measuring its mass could only establish an upper limit of about one-twentieth the electron mass. As the 1950s began, however, this situation was about to change dramatically.

In 1951, following atomic-bomb testing at Eniwetok atoll, Los Alamos physicist Frederick Reines began contemplating experiments in fundamental physics he might attempt [1.8]. The Manhattan Project had provided intense new sources of neutrons and neutrinos that could then be used to ascertain more about these particles. Reines and Clyde Cowan chose to focus on direct detection of neutrinos. They also recognized that recently developed organic scintillating liquids would allow them to build the massive detector required. Together with the intense neutrino fluxes generated by atomic blasts or near a fission reactor, such a large detector might finally overcome the dauntingly minuscule cross section for a neutrino ν to interact with matter [1.8].

Reines and Cowan elected to search for evidence of the interaction

$$\nu + p \longrightarrow n + e^+ \,,$$

which should yield a prompt light flash in the organic scintillator owing to the positron's annihilation, followed several microseconds later by another flash due to neutron capture.[4] After considering and rejecting the idea of placing a detector within 100 meters of an atomic-bomb explosion, they decided instead to put it close to one of the nuclear reactors then in operation. Their initial experiment at the Hanford Engineering Works in Washington State, site of the reactors used to breed plutonium for the Manhattan Project, involved a 300 liter tank of liquid scintillator viewed by 90 phototubes. A marginal increase in signal that they observed with the reactor operating was nearly swamped by a reactor-independent signal later ascribed to cosmic rays [1.8] (see also [1.13]).

Reines, Cowan and their colleagues did a second experiment at the newly built Savannah River reactor, which generated an antineutrino flux of 10^{13} per square centimeter per second at a position 11 m away [1.8] (see also [1.14, 1.15]). (By then it was becoming recognized that the neutrino and antineutrino are distinct, the latter being the one produced along with the electron in beta decay processes.) The detector location was also 12 m underground, permitting much better rejection of cosmic-ray backgrounds. The detector consisted of three large tanks of organic scintillator, each viewed by 110 phototubes; inserted between them were two tanks of water with dissolved cadmium chloride to promote neutron capture (Fig. 1.3).

An antineutrino from the reactor occasionally interacted with a proton in the water, producing a positron and a neutron. The positron annihilated almost immediately with an atomic electron, yielding two 0.51 MeV gamma rays that were detected in the scintillator; about 10 ms later the neutron was captured by a cadmium nucleus, resulting in another burst of gamma rays (Fig. 1.4). A delayed coincidence between the first and second gamma-ray bursts was interpreted as the signature of an antineutrino event; Reines and Cowan observed 3.0 ± 0.2 events per hour with the reactor operating – much greater than the backgrounds due to cosmic rays or accidental coincidences [1.8].

Elated by their discovery, Reines and Cowan sent a telegram to Pauli in Zurich on 14 June 1956: "We are happy to inform you that we have definitely detected neutrinos from fission fragments by observing inverse beta decay of protons." According to Reines, Pauli drank a case of champagne with friends to celebrate the discovery and penned Reines and Cowan a reply: "Thanks for the message. Everything comes to him who knows how to wait" [1.8].[5]

---

[4] Knowledgeable readers will protest that antineutrinos, not neutrinos, participate in this "inverse beta decay" interaction, but the distinction between the two was not clear until after Reines and Cowan had completed their experiments. See, for example, [1.12]. For historical accuracy, I retain the notation used in their earlier papers.

[5] Note that Reines still called these particles "neutrinos," not antineutrinos. According to him, Pauli's letter of reply did not reach them.

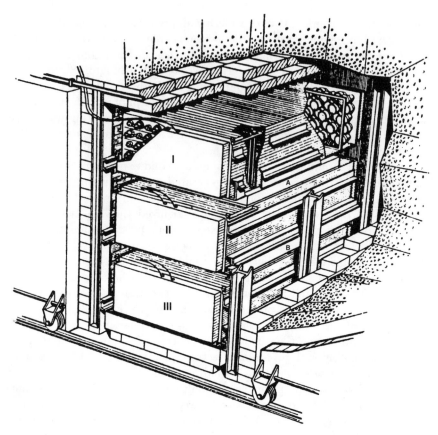

Fig. 1.3. Artist's conception of the detector Reines and Cowan used in their Savannah River experiment. Tanks I, II and III contained liquid scintillator and were viewed on each end by 55 five-inch phototubes. Tanks A and B (between tanks I and II, and tanks II and III), containing 200 liters of water with dissolved cadmium chloride for neutron capture, served as the target volume

1.2 Massless Neutrinos?

Reviewing the status of neutrino physics a year later, Reines and Cowan could cite a variety of major improvements in the understanding of this previously invisible poltergeist [1.16]. Delicate measurements of the electron spectrum in tritium beta decays had by then established that the neutrino's mass was less than $1/2000$ of the electron's, a factor of 100 improvement over the experiments noted by Gamow almost a decade earlier [1.17]. The lack of any evidence for double beta decay indicated that it was most probably a Dirac particle like the electron, with the neutrino and antineutrino distinctly different entities. (The other possibility, first suggested by Ettore Majorana in 1937, was that they are two different polarization states of the same ob-

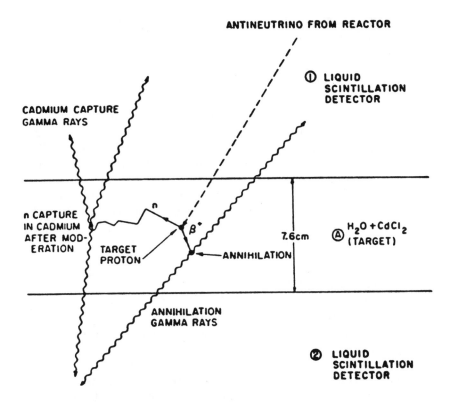

ANTINEUTRINO FROM REACTOR

① **LIQUID SCINTILLATION DETECTOR**

CADMIUM CAPTURE GAMMA RAYS

n CAPTURE IN CADMIUM AFTER MODERATION

TARGET PROTON

n

β⁺

ANNIHILATION

7.6cm

Ⓐ $H_2O + CdCl_2$ (TARGET)

ANNIHILATION GAMMA RAYS

② **LIQUID SCINTILLATION DETECTOR**

Fig. 1.4. Diagram of the antineutrino-detection scheme employed in the Savannah River experiment

ject [1.18].) And the failure of Ray Davis to observe $Cl^{37} \longrightarrow Ar^{37}$ conversions in a tank containing a thousand gallons of carbon tetrachloride (CCl_4) placed near the Savannah River reactor could be most easily explained in the same way: antineutrinos emanating from its fission reactions could not induce such transitions, while neutrinos should have done so [1.16].[6]

The most striking advance in understanding neutrinos had come the previous year in the wake of the earthshaking discovery of parity violation. Tsung-Dao Lee and Chen-Ning Yang, among other theorists, proposed to rescue the deteriorating situation by invoking a peculiarity of the neutrinos emitted in beta-decay and weak-interaction processes [1.19] (see also [1.20]). If they were Dirac particles with absolutely no mass, neutrinos themselves would violate parity because their spin vectors would always be aligned along their direction of motion, while the spins of antineutrinos would point the opposite way. We say that neutrinos have left-handed chirality and antineu-

---

[6] Bruno Pontecorvo apparently suggested doing such an experiment as early as 1946; see [1.2], p. 79.

trinos are right-handed. "Since this new model for the neutrino does not obey the simple parity principle," wrote Reines and Cowan, "no reaction involving such a neutrino can be expected to conserve parity in its restricted sense" [1.16].

A further consequence of this hypothesis was that the cross section for interaction of reactor-produced antineutrinos with protons had to be twice what had been previously calculated – an effect that Reines and Cowan had already begun to observe [1.8]. But the most convincing proof came from a sensitive experiment at Brookhaven National Laboratory by Maurice Goldhaber and colleagues [1.21]. They determined the spin direction of the recoiling samarium-152 nucleus that emerged after europium-152 captured a K shell electron and emitted a neutrino. From this information they concluded that the neutrino is always left-handed, as Lee and Yang had suggested. Thus the neutrino and antineutrino appeared to be distinctly different, completely massless entities.

1.3   Two Kinds of Neutrinos

major mystery in the late 1950s was whether the neutral particles emitted in pion and muon decays were the same neutrino and antineutrino observed in nuclear beta decays – or something else. Because the strengths of these weak interactions were similar, and because they also violated parity, it was widely believed that the very same particles were involved. But if this were the case, then muons should occasionally have been observed decaying into an electron plus a photon, $\mu \longrightarrow e + \gamma$. For example, the neutrino and antineutrino generated in the three-body decay of a muon could have annihilated each other, yielding a photon to pair with the departing electron. Theorists calculated that such processes should occur about once in every 10^4 muon decays,[7] but measurements indicated that nothing like this occurred in 10^7 to 10^8 events. One way to accommodate this apparent discrepancy was to say that two different kinds of neutrinos were involved in muon decay.

Intrigued by these questions and the possibility of working with beams of neutrinos, Melvin Schwartz, Leon Lederman, Jack Steinberger and their colleagues began planning an experiment at Brookhaven [1.23, 1.24]. Spurred by his discussions with Lee, Schwartz recognized that the intense, high-energy beams of protons soon to be available at its Alternating Gradient Synchrotron would allow the generation of neutrino beams with sufficient intensity. Pions and kaons produced by protons colliding with a beryllium target were allowed to decay; charged particles from these decays were absorbed in over 12 m of steel shielding, leaving only neutrinos (and antineutrinos) to penetrate to a detector chamber (Fig. 1.5).

---

[7] See, for example, [1.22].

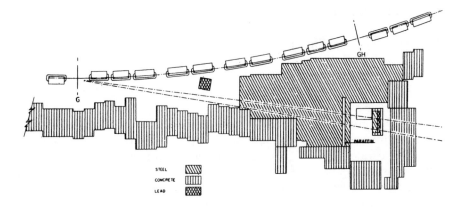

Fig. 1.5. Plan view of the two-neutrino experiment at Brookhaven. Pions and kaons produced in a beryllium target at the *far left* decayed, yielding neutrinos (and antineutrinos) that traveled from *left* to *right* and penetrated the massive steel shielding, striking the detector in the alcove at the *far right*

With energies ranging from hundreds of MeV to several GeV, these neutrinos had interaction cross sections more than a hundred times greater than reactor-born neutrinos, but a large, massive detector was still required to observe a sufficient number of them. Schwartz and his colleagues elected to build a 10 ton spark chamber from aluminum plates. If the neutrinos produced in pion or kaon decays (e.g. $\pi \longrightarrow \mu + \nu$) were distinct from those produced in beta decays, they expected to see only the long, penetrating tracks of muons generated by neutrinos that interacted in the aluminum. However, Schwartz recalled, "If there had been only one kind of neutrino, there should have been as many electron-type as muon-type events" [1.24]. In the initial run of this experiment, which began in late 1961, they recorded 34 events in which there appeared a single muon track originating in the aluminum plates [1.25] (see also [1.24]). There were another 22 events with a muon and other particles, plus six ambiguous events that might have been interpreted as electrons (or positrons). But comparisons with actual electron events from a separate run at Brookhaven's Cosmotron showed little similarity.

Thus the neutrinos produced in tandem with muons in pion and kaon decay are distinct from those produced together with electrons in nuclear beta decay. After Schwarz's pivotal experiment, particle physicists began calling the former "muon neutrinos" (or ν_μ) to distinguish these particles from the latter, called "electron neutrinos" (or ν_e). Whenever a positive muon, for example, decays, it yields a positron, an electron neutrino and a muon antineutrino ($\mu^+ \longrightarrow e^+ + \nu_e + \bar{\nu}_\mu$). There were now four distinct "leptons," or light particles: the electron, the muon and their two respective neutrinos (plus antiparticles).

1.4   The Standard Model

Theoretical and experimental advances that occurred over the ensuing decade resulted in a revolutionary new picture of the subatomic realm that came to be known as the "Standard Model" of particle physics.[8] In this theory, the leptons, and other particles called "quarks" (of which subatomic particles that experience the strong nuclear force – such as protons, neutrons, pions and kaons – are composed) are regarded as elementary, point-like entities. The electromagnetic and weak interactions, previously thought of as distinct forces with widely differing strengths, are now considered to be merely two different aspects of one and the same "electroweak" interaction. The extreme feebleness of the weak interaction arises because it occurs via the exchange of ponderous spin-1 particles known as "gauge bosons." For example, in its beta decay, a neutron coughs up a massive, negatively charged W boson and transforms into a proton ($n \longrightarrow p + W^-$); the W^- immediately converts into an electron plus its antineutrino ($W^- \longrightarrow e^- + \bar{\nu}_e$). Only left-handed electrons and neutrinos (and their right-handed antiparticles) participate in these weak interactions, thereby yielding their characteristic parity-violating property.

An inescapable requirement of this unification of the electromagnetic and weak interactions is the existence of "neutral currents" that occur owing to exchange of another massive, but neutral, boson Z. For example, instead of converting into a muon when it interacts with a nucleus via the exchange of a W boson, a muon neutrino can instead glance away unchanged, *remaining* a muon neutrino and swapping a Z boson. Searches for such neutral currents have had a long and checkered history [1.28]. Schwartz and his colleagues looked for them without success in a follow-up Brookhaven experiment (although they were probably present in the data). Later, his group found several candidate events in a 1970 experiment at the Stanford Linear Accelerator Center (SLAC), in which his massive spark chambers were used to detect neutral particles produced in an electron beam dump 60 m away [1.24].

Neutral currents were finally discovered in 1973 by an international collaboration of physicists working at the European Center for Nuclear Research (CERN) on the Gargamelle bubble chamber [1.29]. Filled with 20 tons of liquid freon (CF_3Br), it was exposed to beams of muon neutrinos and antineutrinos from CERN's proton synchrotron. Initial evidence came from rare events in which these spookinos rebounded elastically from atomic electrons (e.g. $\nu_\mu + e^- \longrightarrow \nu_\mu + e^-$) and imparted lots of energy to them. The electrons left wispy tracks in the bubble chamber (Fig. 1.6), whereas neutrinos or antineutrinos crept away undetected [1.29, 1.30]. Subsequent confirmations came from Gargamelle and experiments at the Fermi National Accelerator Laboratory in which muon neutrinos scattered inelastically from

---

[8] On the establishment of the Standard Model as the dominant theory of particle physics, see [1.26], especially the introductory essay [1.27].

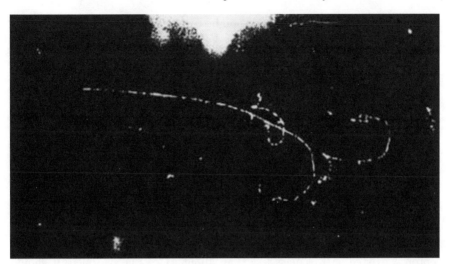

Fig. 1.6. One of the first neutral current events observed in the Gargamelle bubble chamber at CERN. A muon neutrino traveling from *left* to *right* strikes an atomic electron, which makes the gently arcing track in this photo (courtesy of CERN)

nuclei and eluded observation. Such events occurred much more frequently than elastic scattering events, but they were difficult to distinguish at first from background events that had been induced by stray neutrons interacting within the detector [1.28, 1.29, 1.31, 1.32].

By the mid-1970s, it was clear that neutrinos (and, in fact, all the other leptons and quarks) were capable of a new kind of weak interaction in which they maintained their identity instead of transforming into a partner lepton (or quark). The discovery of these neutral currents, together with conclusive evidence for a fourth, or charm, quark c to accompany the initial trio – up u, down d and strange s – provided strong support for the Standard Model [1.27, 1.33]. Quarks and leptons come in pairs: u and d, e and ν_e; c and s, μ and ν_μ. In addition, quark and lepton pairs can be grouped into families of four, often called "generations". Two such families were recognized by 1976: the first includes the up and down quarks plus the electron and electron neutrino, while the second contains the charm and strange quarks plus the muon and muon neutrino. Particle physicists could now discern a highly satisfying symmetry among the elementary entities in their new ontology.

1.5   The Third Family

A few years earlier, two Japanese theorists had suggested there might even be a *third* family of quarks and leptons [1.34]. Working at Kyoto University beginning in 1972, Makoto Kobayashi and Toshihide Maskawa were seeking a way to incorporate the mysterious phenomenon of *CP* violation within the emerging structure of what would soon be recognized as the Standard Model. Discovered in 1964 by James Cronin, Valentine Fitch and their colleagues, this phenomenon indicated that – at least in certain decays of kaons – nature is asymmetric under the combined operations of charge conjugation (C) and parity inversion (P) [1.35]. They observed that kaons behave differently if one replaces particles by their antiparticles and views the interactions in a mirror. Kobayashi and Maskawa could not obtain CP violation using two families of quarks and leptons. But if they added a third family to the mix, including two more quarks plus another charged lepton and its neutrino, they discovered CP violation arose naturally [1.36]. The paper did not attract much attention initially, however, because it had been published in a relatively obscure journal.

At about the same time, Martin Perl and his colleagues were beginning their search for another heavy, charged lepton using the SLAC-LBL detector at the new electron–positron collider SPEAR [1.37, 1.38]. If such a heavy lepton λ existed with a mass less than about 4 GeV, he reasoned, it should be pair-produced in high-energy electron–positron collisions on this facility:

$$e^+ + e^- \longrightarrow \lambda^+ + \lambda^- .$$

Depending on their mass, such heavy leptons could have a variety of decays, two of which would be similar (by analogy) to the muon's familiar decay:

$$\lambda^- \longrightarrow e^- + \nu_\lambda + \bar{\nu}_e ,$$
$$\lambda^- \longrightarrow \mu^- + \nu_\lambda + \bar{\nu}_\mu .$$

By searching for events in which one of these hypothetical leptons decayed into an electron and the other into an oppositely charged muon (plus unseen neutrinos and antineutrinos), they hoped to find evidence for the existence of a new heavy lepton – and (again by analogy) a third neutrino.

By 1974 Perl's group had begun to find such "anomalous $e\mu$ events" in the data samples being collected on the SLAC-LBL detector, and by 1975 they had dozens. But convincing their colleagues in the collaboration and the rest of the particle physics community that these events gave conclusive evidence for another heavy lepton took a few more years [1.37]. At issue was whether these events had arisen owing to background processes, such as the misidentification of hadrons or the production of additional decay particles that had escaped detection. These ambiguities were gradually resolved by the addition of better electron and muon detection capabilities to the detector, the accumulation of more anomalous events (Fig. 1.7) and more sophisticated data analysis.

A 1975 paper by the SLAC-LBL collaboration claimed "evidence" for a new lepton with a mass in the range 1.6–2.0 GeV but stopped short of saying it had made a discovery [1.39]. After two years of "confusion and uncertainty" over these events and their interpretation, confirmation of the SLAC results began to come in from the DORIS electron–positron collider

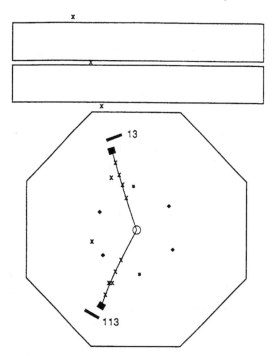

Fig. 1.7. One of the first anomalous $e\mu$ events observed after a "muon tower" had been added to the SLAC-LBL detector in 1975 [1.37]. The muon travels upwards through the tower in this computer reconstruction of the detector's cross section, while the electron moves downwards; the numbers 13 and 113 represent the amounts of energy deposited by the muon and electron in electromagnetic shower counters

in Hamburg [1.37, 1.40]. At the major summer conference that year, Perl concluded that there was no other way to account for all the anomalous events than by postulating the existence of a new lepton with a mass of 1.8–2.0 GeV. Further experiments on SPEAR and DORIS narrowed this mass range to 1.777–1.786 GeV and identified the decay of this "tau lepton" τ into a pion and a tau neutrino ($\tau^- \longrightarrow \pi^- + \nu_\tau$). By the summer of 1978, little doubt remained among particle physicists about the existence of the tau and its neutrino.

Two decades later, physicists still had not achieved the direct detection of a free tau neutrino, well separated from its point of production, despite a number of attempts to do so.[9] The problem comes from the difficulties of making a sufficiently intense beam of tau neutrinos and of detecting them unambiguously. Still, there is essentially no doubt today that the tau neu-

---

[9] The Fermilab DONUT experiment has reported several candidate tau neutrino events but has yet to make an unambiguous claim for its discovery, owing to difficulties in estimating the backgrounds. See, for example, [1.41].

trino exists. The two quarks expected in the third family – the top τ and bottom b quarks – have since been isolated [1.42, 1.43], leaving only the ν_τ remaining to be found. So dominant has the Standard Model become that no particle physicist now seriously questions that the ν_τ eventually will be discovered.[10]

And cosmological arguments about nucleosynthesis of light elements during the first few minutes of the Big Bang require that a third neutrino – and perhaps even a fourth – ought to exist [1.44].[11] The abundance of primordial helium-4 synthesized during this process is determined by the expansion rate of the Universe at that moment, which in turn is sensitively related to the number of different kinds of light neutrinos. As measurements became more and more accurate during the 1980s, primordial helium-4 was shown to contribute about a quarter of the visible mass in the Universe, suggesting a third (and a remotely possible fourth) kind of neutrino.

The issue was settled in 1989 by terrestrial experiments at new electron–positron colliders – the SLC at SLAC and LEP at CERN – that were capable of producing the ultraheavy Z boson. Within the structure of the Standard Model, additional kinds of neutrinos give this particle additional channels for its decay, thereby shortening its half-life. By making precision measurements of the height and width of the Z resonance peak, physicists at the two facilities concluded that there must be three, and *only* three, kinds of "conventional", weakly interacting neutrinos with masses less than 45.5 GeV, half the Z boson mass.[12] These experiments confirmed the cosmological predictions and put a firm upper limit on the complexity of the Universe. Its table of fundamental entities could contain only three "conventional" families of quarks and leptons. Any further families must be be weird indeed, harboring a neutrino more massive than a titanium atom!

1.6 Summary

From this account it should be clear that the idea of the neutrino has evolved substantially since its conception in the early 1930s. Pauli's tenuous hypothesis was just the starting point for a lengthy process of theoretical and experimental elaboration that still continues today. Where Pauli at first regarded

---

[10] *Note Added in Proof:* On July 21, 2000, the DONUT collaboration announced the discovery of direct evidence for the tau neutrino in a colloquium at Fermilab, based on four events in which a tau lepton was created in collisions of neutrinos with iron nuclei. This discovery was subsequently reported at the 30th International Conference on High Energy Physics, Osaka, Japan.

[11] For a more general account, see [1.45].

[12] The initial results on the decay of Z bosons produced by the SLC and LEP colliders excluded a fourth light neutrino at the 2–3 standard deviation level. See, for example, [1.46, 1.47]. This kind of experiment does not rule out, however, the possible existence of "sterile" neutrinos, which do not experience the weak interaction. See Chap. 10 for further discussion.

them as constituents of nuclei, Fermi showed how neutrinos can instead be created in nuclear transformations. Where Pauli mainly sought a minimalist way to preserve the conservation of energy and angular momentum in individual beta decay events, physicists have recently established that at least one – and possibly all three – of the neutrinos have a tiny bit of mass [1.48].[13] Where he speculated that this mass might be greater than the electron's, physicists now think it is at least a million times smaller. This intriguing question of neutrino mass, and its implications for physics beyond the Standard Model, is the subject of many of the succeeding chapters in this volume.

But, despite the great transformations that have occurred since 1930 in the idea of the neutrino, Pauli deserves due credit for being the one daring enough to take the great conceptual leap of introducing another fundamental entity into the minimalist ontology of his day. No doubt the enduring success of his bold scheme has encouraged theorists of later decades to repeat this exercise whenever a cherished symmetry or conservation law appears to be violated. In this subtle fashion, the ghost of Wolfgang Pauli still haunts the way particle physics is practiced today.

References

1.1 W. Pauli, transl. in L. M. Brown, Phys. Today, Sept. 1978, 23.

1.2 C. Sutton, *Spaceship Neutrino* (Cambridge University Press, Cambridge, 1992).

1.3 C. D. Ellis and W. A. Wooster, Proc. R. Soc. (Lond.) A **117**, 109 (1927); L. Meitner and W. Orthmann, Z. Phys. **60**, 143 (1930).

1.4 N. Bohr, J. Chem. Soc. (Lond.) **135**, 349 (1932).

1.5 J. Chadwick, Nature 129, 312 (1932); Proc. R. Soc. (Lond.) A **136**, 692 (1932).

1.6 E. Fermi, Z. Phys. **88**, 161 (1934) [transl. in F. L. Wilson, Am. J. Phys. **36**, 1150 (1960)].

1.7 H. A. Bethe and R. E. Peierls, Nature **133**, 532 (1934).

1.8 F. Reines, Rev. Mod. Phys. **68**, 317 (1996).

1.9 A. S. Eddington, *The Philosophy of Physical Science* (Cambridge University Press, Cambridge, 1939).

1.10 G. Gamow, Phys. Today, July 1948, 6.

1.11 J. Allen, Phys. Rev. **61**, 692 (1942).

1.12 C. L. Cowan and F. Reines, Phys. Rev. **106**, 825 (1957).

1.13 F. Reines and C. L. Cowan Jr., Phys. Rev. **92**, 830 (1953).

1.14 C. L. Cowan Jr. et al., Science **124**, 103 (1956).

1.15 F. Reines et al., Phys. Rev. **117**, 159 (1960).

1.16 F. Reines and C. Cowan, Phys. Today, Aug. 1957, 12.

1.17 L. M. Langer and R. J. D. Moffat, Phys. Rev. **88**, 689 (1952).

1.18 E. Majorana, Nuovo Cimento **14**, 171 (1937).

1.19 T. D. Lee and C. N. Yang, Phys. Rev. **105**, 1671 (1957).

1.20 L. Landau, Nucl. Phys. **3**, 127 (1957); A. Salam, Nuovo Cimento **5**, 299 (1957).

1.21 M. Goldhaber, L. Grodzins and A. W. Sunyar, Phys. Rev. **106**, 826 (1957).

---

[13] For more recent results, see Chap. 5.

1.22 G. Feinberg, Phys. Rev. **110**, 1482 (1958).

1.23 M. Schwartz, Rev. Mod. Phys. **61**, 527 (1989).

1.24 M. Schwartz, "The early history of high-energy neutrino physics", in *The Rise of the Standard Model: Particle Physics in the 1960s and 1970s*, ed. by L. Hoddeson et al. (Cambridge University Press, New York, 1997), p. 411.

1.25 G. Danby et al., Phys. Rev. Lett. **9**, 36 (1962).

1.26 L. Hoddeson et al. (eds.), *The Rise of the Standard Model: Particle Physics in the 1960s and 1970s* (Cambridge University Press, New York, 1997).

1.27 L. Brown et al., "The rise of the Standard Model: 1964–1979", in *The Rise of the Standard Model: Particle Physics in the 1960s and 1970s*, ed. by L. Hoddeson et al. (Cambridge University Press, New York, 1997), p. 3.

1.28 P. Galison, Rev. Mod. Phys. **55**, 477 (1983).

1.29 D. Perkins, "Gargamelle and the discovery of neutral currents", in *The Rise of the Standard Model: Particle Physics in the 1960s and 1970s*, ed. by L. Hoddeson et al. (Cambridge University Press, New York, 1997), p. 428.

1.30 F. J. Hasert et al., Phys. Lett. B **46**, 121 (1973).

1.31 F. J. Hasert et al., Phys. Lett. B **46**, 138 (1973).

1.32 A. Benvenuti et al., Phys. Rev. Lett. **32**, 1457 (1974).

1.33 G. Goldhaber, "From the psi to charmed mesons: three years with the SLAC–LBL detector at SPEAR", in *The Rise of the Standard Model: Particle Physics in the 1960s and 1970s*, ed. by L. Hoddeson et al. (Cambridge University Press, New York, 1997), p. 57.

1.34 M. Kobayashi, "Flavor mixing and *CP* violation", in *The Rise of the Standard Model: Particle Physics in the 1960s and 1970s*, ed. by L. Hoddeson et al. (Cambridge University Press, New York, 1997), p. 137.

1.35 J. Cronin, "The discovery of *CP* violation", in *The Rise of the Standard Model: Particle Physics in the 1960s and 1970s*, ed. by L. Hoddeson et al. (Cambridge University Press, New York, 1997), p. 114.

1.36 M. Kobayashi and T. Maskawa, Prog. Theor. Phys. **49**, 652 (1973).

1.37 M. Perl, "The discovery of the tau lepton", in *The Rise of the Standard Model: Particle Physics in the 1960s and 1970s*, ed. by L. Hoddeson et al. (Cambridge University Press, New York, 1997), p. 79.

1.38 G. Feldman, "The discovery of the τ, 1975–1977: a tale of three papers", in *Proceedings of the 20th SLAC Summer Institute of Particle Physics*, ed. by J. Hawthorne (Stanford University, 1992), p. 631.

1.39 M. Perl et al., Phys. Rev. Lett. **35**, 1489 (1975).

1.40 J. Burmeister et al., Phys. Lett. B **58**, 297 (1977).

1.41 M. Nakamura, Nucl. Phys. B, Proc. Suppl. **77**, 259 (1999).

1.42 L. Lederman, "The discovery of the upsilon, bottom quark, and B mesons", in *The Rise of the Standard Model: Particle Physics in the 1960s and 1970s*, ed. by L. Hoddeson et al. (Cambridge University Press, New York, 1997), p. 101.

1.43 F. Abe et al., Phys. Rev. D **50**, 2966 (1994).

1.44 J. Yang et al., Astrophys. J. **281**, 493 (1984).

1.45 D. N. Schramm and G. Steigman, Sci. Am., June 1988, 66.

1.46 G. S. Abrams et al., Phys. Rev. Lett. **63**, 2173 (1989).

1.47 D. Decamp et al., Phys. Lett. **231**, 519 (1989).

1.48 Y. Fukuda et al., Phys. Rev. Lett. **81**, 1562 (1998).

2 The Nature of Massive Neutrinos

Boris Kayser and Rabindra N. Mohapatra

2.1 Introduction

Neutrinos are spin-half, electrically neutral particles introduced by Pauli in 1930 to explain the energy crisis that had plagued the understanding of beta decay processes since the time of their discovery. Soon after Pauli's suggestion, Fermi wrote down the Hamiltonian describing the interactions of the neutrino with other elementary-particles required to explain the nuclear beta decay observations, and neutrinos became an inseparable part of elementary particle physics. The discussion of their discovery and the rest of the history of neutrinos is the subject of the previous chapter.

Since beta decay is a weak process, the interactions of neutrinos with matter are weak. This helps neutrinos to pass through most matter in the Universe with very little interaction. On the other hand, it is also their weak interaction that helps them to play a crucial role in the evolution of the Universe, e.g. in determining the present matter content of the Universe. In fact, next to radiation, neutrinos are believed to be the most abundant constituent of the Universe.

At the time when Pauli postulated the neutrino and for almost half a century after that, most people believed that neutrinos were massless particles. This feature was embodied in the construction of the Standard Model of the electroweak and strong interactions by Glashow, Weinberg and Salam. However, recent neutrino oscillation data, discussed elswhere in this book, have provided compelling evidence that one or more neutrino species have nonzero masses. This raises a number of theoretical and phenomenological issues that were not present for the case of massless neutrinos predicted by the Standard Model. A study of these questions is therefore important not only for a better understanding of the true nature of neutrinos, but also to gain insight into the nature of the new physics present at high mass scales that could explain their properties. With this in mind, we study several topics related to massive neutrinos in a model-independent manner in this chapter.

We address three topics. The first is the key question of the nature of the neutrino mass, i.e. whether it is a Majorana or Dirac type of mass, the first being the case when a neutrino is its own antiparticle and the second when it is not. In the second case, one can define a lepton number symmetry under which the theory is invariant. Such questions do not arise for other known fermions in nature, owing to the fact that they are electrically charged, unlike the neutrino. Mass, of course, is not the only property that can be used to decide whether a neutrino is its own antiparticle; interactions can also determine the answer to this question, if they violate lepton number symmetry. However, at a fundamental level, when interactions violate lepton number symmetry, they always lead at some level of perturbation theory to a Majorana mass for the neutrino.

A second topic of great interest is neutrino mixing and the oscillations that occur during the time evolution of the neutrinos. This behavior provides another way to distinguish between massive neutrinos and massless ones. In fact, the evidence for massive neutrinos at present comes only from oscillation experiments.

Finally, another distinguishing feature of massive neutrinos is that a member of one species could decay to another one, giving rise to many interesting new phenomena in the laboratory as well as in the cosmological domain.

It is the goal of this chapter to discuss these properties of massive neutrinos and note some experimental tests. We have arranged the chapter as follows: in Sect. 2.2, we discuss the distinction between Dirac and Majorana neutrinos; Sect. 2.3 is devoted to a discussion of neutrino mixings and oscillations; in Sect. 2.4, we discuss neutrino decays.

2.2   Dirac and Majorana Masses for Neutrinos

In this section, we discuss the difference between Dirac and Majorana neutrinos. A Majorana neutrino is its own antiparticle, whereas a Dirac neutrino is not. This apparently simple diffrence between them leads to a large number of profound and distinguishing physical implications which can be used to test whether the neutrino is a Majorana or a Dirac fermion. Clearly, a full understanding of the nature of the massive neutrino requires that we know whether it is its own antiparticle or not. In this section, we give the mathematical formulation for the Majorana neutrino and some ways to test for its Majorana character.

As is well known from the study of the Dirac equation in quantum mechanics, a spin-half massive fermion like the electron is described by a four-component wave function. Since, for an electron, there are two spin states for the particle and two for the antiparticle, all encoded in the same wave function, the four components needed to describe the electron are accounted for. On the other hand, in the case of a Majorana neutrino, since it is its own antiparticle, naive arguments would suggest that we should need only two components to describe it. This is indeed true; however, for many calculational purposes, it is convenient to use the four-component wave function. Therefore, for completeness, we present both formulations [2.1]. We first present the four-component notation and follow it up in the next subsection with a discussion of the two-component one. We shall see that the two-component notation is particularly suitable for discussion of the mass matrices, whereas the four-component one makes the manipulations involving the Feynman diagrams a lot easier. The basic physics is of course the same, as would be expected.

2.2.1 Four-Component Notation

The four-component notation for massive spin-half particles is familiar from the discussion of the Dirac equation for the electron. In analogy with the electron, we can denote the wave function for a free neutrino by a wave function ψ which satisfies the equation

$$i\gamma^\lambda \partial_\lambda \psi - m\psi = 0\,. \tag{2.1}$$

This equation follows from the free Lagrangian

$$\mathcal{L} = i\bar{\psi}\gamma^\lambda \partial_\lambda \psi - m\bar{\psi}\psi \tag{2.2}$$

and leads to the relativistic energy–momentum relation $p^\lambda p_\lambda = m^2$ for a spin-half particle only if the four γ_λs anticommute. If we take the γ_λs to be $n \times n$ matrices, the smallest value of n for which four anticommuting matrices exist is four and hence the minimal number of components in ψ must be four.

A spin-half particle is said to be a Majorana particle if the spinor field ψ satisfies the condition of being self-charge-conjugate, i.e.

$$\psi = \psi^c \equiv C\bar{\psi}^{\mathrm{T}}, \tag{2.3}$$

where C is the charge conjugation matrix and has the property $C\gamma_\lambda C^{-1} = -\gamma^{\lambda\mathrm{T}}$. Using this condition, the mass term in the Lagrangian in (2.2) can be written as $\psi^{\mathrm{T}}C^{-1}\psi$, where we have used the fact that C is a unitary matrix. Writing the mass term in this way makes it clear that if a field carries a $U(1)$ charge and the theory is invariant under those $U(1)$ transformations, then the mass term is forbidden. This means that one cannot impose the Majorana condition on this particle. In other words, a particle carrying $U(1)$ charge (such as electric charge) cannot be a Majorana particle if we want to maintain the invariance of the theory under that $U(1)$ symmetry. Since neutrinos do not have electric charge, they can be Majorana particles, unlike the quarks, the electron and the muon. It is of course well known that the gauge boson contributions to weak interactions in a gauge theory Lagrangian conserve a global $U(1)$ symmetry known as lepton number, with the neutrino and electron carrying the same lepton number. If lepton number were to be established as an exact symmetry of nature, a Majorana mass for the neutrino would be forbidden, and the neutrino, like the electron, would be a Dirac particle.

It is worth pointing out that the Majorana condition on the field ψ can be generalized to the form

$$\psi = e^{i\alpha} C\bar{\psi}^{\mathrm{T}}, \tag{2.4}$$

if the mass term in the Lagrangian has the form $me^{i\alpha}\psi^T C^{-1}\psi$. In what follows, we shall make the choice $\alpha = 0$.

These properties of a Majorana fermion can also be seen in its free field expansion in terms of creation and annihilation operators:

$$\psi(x) = \int \frac{d^3p}{\sqrt{(2\pi)^3 2E_p}} \sum_s \left[a_s(\boldsymbol{p}) u_s(\boldsymbol{p}) e^{-ip\cdot x} + a_s^\dagger v_s(\boldsymbol{p}) e^{ip\cdot x} \right]. \qquad (2.5)$$

In the gamma matrix convention, where $\gamma_i = \begin{pmatrix} 0 & \sigma_i \\ -\sigma_i & 0 \end{pmatrix}$ and $\gamma_0 = \begin{pmatrix} 0 & \mathbf{I} \\ \mathbf{I} & 0 \end{pmatrix}$, the u_s and v_s are given by

$$u_s(\boldsymbol{p}) = \frac{m}{\sqrt{E}} \begin{pmatrix} \alpha_s \\ [(E - \sigma \cdot \boldsymbol{p}/m)]\alpha_s \end{pmatrix} \qquad (2.6)$$

and

$$v_s(\boldsymbol{p}) = \frac{m}{\sqrt{E}} \begin{pmatrix} -[(E + \sigma \cdot \boldsymbol{p})/m]\alpha_s' \\ \alpha_s' \end{pmatrix}. \qquad (2.7)$$

If we choose $\alpha_s' = \sigma_2 \alpha_s$, we obtain the relation $C\gamma_0 u_s^*(\boldsymbol{p}) = v_s(\boldsymbol{p})$ between the spinors $u_s(\boldsymbol{P})$ and $v_s(\boldsymbol{P})$, and the Majorana condition follows. Note that if ψ were to describe a Dirac spinor, then we would have a different creation operator $b^\dagger$ in the second term in the free-field expansion above.

Using the anticommutation property of spinor fields and the property $C\gamma_\mu C^{-1} = -\gamma^{\mu T}$, it is easy to verify that, for Majorana spinors satisfying the Majorana condition in (2.3), the vector current $\bar{\psi}\gamma_\mu\psi$ and the tensor current $\bar{\psi}\sigma_{\mu\nu}\psi$ are identically zero. This has important physical implications, i.e. a Majorana neutrino cannot have a magnetic moment and cannot support a vector charge. It can, however, support an axial-vector charge, as can also be checked the same way.

The propagator for the Majorana neutrino is the same as that for the Dirac neutrino. The Feynman rules are, however, somewhat different, especially when they involve electron lines connecting to the internal neutrino lines. This is beyond the scope of the present article.

2.2.2   Two-Component Notation

To see why the two-component notation is as appropriate for discussion of fermions as the four-component notation, it is useful to discuss the $SL(2, C)$ group, its connection to the Lorentz group and its representations.

The first point to note is that the proper Lorentz group is locally isomorphic to the $SL(2, C)$ group, which is the group of transformations on a two-dimensional complex linear space with unit determinant. If we denote the elements of the $SL(2, C)$ transformation group by complex 2×2 matrices

$$S \equiv \begin{pmatrix} S_{11} & S_{12} \\ S_{21} & S_{22} \end{pmatrix},$$

then the elements S_{ij} satisfy the additional constraint $S_{11}S_{22} - S_{12}S_{21} = 1$. Simple counting then shows that there are six independent parameters describing the $SL(2,C)$ transformations, which is the same as the number of parameters for the Lorentz group. The fundamental representations of the $SL(2,C)$ group are two-dimensional; these we denote by χ. Therefore, under a Lorentz transformation, $\chi \to S\chi$. The matrix S can also be written as $S = e^{i\sigma \cdot n} \cdot e^{\sigma \cdot m}$, where the σ's are the Pauli matrices and the components of m, n are the six parameters corresponding to the Lorentz transformations. Note that χ^* transforms differently from χ, i.e. χ and χ^* are two inequivalent representations of the $SL(2,C)$ group. It is therefore clear that for a Lorentz-invariant description, it is sufficient to use the two-component wave functions χ to denote a spin-half fermion;

$$\chi \equiv \begin{pmatrix} \chi_1 \\ \chi_2 \end{pmatrix} . \tag{2.8}$$

To be more explicit, under Lorentz transformation, $\chi \to \chi' = S\chi$. Note that χ^* is inequivalent to χ since it transforms as $\chi^* \to \chi^{*'} = S^*\chi^*$, and S and S^* cannot be related to each other by a similarity transformation. We shall call χ a (2,1) representation and χ^* a (1,2) representation. Then $\sigma_\mu P^\mu$ transforms as a (2,2) representation under $SL(2,C)$. The components of χ will be denoted by χ_a with a subscript index, and those of χ^* by $\bar{\chi}^p$ with a superscript index.

We can now proceed to construct Lorentz-invariant bilinears, which will enable us to define Majorana and Dirac neutrinos. The unit determinant property of S implies that, if χ and ϕ are two two-component spinors, then $\epsilon^{\alpha\beta}\chi_\alpha\phi_\beta$ is invariant under $SL(2,C)$, where ϵ is the antisymmetric tensor defined by $\epsilon^{11} = \epsilon^{22} = 0$, $\epsilon^{12} = -\epsilon^{21} = 1$. Note that the anticommuting property of spinor fields allows one to write $\epsilon^{\alpha\beta}\chi_\alpha\chi_\beta$ also as a nonvanishing Lorentz-invariant bilinear. These invariants can be written as $i\chi^T\sigma_2\phi$ or $i\chi^T\sigma_2\chi$ in matrix notation. Similar invariants exist for $\bar{\chi}$ and $\bar{\phi}$.

A four-component Dirac spinor can in general be written as

$$\psi = \begin{pmatrix} \chi \\ -i\sigma_2\phi^* \end{pmatrix} . \tag{2.9}$$

Thus, if we choose

$$\gamma_0 = \begin{pmatrix} 0 & 1 \\ 1 & 0 \end{pmatrix} ,$$

a conventional Dirac mass term $\bar{\Psi}\Psi$ becomes

$$\bar{\psi}\psi = i\phi^T\sigma_2\chi - i\chi^\dagger\sigma_2\phi^* . \tag{2.10}$$

Note that such a mass term is invariant under the $U(1)$ transformation under which

$$\phi \to e^{i\theta}\phi, \qquad \chi \to e^{-i\theta}\chi . \tag{2.11}$$

This $U(1)$ can be identified with the lepton number if χ represents a neutrino. The field ϕ will then represent an antilepton or perhaps a different fermion with opposite lepton number.

A general mass term involving many two-component spinors can be written as

$$-\mathcal{L}_{\text{mass}} = \frac{i}{2} \sum_{a,b} M_{ab} \phi_a^{\mathrm{T}} \sigma_2 \chi_b + \text{H.c.}, \tag{2.12}$$

where now the indices a and b label different spinors. It is then easy to check that $M = M^{\mathrm{T}}$. To discuss the neutrino wave function, let us assume that a free two-component neutrino is described by the following Lagrangian:

$$\mathcal{L} = \chi^\dagger \sigma^\lambda i \partial_\lambda \chi - \frac{im}{2} \chi^{\mathrm{T}} \sigma_2 \chi + \frac{im}{2} \chi^\dagger \sigma_2 \chi^*. \tag{2.13}$$

This leads to the following equation of motion for the field χ:

$$i\sigma^\lambda \partial_\lambda \chi - im\sigma_2 \chi^* = 0. \tag{2.14}$$

As is conventionally done in field theories, we can now give a free-field expansion of the two-component Majorana field in terms of the creation and annihilation operators:

$$\chi(x,t) = \sum_{p,s} [a_{p,s} \alpha_{p,s} e^{-ip.x} + a_{p,s}^\dagger \beta_{p,s} e^{ip.x}], \tag{2.15}$$

where the sum over s goes over the spin-up and spin-down states; α and β are two-component spinors that must satisfy the field equations that follow from the (2.14). Solving those equations, we obtain the following form for the χ field:

$$\chi(x,t) = \sum_p [a_{p,+} e^{-ip.x} - a_{p,-}^\dagger e^{ip.x}] \alpha \sqrt{E+p}$$
$$+ \sum_p [a_{p,-} e^{-ip.x} + a_{p,+}^\dagger e^{ip.x}] \beta \sqrt{E-p}, \tag{2.16}$$

where choosing the momentum along the z direction leads to

$$\alpha = \begin{pmatrix} 1 \\ 0 \end{pmatrix}$$

and

$$\beta = \begin{pmatrix} 0 \\ 1 \end{pmatrix}.$$

For small masses, only the first terms are important; i.e. the field χ creates the "down" helicity state and destroys the "up" one to zeroth order in m/E, and the $\chi^\dagger$ field will dominantly create the "up" and destroy the "down" state, correspondingly. The two-component field behaves like a Weyl fermion.

Let us now turn to the definition of parity and charge conjugation for the two-component fermion. Note that the two-component Weyl equation (i.e. (2.14) with the mass term set to zero) does not respect parity or charge conjugation invariance defined in the conventional way, i.e. parity as space inversion and C such that it transforms the field χ to χ^*. However, for the massive Majorana spinor, one can define both P and C so that they match the corresponding definitions in the four-component case or, alternatively, one can include additional fields in the theory and define Z_2 symmetries relating them, which can be called P and/or C. For the first case one can have

$$
\begin{aligned}
P &: \chi \to -\sigma_2 \chi^*, \\
C &: \chi \to \sigma_2 \chi^*.
\end{aligned}
\tag{2.17}
$$

It is important to point out that the above transformations are true only if the mass of the particle is real. For a complex Majorana mass parameter, with a phase $e^{i\delta}$, the C transformations change and acquire an additional phase. If there are two 2-component spiners, we can define

$$
\begin{aligned}
P &: \chi \to -\sigma_2 \phi^* \\
C &: \chi \to \sigma_2 \phi^*.
\end{aligned}
\tag{2.17a}
$$

Before closing this subsection, let us discuss under what conditions two two-component neutrinos (say χ and ϕ) form a four-component Dirac neutrino. This is determined by two conditions: (i) their mass term must be given by $\mathcal{L} = im\chi^T \sigma_2 \phi + \text{h.c.}$, which allows the definition of a conserved $U(1)$ quantum number under which χ and ϕ have opposite charges, and (ii) all interactions in the theory must conserve the above $U(1)$. The mass term given in condition (i) implies that the mass eigenstates are $(\phi \pm \chi)/\sqrt{2}$. Using the definition of C and P in (2.17a), we note that under CP $\chi \to \phi$ and $\phi \to \chi$. Therefore, $\chi \pm \phi$ have $CP \pm 1$. Thus a Dirac four-component neutrino can be thought of as consisting of two opposite-CP Majorana neutrinos. However, an arbitrary pair of two-component spinors of opposite CP does not necessarily define a Dirac fermion, since it may not satisfy the second condition. A simple example that violates condition (ii) is $\mathcal{L} = W_\lambda \bar{e} \bar{\sigma}^\lambda (\phi + \chi) + m\phi^T \sigma_2 \chi + \text{h.c.}$, where $\bar{\sigma}^\mu = (1, -\sigma)$. A more physically relevant example is that of a Majorana ν_e and ν_μ with the same mass but opposite CP. An example of such a mass term is $\mathcal{L} = im(\nu_e^T \sigma_2 \nu_e - \nu_\mu^T \sigma_2 \nu_\mu)$. Clearly this pair does not form a Dirac fermion – they are simply two two-component neutrinos with opposite CP. Lepton-number-violating processes such as $(Z, A) \to (Z + 2, A) + e^- e^-$, the neutrinoless double beta decay denoted later by $\beta\beta 0\nu$, is allowed in this case. These two neutrinos, however, form a Dirac fermion if their mass term connects them, i.e. it is of the form $\nu_e^T \sigma_2 \nu_\mu$. This mass term conserves $L_e - L_\mu$ and thus forbids the neutrinoless double beta decay process mentioned above.

2.2.3 Electromagnetic Properties of a Majorana Neutrino

In this section, we study the electromagnetic properties of a Majorana neutrino in the two component notation.

Magnetic Moment of a Majorana Neutrino. The well-known property that a Majorana neutrino cannot have a magnetic moment can be seen in the two-component notation as follows. Using the expansion of γ matrices in terms of the two-component σ matrices, it is easy to show that $\bar{\psi}\sigma_{\lambda\lambda'}\psi = -i\nu^T\sigma_2(\bar{\sigma}_\lambda\sigma_{\lambda'} - \bar{\sigma}_{\lambda'}\sigma_\lambda)\nu + \text{h.c.}$ Let us now look at the components of this tensor. It is easy to see that the the nonzero components are all in the form $\nu^T\sigma_2\sigma_i\nu$. It is easily checked that $\sigma_2\sigma_i$ is a symmetric matrix, and the fermionic anticommutation implies that for two identical fermions ν, the above operator vanishes identically.

Vector versus Axial Vector Charge of a Majorana Fermion. In discussing the question of vector versus axial vector charge of a Majorana neutrino, it is important to remember that a Lorentz vector transforms under $SL(2, C)$ like a (2,2) representation but has no definite parity. We shall therefore use the definition of parity given above to write the vector and axial-vector currents (denoted by J_μ and $J_{5,\mu}$, respectively). We first observe that under parity, $\chi^\dagger\sigma^\lambda\chi \to \chi^T\sigma_2\bar{\sigma}_\lambda\sigma_2\chi^*$. Therefore, we can write

$$J_\lambda = \chi^\dagger\sigma^\lambda\chi - \chi^T\sigma_2\bar{\sigma}_\lambda\sigma_2\chi^*,$$
$$J_{5,\lambda} = \chi^\dagger\sigma^\lambda\chi + \chi^T\bar{\sigma}_2\bar{\sigma}_\lambda\sigma_2\chi^*. \qquad (2.18)$$

It is then a matter of simple algebra to see that $J_\mu = 0$, from which the well-known conclusion that Majorana spinors cannot support a vector charge emerges. (One simple way to see this is to note that $\sigma_2\bar{\sigma}_\lambda^T\sigma_2 = -\sigma_\lambda$ and, the ν fields being fermionic, its components anticommute.) On the other hand, the pseudo-vector current is nonvanishing, and it is also even under charge conjugation, in agreement with the result from the four-component notation, where only the $\bar{\psi}\gamma_5\gamma_\lambda\psi$ part is nonvanishing for a Majorana field.

2.3 Leptonic Mixing and Neutrino Oscillation

2.3.1 Leptonic Mixing

We shall assume that the interaction between neutrinos and other particles is correctly described by the very well-confirmed Standard Model. In the Standard Model, the coupling between neutrinos, charged leptons, and the W boson is given by

$$\mathcal{L}_{\ell\nu W} = -\frac{g}{\sqrt{2}}\overline{\ell^0_L}\gamma_\lambda\nu^0_L W^\lambda + \text{H.c.} \qquad (2.19)$$

Here, g is the semiweak coupling constant. If nature contains N generations, each comprising a charged lepton and a neutrino, then ℓ_L^0 and ν_L^0 are N-component vectors. The ath component of ℓ_L^0 (where the subscript L denotes left-handedness) is the field ℓ_{La}^0 for the charged lepton in generation a, while the ath component of ν_L^0 is the field ν_{La}^0 for the corresponding neutrino.

In general, the "particles" ℓ_{La}^0 and ν_{La}^0, which have the simple weak coupling of (2.19), do not have definite mass. Rather, each of them is a coherent superposition of particles with definite mass. For the charged leptons, we have a mass term

$$\mathcal{L}_{m_\ell} = -\overline{\ell_R^0} \mathcal{M}_\ell \ell_L^0 + \text{H.c.} \tag{2.20}$$

Here, ℓ_R^0 is the right-handed counterpart of the vector ℓ_L^0, and $\mathcal{M}_\ell$ is the $N \times N$ charged-lepton mass matrix. The matrix $\mathcal{M}_\ell$ can be diagonalized by the biunitary transformation [2.1]

$$V_{R\ell}^\dagger \mathcal{M}_\ell V_{L\ell} = \mathcal{D}_\ell , \tag{2.21}$$

where $V_{R\ell}$, $V_{L\ell}$ are two distinct unitary matrices, and

$$\mathcal{D}_\ell = \begin{bmatrix} m_e & & & 0 \\ & m_\mu & & \\ & & m_\tau & \\ 0 & & & \ddots \end{bmatrix} \tag{2.22}$$

is a diagonal matrix whose diagonal elements are the masses of the charged leptons $e, \mu, \tau, \ldots$ of definite mass.

The vector

$$\ell = \begin{pmatrix} e \\ \mu \\ \tau \\ \vdots \end{pmatrix} , \tag{2.23}$$

containing the fields of the charged leptons of definite mass, is related to $\ell_{L,R}^0$ by

$$\ell_{L,R}^0 = V_{L\ell,R\ell} \ell_{L,R} . \tag{2.24}$$

For the neutrinos, there can be both Dirac and Majorana masses, as discussed in Sect. 2.1. Thus, the most general neutrino mass term, in four-component notation, is

$$\mathcal{L}_{M_\nu} = -\frac{1}{2} [\overline{(\nu_L^0)^c}, \overline{\nu_R^0}] \mathcal{M}_\nu \begin{bmatrix} \nu_L^0 \\ (\nu_R^0)^c \end{bmatrix} + \text{H.c.} \tag{2.25}$$

Here, ν_R^0 is the right-handed counterpart of the vector ν_L^0, and $\mathcal{M}_\nu$ is a $2N \times 2N$ matrix. In two-component notation, this mass term takes the form indicated by (2.12).

The matrix $\mathcal{M}_\nu$ may be decomposed as

$$\mathcal{M}_\nu = \begin{bmatrix} M_{\mathrm{L}} & M_{\mathrm{D}}^{\mathrm{T}} \\ M_{\mathrm{D}} & M_{\mathrm{R}} \end{bmatrix} , \qquad (2.26)$$

in which each of M_{L}, M_{R}, and M_{D} is an $N \times N$ submatrix. The submatrices M_{L} and M_{R} contain left-handed and right-handed Majorana mass terms of the kind defined in Sect. 2.2, while M_{D} contains Dirac mass terms. Owing to relations such as $(\nu_{La}^0)^c \, \nu_{Lb}^0 = \overline{(\nu_{Lb}^0)^c} \, \nu_{La}^0$, M_{L} and M_{R} may be taken to be symmetric, so that $\mathcal{M}_\nu$ is a symmetric matrix. As a result, in the neutrino analogue of the diagonalizing transformation (2.21), the matrix preceding the mass matrix can be chosen as the transpose of the one following it. Thus, $\mathcal{M}_\nu$ may be diagonalized by the transformation

$$Y_\nu^{\mathrm{T}} \mathcal{M}_\nu Y_\nu = \mathcal{D}_\nu . \qquad (2.27)$$

Here, Y_ν is a $2N \times 2N$ unitary matrix, and $\mathcal{D}_\nu$ is a diagonal matrix whose diagonal elements are the real, positive-definite neutrino mass eigenalues m_i. Let us define the $2N$-component left-handed vector ν_{L} by

$$\nu_{\mathrm{L}} = Y_\nu^\dagger \begin{bmatrix} \nu_{\mathrm{L}}^0 \\ (\nu_{\mathrm{R}}^0)^c \end{bmatrix} , \qquad (2.28)$$

the corresponding right-handed vector ν_{R} by

$$\nu_{\mathrm{R}} = (\nu_{\mathrm{L}})^c \qquad (2.29)$$

and a vector ν by

$$\nu = \nu_{\mathrm{L}} + \nu_{\mathrm{R}} . \qquad (2.30)$$

Then, from (2.25) and (2.27), we have

$$\mathcal{L}_{m_\nu} = -\frac{1}{2} \sum_{i=1}^{2N} m_i \overline{\nu_i} \nu_i , \qquad (2.31)$$

where ν_i is the ith component of the vector ν. Each term on the right-hand side of this equation is just the mass term for a neutrino ν_i, so we see that the ν_i are the neutrinos of definite mass. Furthermore, from (2.29) and (2.30), we see that the fields ν_i satisfy

$$\nu_i^c = \nu_i . \qquad (2.32)$$

Thus, the neutrinos ν_i are Majorana particles.[1] Note that in defining ν_{R} and ν by (2.29) and (2.30), we have chosen to set the arbitrary phase α defined by (2.4) equal to zero for all the neutrinos ν_i.

---

[1] Our discussion of the neutrino mass matrix and its diagonalization is based on work done in collaboration with S. Petcov, whom we thank for this very instructive collaboration.

It is convenient to write

$$Y_\nu = \begin{bmatrix} V_\nu \\ W_\nu \end{bmatrix} , \tag{2.33}$$

in which V_ν and W_ν each contains N rows and $2N$ columns. Then, from (2.28),

$$\nu_L^0 = V_\nu \nu_L . \tag{2.34}$$

In terms of the charged leptons of definite mass (e, μ, τ and any others that may exist), the weak interaction $\mathcal{L}_{\ell\nu W}$ of (2.19) may be written as

$$\begin{aligned}
\mathcal{L}_{\ell\nu W} &= -\frac{g}{\sqrt{2}} \overline{\ell_L} \gamma_\lambda (V_{L\ell}^\dagger \nu_L^0) W^\lambda + \text{H.c.} \\
&\equiv -\frac{g}{\sqrt{2}} \overline{\ell_L} \gamma_\lambda \nu_L' W^\lambda + \text{H.c.} \\
&= -\frac{g}{\sqrt{2}} (\overline{e_L} \gamma_\lambda \nu_{Le}' + \overline{\mu_L} \gamma_\lambda \nu_{L\mu}' + \overline{\tau_L} \gamma_\lambda \nu_{L\tau}' + \ldots) W^\lambda + \text{H.c.}
\end{aligned} \tag{2.35}$$

Here, we have used (2.24) and introduced the vector

$$\nu_L' = V_{L\ell}^\dagger \nu_L^0 , \tag{2.36}$$

whose N components are the neutrinos $\nu_{L\alpha}'$ of definite "flavor". The neutrino of flavor α, $\nu_{L\alpha}'$, is coupled by the weak interaction (2.35) to the definite-mass charged lepton with the same flavor: ν_e' to e, ν_μ' to μ and so forth. However, like the neutrinos ν_{La}^0 that have simple couplings to the ℓ_{La}^0, the neutrinos $\nu_{L\alpha}'$ of definite flavor do not have definite masses. To express the weak interaction of (2.19) in terms of the neutrinos which do have definite masses, we use (2.24) and (2.34), and obtain

$$\begin{aligned}
\mathcal{L}_{\ell\nu W} &= -\frac{g}{\sqrt{2}} \overline{\ell_L} \gamma_\lambda (V_{L\ell}^\dagger V_\nu) \nu_L W^\lambda + \text{H.c.} \\
&\equiv -\frac{g}{\sqrt{2}} \overline{\ell_L} \gamma_\lambda U \nu_L W^\lambda + \text{H.c.} \\
&= -\frac{g}{\sqrt{2}} \sum_{\substack{\alpha=1,\ldots,N \\ i=1,\ldots,2N}} \overline{\ell_{L\alpha}} \gamma_\lambda U_{\alpha i} \nu_{Li} W^\lambda + \text{H.c.}
\end{aligned} \tag{2.37}$$

Here, we have introduced the $N \times 2N$ "leptonic mixing matrix",

$$U \equiv V_{L\ell}^\dagger V_\nu . \tag{2.38}$$

The leptonic mixing matrix is increasingly often called the "Maki–Nakagawa–Sakata matrix", to honor those three people for their very insightful early work on neutrino mixing and oscillation [2.2].

If all the Majorana mass terms in $\mathcal{M}_\nu$ vanish, then the $2N$ Majorana mass eigenstates of $\mathcal{M}_\nu$ pair up to make N Dirac neutrinos, in the manner explained in Sect. 2.2.

The Standard Model weak interaction of (2.19) will only absorb a neutrino to make a negatively charged lepton or (through the Hermitian-conjugate term) absorb an antineutrino to make a positively charged lepton. That is, this interaction conserves the lepton number L, which distinguishes a neutrino or negatively charged lepton on the one hand from an antineutrino or positively charged lepton on the other hand. If this interaction, plus the Standard Model couplings of the neutral Z boson, which are also L-conserving, are the only interactions which need to be taken into account, then the violation of L conservation which is required before $\beta\beta 0\nu$ decay can occur can come only from the neutrino mass terms. As explained in Sect. 2.1, Dirac mass terms conserve L. However, Majorana mass terms such as $\overline{(\nu_{La}^0)^c}\, \nu_{La}^0$ or $\overline{\nu_{Ra}^0}\, (\nu_{Ra}^0)^c$, which convert a neutrino into an antineutrino and vice versa, respectively, clearly break L conservation.

2.3.2   Neutrino Oscillation

One of the most interesting consequences of neutrino mass and mixing is neutrino oscillation in vacuum. This is the phenomenon in which a neutrino is born in association with a charged lepton of one flavor, such as an e, and then interacts to make a charged lepton of a different flavor, such as a μ. Since the neutrino born in association with an e is, by definition, a ν_e' (see (2.35)), while the one whose interaction yields a μ is, by definition, a ν_μ', this sequence of events is usually described as the metamorphosis, or oscillation, of a ν_e' into a ν_μ'. Or, omitting the primes from now on without fear of confusion, it is described as the oscillation $\nu_e \to \nu_\mu$. We shall see that the probability for this neutrino flavor transformation does have an oscillatory behavior, which is the reason the transformation is referred to as "oscillation".

From the expression for the weak interaction in terms of leptonic mass eigenstates (2.37), we see that the neutrino created in association with the charged lepton ℓ_α can be any of the mass eigenstates ν_i. We also see that, when it interacts, this mass eigenstate ν_i can produce any charged lepton ℓ_β. Thus, the amplitude for the neutrino oscillation $\nu_\alpha \to \nu_\beta$ is given by

$$A(\nu_\alpha \to \nu_\beta) = \sum_i [A(\text{neutrino born with } \ell_\alpha \text{ is a } \nu_i)$$

$$\times\, A(\nu_i \text{ propagates})\, A(\text{when } \nu_i \text{ interacts it makes } \ell_\beta)]\,. \qquad (2.39)$$

Here, A denotes an amplitude. From the Hermitian-conjugate term in (2.37), $A(\text{neutrino born with } \ell_\alpha \text{ is a } \nu_i) = U_{\alpha i}^*$. Similarly, from (2.37) again, $A(\text{when } \nu_i \text{ interacts it makes } \ell_\beta) = U_{\beta i}$. Finally, to find $A(\nu_i \text{ propagates})$, we note that in the rest frame of ν_i, where the proper time is τ_i, Schrödinger's equation states that

$$i\frac{\partial}{\partial \tau_i}|\nu_i(\tau_i)\rangle = m_i|\nu_i(\tau_i)\rangle\,. \qquad (2.40)$$

This implies that [2.3]

$$|\nu_i(\tau_i)\rangle = e^{-im_i\tau_i}|\nu_i(0)\rangle . \tag{2.41}$$

Thus, for propagation over a proper-time interval τ_i,

$$A(\nu_i \text{ propagates}) \equiv \langle\nu_i(0)|\nu_i(\tau_i)\rangle = e^{-im_i\tau_i} . \tag{2.42}$$

Now, in terms of the time t and position L in the laboratory frame, the Lorentz-invariant phase factor $\exp(-im_i\tau_i)$ is

$$e^{-i(E_it-p_iL)} . \tag{2.43}$$

Here, E_i and p_i are the energy and momentum, respectively, of ν_i in the laboratory frame. Since, in practice, our neutrino will be highly relativistic, we shall be interested in evaluating the phase factor (2.43) where $t \cong L$, where it becomes

$$e^{-i(E_i-p_i)L} . \tag{2.44}$$

Suppose that our neutrino is born with a definite momentum p, so that it has this momentum regardless of which mass eigenstate it is. Then, if it is the particular mass eigenstate ν_i, it has an energy $E_i = \sqrt{p^2 + m_i^2} \cong p + m_i^2/2p$, assuming that all neutrino masses are small compared with p. From (2.42) and (2.44), we then have

$$A(\nu_i \text{ propagates}) \cong e^{-i(m_i^2/2p)L} . \tag{2.45}$$

Alternatively, imagine that our neutrino is produced with a definite energy E, so that it has this energy regardless of which mass eigenstate it is. [2.4] Then, if it is the particular mass eigenstate ν_i, it has a momentum $p_i = \sqrt{E^2 - m_i^2} \cong E - m_i^2/2E$. We then have

$$A(\nu_i \text{ propagates}) \cong e^{-i(m_i^2/2E)L} . \tag{2.46}$$

Since highly relativistic neutrinos have $E \cong p$, the propagation amplitudes given by (2.45) and (2.46) are essentially identical. Thus, it makes no difference whether our neutrino is created with definite momentum or definite energy.

Putting together the various factors in (2.39), we find that the amplitude $A(\nu_\alpha \to \nu_\beta)$ for a neutrino with energy E to undergo the oscillation $\nu_\alpha \to \nu_\beta$ while traveling a distance L is given by

$$A(\nu_\alpha \to \nu_\beta) = \sum_i U_{\alpha i}^* e^{-i(m_i^2L/2E)} U_{\beta i} . \tag{2.47}$$

The probability $P(\nu_\alpha \to \nu_\beta)$ for this oscillation is then given by [2.5]

$$P(\nu_\alpha \to \nu_\beta) = |A(\nu_\alpha \to \nu_\beta)|^2$$

$$= \delta_{\alpha\beta} - 4 \sum_{i>j} \text{Re}(U_{\alpha i}^* U_{\beta i} U_{\alpha j} U_{\beta j}^*) \sin^2\left(\delta m_{ij}^2 \frac{L}{4E}\right)$$

$$+ 2 \sum_{i>j} \text{Im}(U_{\alpha i}^* U_{\beta i} U_{\alpha j} U_{\beta j}^*) \sin\left(\delta m_{ij}^2 \frac{L}{2E}\right) . \tag{2.48}$$

Here, $\delta m_{ij}^2 \equiv m_i^2 - m_j^2$. In deriving (2.48), we have used the fact that the mixing matrix U is "unitary in the horizontal direction":

$$\sum_{i=1}^{2N} U_{\alpha i}^* U_{\beta i} = \delta_{\alpha\beta} \ . \qquad (2.49)$$

This follows from the definition of U (2.38) and the unitarity of $V_{L\ell}$ and Y_ν. Note that, in general, the matrix U may not be fully unitary, but it always obeys the constraint (2.49). Thus, we see from (2.47) that neutrino oscillation from one flavor to a different one requires neutrino mass. If all the masses m_i vanish, then $A(\nu_\alpha \to \nu_\beta) = \delta_{\alpha\beta}$.

The quantum mechanics of neutrino oscillation is both intriguing and subtle, and continues to be investigated [2.3, 2.4, 2.6].

We note from (2.48) for $P(\nu_\alpha \to \nu_\beta)$ that this "oscillation" probability does indeed have oscillatory behavior. If we include the so-far omitted factors of $\hbar$ and c in the quantity $\delta m_{ij}^2 L/4E$ on which the oscillating probabilities $P(\nu_\alpha \to \nu_\beta)$ depend, it becomes

$$1.27 \, \delta m_{ij}^2 \, (\text{eV}^2) \frac{L \, (\text{km})}{E \, (\text{GeV})} \ . \qquad (2.50)$$

Imagine, then, that some experiment is characterized by a certain value of $L \, (\text{km}) / E \, (\text{GeV})$. From (2.48), we see that this experiment will not observe oscillation probabilities departing appreciably from the "no-oscillation limit", $P(\nu_\alpha \to \nu_\beta) = \delta_{\alpha\beta}$, unless there is at least one $\delta m_{ij}^2 \, (\text{eV}^2)$ which is not small compared with $[L \, (\text{km}) / E \, (\text{GeV})]^{-1}$. However, by building experiments involving very large values of $L \, (\text{km}) / E \, (\text{GeV})$, we can achieve sensitivity to very small neutrino mass splittings. Indeed, some searches for neutrino oscillation are sensitive to much smaller neutrino masses than those to which any other experiments of any kind are sensitive.

Several simple examples of neutrino oscillation merit special mention. The simplest situation of all is two-neutrino oscillation. This occurs when the weak interaction of (2.37) couples two charged leptons of definite mass, say e and μ, to two neutrinos of definite mass, ν_1 and ν_2, but only negligibly to any other neutrinos of definite mass. It is then trivial to show that the 2×2 submatrix

$$\hat{U} = \begin{bmatrix} U_{e1} & U_{e2} \\ U_{\mu1} & U_{\mu2} \end{bmatrix} \qquad (2.51)$$

of the mixing matrix U must be unitary all by itself. This means that the neutrinos of definite flavor ν_e and ν_μ are made up exclusively of the mass eigenstates ν_1 and ν_2 and do not mix with neutrinos of any other flavor. Using the unitarity of $\hat{U}$, we find immediately from (2.48) and (2.50) that, for $\alpha\beta = e\mu$ or μe,

$$P(\nu_\alpha \to \nu_{\beta \neq \alpha}) = 4|U_{\alpha2}|^2|U_{\beta2}|^2 \sin^2 \left(1.27 \, \delta m_{21}^2 \, (\text{eV}^2) \frac{L \, (\text{km})}{E \, (\text{GeV})} \right) \ , \qquad (2.52)$$

and

$$P(\nu_\alpha \to \nu_\alpha) = 1 - 4|U_{\alpha 2}|^2(1 - |U_{\alpha 2}|^2)\sin^2\left(1.27\,\delta m_{21}^2\,(\text{eV}^2)\frac{L\,(\text{km})}{E\,(\text{GeV})}\right)$$
$$= 1 - P(\nu_\alpha \to \nu_\beta)\,. \tag{2.53}$$

The unitarity constraints on $\hat{U}$ imply that it can be written in the form

$$\hat{U} = \begin{bmatrix} e^{i\phi_1}\cos\theta & e^{i\phi_2}\sin\theta \\ -e^{i\phi_3+i\phi_1}\sin\theta & e^{i\phi_3+i\phi_2}\cos\theta \end{bmatrix}, \tag{2.54}$$

where θ is a mixing angle and the ϕ_i are phases. Thus, in (2.52) we may write $4|U_{\alpha 2}|^2|U_{\beta 2}|^2 = \sin^2 2\theta$, whereupon $P(\nu_\alpha \to \nu_{\beta \neq \alpha})$ takes the well-known form

$$P(\nu_\alpha \to \nu_{\beta \neq \alpha}) = \sin^2 2\theta \, \sin^2\left(1.27\,\delta m_{21}^2\,(\text{eV}^2)\frac{L\,(\text{km})}{E\,(\text{GeV})}\right)\,. \tag{2.55}$$

A second simple example of oscillation is one that may prove to be of great relevance in the real world. This is the case where the charged leptons e, μ, and τ are coupled by the weak interaction to three mass-eigenstate neutrinos ν_1, ν_2 and ν_3, but the mass-squared splitting between ν_1 and ν_2 is very small compared with that between the ν_1–ν_2 pair and ν_3 [2.7]. That is,

$$|\delta m_{21}^2| \equiv \delta m_{\text{Small}}^2 \ll |\delta m_{32}^2| \cong |\delta m_{31}^2| \equiv \delta m_{\text{Big}}^2\,. \tag{2.56}$$

Now, suppose an oscillation experiment has L/E such that $\delta m_{\text{Big}}^2\,L/E$ is of order unity, so that $\delta m_{\text{Small}}^2\,L/E \ll 1$. In this experiment, the $ij = 21$ term in (2.48) for $P(\nu_\alpha \to \nu_\beta)$ is negligible. In addition, the $ij = 32$ and 31 terms have identical L/E dependence, so they can be combined. Then, assuming that the 3×3 mixing matrix U that describes the $e\mu\tau$–$\nu_1\nu_2\nu_3$ coupling is unitary, we immediately find from (2.48) and (2.50) that, for $\alpha, \beta = e, \mu, \tau$,

$$P(\nu_\alpha \to \nu_{\beta \neq \alpha}) = 4|U_{\alpha 3}|^2|U_{\beta 3}|^2\sin^2\left(1.27\,\delta m_{\text{Big}}^2\,(\text{eV}^2)\frac{L\,(\text{km})}{E\,(\text{GeV})}\right)\,, \tag{2.57}$$

and

$$P(\nu_\alpha \to \nu_\alpha) = 1 - 4|U_{\alpha 3}|^2(1 - |U_{\alpha 3}|^2)\sin^2\left(1.27\,\delta m_{\text{Big}}^2\,(\text{eV}^2)\frac{L\,(\text{km})}{E\,(\text{GeV})}\right)$$
$$= 1 - \sum_{\beta \neq \alpha} P(\nu_\alpha \to \nu_\beta)\,. \tag{2.58}$$

We see that the probabilities for three-neutrino oscillation when two of the neutrinos are nearly degenerate, given by (2.57) and (2.58), are the same as those for two-neutrino oscillation, given by (2.52) and (2.53), except that the role of ν_2 in the two-neutrino case is played by ν_3 in the three-neutrino case. This behavior is very easy to understand. When the three-neutrino case is studied in an experiment with $\delta m_{\text{Small}}^2\,L/E \ll 1$, the mass splitting between ν_1 and ν_2 is invisible. Thus, in this experiment, there appear to be only two mass-eigenstate neutrinos: the effectively degenerate ν_1–ν_2 pair, which behaves like a single neutrino, and ν_3. Thus, it is no surprise that the probabilities for neutrino oscillation are the same as in the two-neutrino case.

2.4   Experimental Tests of the Majorana Nature of the Neutrino

Let us now discuss the key experimental tests of the Majorana nature of the neutrino. Any process that violates the total lepton number $(L_e + L_\mu + L_\tau)$ will be evidence in favor of the Majorana nature of the neutrino. The typical process that provides experimentally accessible evidence for the Majorana nature of ν_e is the nuclear process neutrinoless double beta decay (see Chap. 8 by P. Vogel in this book). At the nuclear level, the process involves the decay of a nucleus (A, Z) to the daughter nucleus $(A, Z \pm 2)$ accompanied by only two electrons (or positrons). At the level of nucleons, the process is $2n \rightarrow 2p+2e^-$, and at the quark level one has $d+d \rightarrow u+u+e^- +e^-$. The Feynman diagram responsible for this process involves two weak-interaction beta decay processes with the neutrinos annihilating each other via their mass. In the presence of general neutrino mixing, the light-neutrino contribution to the amplitude for neutrinoless double beta decay is given by

$$M_{\beta\beta 0\nu} \simeq G_F^2 \langle \frac{1}{k^2} \rangle \langle m_\nu \rangle_e \ , \tag{2.59}$$

where $\langle m_\nu \rangle_e = \sum_i U_{ei}^2 m_i$ and the sum over i covers the mass eigenstates. Clearly, if the electron neutrino is not mixed with any other neutrino, then $\langle m_\nu \rangle_e = m_{\nu_e}$. Furthermore, it is also easy to check that in the basis in which the charged-lepton mass matrix is diagonal, $\langle m_\nu \rangle_e$ is equal to the $\nu_e \nu_e$ entry of the mass matrix of the weak-eigenstate neutrinos; therefore, for any model for which $m_{\nu_e \nu_e} = 0$ in the diagonal charged-lepton basis, the light-neutrino contribution to the $\beta\beta 0\nu$ decay vanishes even though, in the mass eigenbasis, the eigenstates may be mixed with each other. The only exception to this rule arises when some of the mass eigenstates are heavier than the allowed energy in the double beta decay process, so that their contributions are exponentially suppressed. In this case, the sum above takes the form

$$\langle m_\nu \rangle_e = \sum_i U_{ei}^2 m_i \langle e^{-m_i r} \rangle_{\text{nucl.}} \ , \tag{2.60}$$

where the subscript "nucl." stands for averaging over nuclear states.

An analogous process for muons can test the Majorana nature of the muon neutrino, although in conjunction with other observations, neutrinoless double beta decay by itself may also shed light on the Majorana character of muon and tau neutrinos. The muonic analogue of neutrinoless double beta decay is the process $\mu^- + (A, Z) \rightarrow \mu^+ + (A, Z - 2)$. The experimental feasibility and theoretical expectations for this process were investigated in [2.8], where it was noted that the most suitable nucleus from considerations of energy and momentum conservation is ^{44}Ti, which has a half-life of 47 years. The expectation based on known extensions of the Standard Model, however, appear to give $(\mu^- \rightarrow \mu^+)/(\mu^- \rightarrow \nu_\mu) \sim 10^{-16}$, which may not be easily accessible.

Other processes that can also in principle test for the Majorana nature of the muon neutrinos are processes such as $K^+ \to \pi^- \mu^+ \mu^+$. The simplest Feynman diagram contributing to this process is the one mediated by the Majorana mass of the ν_μ, and it leads to an amplitude of order

$$M(K \to \pi\mu\mu) \sim G_{\rm F}^2 \sin\theta_{\rm c}\langle m_{\nu_\mu}\rangle_\mu \left\langle \frac{1}{k^2} \right\rangle \tag{2.61}$$

and therefore has a negligible branching ratio $\leq 10^{-20}$ for $m_{\nu_\mu} \simeq 100$ keV. On the other hand, if the muon neutrino is light, as is generally assumed in fitting neutrino data, and there are heavy sterile Majorana neutrinos that mix with the muon neutrino, then the limit on the branching ratio goes down to the level of 10^{-25} [2.9].

A point worth remembering is that any theory that leads to lepton-number-violating processes will necessarily lead to a Majorana mass for neutrinos regardless of whether the process receives a dominant contribution from Feynman diagrams involving the Majorana mass of the corresponding neutrino [2.10].

2.5   Neutrino Decays

Once neutrinos are known to be massive, a natural question that arises is, are they stable, and if they are not, what are their lifetimes and what are the implications in other areas of particle physics?

Massive neutrinos of one species can decay into another, lighter species plus a light (or massless) boson, if an appropriate flavor-changing interaction is present in the Lagrangian. While there is no evidence for any such decay processes, they have been invoked in many particle physics as well as cosmological contexts. There are two choices for the lighter boson: a photon and a hypothetical particle called a "majoron" [2.11]. In this section, we give a brief outline of some of these possibilities, accompanied by an overview of possible consequences.

2.5.1   Radiative Neutrino Decays

The simplest decay process that involves only known particles is $\nu_i \to \nu_j + \gamma$. Such a process can actually be generated as a loop effect if the neutrinos of the Standard Model are given a mass and they are allowed to mix [2.12]:

$$T = \bar{u}'(\boldsymbol{p}')\Gamma_\lambda u(\boldsymbol{p})\epsilon^{*\lambda}\,, \tag{2.62}$$

where ϵ^λ is the photon polarization. For off-shell photons, the most general form for Γ_λ contains four form factors. However, for physical photons the on-shell condition and the Lorentz gauge condition read

$$q^2 = 0\,, \quad \epsilon^\lambda q_\lambda = 0\,, \tag{2.63}$$

so that the most general form for Γ_λ is given by

$$\Gamma_\lambda = \left[F(q^2) + F_5(q^2)\gamma_5\right]\sigma_{\lambda\rho}q^\rho , \tag{2.64}$$

where F and F_5 are Lorentz-invariant form factors, whose values are model-dependent. They are often called the transition magnetic and electric dipole moments between the two neutrinos involved. If the initial and final neutrinos are Majorana particles and the neutrino interactions contributing to this decay conserve CP, then two interesting constraints follow: (i) if the initial and final states have the same CP, one has $F = 0$, and if they have opposite CP, then $F_5 = 0$.

The decay rate of ν_i can be obtained in a straightforward manner from the matrix element given above. In the rest frame of the decaying neutrino, one obtains

$$\Gamma = \frac{(m_i^2 - m_j^2)^3}{8\pi m_i^3}\left(|F|^2 + |F_5|^2\right) . \tag{2.65}$$

One can express the effective coupling of (2.65) in units of the Bohr magneton as follows:

$$\kappa_{ij} \equiv \left(\frac{e}{m_{\nu_i} + m_{\nu_j}}\right)\sqrt{|F|^2 + |F_5|^2} \equiv \kappa_{0,ij}\mu_B . \tag{2.66}$$

Upper limits on $\kappa_{0,ij}$ can be derived from neutrino scattering experiments (for a recent review, see [2.13]):

$$\kappa_{0,e} \leq 10^{-10} ,$$
$$\kappa_{0,\mu} \leq 7.4 \times 10^{-10} ,$$
$$\kappa_{0,\tau} \leq 5.4 \times 10^{-7} . \tag{2.67}$$

The above upper limits, in turn, imply lower limits on the radiative lifetimes of heavy neutrinos. Assuming $m_{\nu_H} \sim \text{eV} \gg m_{\nu_L}$, they imply the limits

$$\tau_{\nu_e} \geq 5 \times 10^{18} \text{ s} ,$$
$$\tau_{\nu_\mu} \geq \times 10^{17} \text{ s} ,$$
$$\tau_{\nu_\tau} \geq 2 \times 10^{11} \text{ s} . \tag{2.68}$$

The limits in (2.67) and (2.68) are based on the assumption of small neutrino mixing so that the flavor and mass eigenstates are nearly identical, although more general statements can be easily derived. The theoretical expectations for F and F_5 are model-dependent. For instance, in the Standard Model with Majorana neutrinos,

$$\Gamma_\lambda =$$
$$-\frac{eG_F}{4\sqrt{2}\pi^2}\sigma_{\lambda\rho}q^\rho \sum_\alpha \left[U_{\alpha i}U_{\alpha j}^*(m_i\mathbf{R} + m_j\mathbf{L}) - U_{\alpha i}^*(m_i\mathbf{L} + m_j\mathbf{R})\right] f(r_\alpha) ,$$
$$\tag{2.69}$$

where $r_\alpha = (m_\alpha/m_W)^2$ (m_α is the mass of the charged lepton of the αth generation) and $f(r_\alpha) \sim -3/2$ for small values of r_i, as is the case. Clearly, if $i = j$, the two terms in the square brackets cancel each other, showing that the dipole moments of a Majorana neutrino vanish, as already discussed. The radiative decay width of the neutrino can be written as

$$\Gamma(\nu_i \to \nu_j + \gamma) \simeq \frac{\alpha G_F^2}{64\pi^4} \left(\frac{m_i^2 - m_j^2}{m_i}\right)^3 (m_i - m_j)^2 \left|\sum_\alpha U_{\alpha j} U_{\alpha i}^* f(r_\alpha)\right|^2 .$$

(2.70)

Clearly, for neutrino masses in the eV range, this leads to a lifetime of order 10^{28} years. The lifetimes become considerably shorter in the context of left–right models with right-handed W's in the TeV range [2.1]. One generally expects an enhancement of the lifetime by a factor of $m_\alpha^2/m_{\nu_i}^2 \, (M_{W_L}/M_{W_R})^4$. This can lower the lifetimes by eight to ten orders of magnitude depending on the choice of neutrino generation.

2.5.2   Invisible Decays

In extensions of the Standard Model, it is possible to contemplate invisible decays of neutrinos. There are two distinct invisible decays of interest: (i) $\nu_i \to \nu_p \nu_j \nu_k$ and (ii) $\nu_i \to \nu_j + \phi$ where ϕ is an ultralight scalar such as the majoron or the familon. We discuss these decays below.

Three-Neutrino Decays. These decays could arise in the presence of flavor-changing leptonic neutral currents or in the presence of "exotic" Higgs bosons such as those present in the left–right-symmetric models of weak interactions [2.14]. Typically, the decay rate for such processes is given by the formula

$$\Gamma(\nu_i \to 3\nu) \simeq \frac{G^2}{192\pi^3} m_{\nu_i}^5 ,$$

(2.71)

where $G \simeq G_F \sin\theta$ for models with flavor-changing Z-mediated neutral leptonic currents (θ denotes the neutrino mixing angle), and $G = f_\Delta^2/M_\Delta^2$ in models with B − L-carrying Higgs bosons such as the ones used in the left–right-symmetric models. For an MeV heavy neutrino, the first mechanism leads to a neutrino lifetime of the order of 1000 s. for maximal values of α. From a cosmological point of view, such long lifetimes are problematic for understanding the success of Big Bang nucleosynthesis. However, the fifth-power dependence of the decay rate makes it clear that a 10 MeV tau neutrino will have a lifetime of order 10^{-2} s, making it cosmologically viable. This is of practical interest, considering the collider upper bounds on the ν_τ mass of 18 MeV [2.15]. In this discussion, we are disregarding the prevalent interpretation of the atmospheric neutrino data.

As far as the second mechanism is concerned, if we choose the mass of the Δ Higgs boson to be less than 100 GeV, which is completely consistent with all known collider data, similar arguments imply that consistency with Big Bang nucleosynthesis can be satisfied for $m_{\nu_\tau} \geq 2$ MeV.

Majoron Decay Modes. If the $B-L$ symmetry is a spontaneously broken global symmetry [2.11, 2.16], this implies the existence of a massless particle, the majoron, which is naturally coupled to the neutrinos. With simple modification of the simplest models [2.17], one can have flavor-changing couplings of type $\nu_i \to \nu_j + J$ (where J denotes the majoron). The decay rate for this process is approximately

$$\Gamma(\nu_i \to \nu_j + J) \simeq \frac{m_{\nu_H}^3}{16\pi v_{B-L}^2} , \tag{2.72}$$

where the scale v_{B-L} corresponds to the scale at which $B-L$ symmetry breaks. If this scale is in the TeV range, it can lead to lifetimes of less than a second for $m_{\nu_H} \sim 1$ keV or so. This considerably extends the range of masses which can be cosmologically acceptable.

In the case of fast neutrino decays, the neutrino oscillation formulae undergo an interesting modification. To illustrate the nature of this modification, let us consider a two-generation model with the mixing angle given by θ. Let us express the weak (flavor) eigenstates $\nu_{\alpha,\beta}$ in terms of the mass eigenstates $\nu_{1,2}$ as follows:

$$\nu_\alpha = \nu_1 \cos\theta + \nu_2 \sin\theta ,$$
$$\nu_\beta = -\nu_1 \sin\theta + \nu_2 \cos\theta . \tag{2.73}$$

Then, if ν_2 is unstable, the survival probability of ν_α is given by [2.18]:

$$P_{\alpha\alpha} = \sin^4\theta + \cos^4\theta e^{-m_2 L/\tau E} + 2\sin^2\theta \cos^2\theta e^{-m_2 L/2\tau E} \cos(\delta m^2 L/2E). \tag{2.74}$$

It is therefore clear that the generic oscillation formulae are modified in a way that allows for new neutrino phenomenology. In the limit of large δm^2, the last term in the above equation averages out to zero and one has

$$P_{\alpha\alpha} = \sin^4\theta + \cos^4\theta e^{-m_2 L/\tau E} , \tag{2.75}$$

whereas, in the limit of small δm^2, one has

$$P_{\alpha\alpha} = (\sin^2\theta + \cos^2\theta e^{-m_2 L/2\tau E})^2 . \tag{2.76}$$

Similarly, the transition probability to another species of neutrino is given by

$$P_{\alpha\beta} = \cos^2\theta \sin^2\theta \left(1 + e^{\frac{-m_2 L}{2E\tau}} - 2\cos\frac{\delta m^2 L}{2E} e^{\frac{-m_2 L}{2E\tau}} \right) . \tag{2.77}$$

Armed with these modifications of the conventional oscillation formulae, several groups have attempted to see if decaying neutrinos could be playing a role in neutrino observations. In particular, could a decaying neutrino be responsible for the missing neutrinos in solar [2.19] and atmospheric [2.20] data? One of the clear-cut statements that can be made is that, in the absence of mixings, solar-neutrino data cannot be explained by neutrino decay, because if neutrinos did decay within distances of order of the earth–Sun distance, they should have certainly decayed on their way from the supernova 1987A. But, apparently, the full spectrum of the ν_es from SN1987A was observed in the Kamiokande and IMB detectors. In the presence of mixings, however, the situation changes, and attempts have been made to fit observed solar, atmospheric and LSND data using three unstable neutrinos. However, it appears that it is impossible to fit all data with three neutrinos; on the other hand, if one includes a sterile neutrino, one can provide new ways to understand the neutrino data without ascribing all neutrino deficits to conventional oscillations [2.20].

Before closing this section, we note that the final state in an invisible decay could involve another hypothetical particle, the familon [2.21]. However, in the simplest familon models, the familon arises from an iso-singlet Higgs boson; therefore, one expects flavor-changing invisible decays involving charged leptons such as $\mu \to e + f$, $\tau \to (e, \mu) + f$, which are very strongly constrained by present experiments; for example, current bounds on the branching ratios for these processes are at the level of 2–7×10^{-6}. This implies familon decay lifetimes for eV mass neutrinos higher than 10^{24} s for ν_μ and 10^{27} s for ν_τ. These lifetimes are clearly much longer than those for the majoron mode and of less interest for cosmological discussions.

To summarize, we have presented an overview of some of the properties of massive neutrinos and related experimental tests which can not only help us better understand the nature of the neutrino but can also provide important clues to new physics beyond the Standard Model.

The work of R. N. M. is supported by the National Science Foundation grant no. PHY-9802551.

References

2.1 P. Langacker, Phys. Rep. **72**, 185 (1981); B. Kayser, F. Gibrat-Debut and F. Perrier, *Massive Neutrinos* (World Scientific, Singapore, 1989); R. N. Mohapatra and P. B. Pal, *Massive Neutrinos in Physics and Astrophysics* (World Scientific, Singapore, 1998, second edition); M. Zralek, hep-ph/9711506.

2.2 Z. Maki, M. Nakagawa and S. Sakata, Prog. Theor. Phys. **28**, 870 (1962).

2.3 B. Kayser and L. Stodolsky, Phys. Lett. B **359**, 343 (1995); Y. Srivastava, A. Widom and E. Sassaroli, Z. Phys. C **66**, 601 (1995).

2.4 Y. Grossman and H. Lipkin, Phys. Rev. D **55**, 2760 (1997); H. Lipkin, Phys. Lett. B **348**, 604 (1995).

2.5 C. Albright et al., Fermilab preprint FERMILAB-FN-692.

2.6 B. Kayser, Phys. Rev. D **24**, 110 (1981); C. Giunti, C. Kim and U. Lee, Phys. Rev. D **44**, 3635 (1991); T. Goldman, hep-ph/9604357; F. Boehm and P. Vogel, *Physics of Massive Neutrinos* (Cambridge University Press, Cambridge, 1987) p. 87.

2.7 P. Fisher, B. Kayser and K. McFarland, Annu. Rev. Nucl. Part. Sci. **49**, 481; B. Kayser, a mini-review of neutrino mass in D. Groom et al. (Particle Data Group) Euro. Phys. J. C **15**, 1 (2000).

2.8 J. Missimer, R. N. Mohapatra and N. Mukhopadhyay, Phys. Rev. D **50**, 2067 (1997); M. Flanz, J. Rodejohan and K. Zuber, Phys. Lett. B **473**, 321 (2000); K. Zuber, hep-ph/0003160.

2.9 L. Littenberg and R. Schrock, hep-ph/0005285.

2.10 J. Schecter and J. W. F. Valle, Phys. Rev. D **25**, 2951 (1982); B. Kayser, S. Petcov and P. Rosen, unpublished.

2.11 Y. Chikashige, R. N. Mohapatra and R. D. Peccei, Phys. Rev. Lett. **45**, 1926 (1980).

2.12 P. B. Pal and L. Wolfenstein, Phys. Rev. D **26**, 766 (1982).

2.13 S. Pakvasa, hep-ph/0004077.

2.14 J. C. Pati and A. Salam, Phys. Rev. D **10**, 275 (1974); R. N. Mohapatra and J. C. Pati, Phys. Rev. D **11**, 566, 2558 (1975); G. Senjanović and R. N. Mohapatra, Phys. Rev. D **12**, 1502 (1975); R. N. Mohapatra and G. Senjanović, Phys. Rev. Lett. **44**, 912 (1980).

2.15 Particle Data Group, Eur. Phys. J. C **3**, 1 (1998).

2.16 G. Gelmini and M. Roncadelli, Phys. Lett. B **99**, 411 (1981).

2.17 J. Valle, Phys. Lett. B **131**, 87 (1983); A. Kumar and R. N. Mohapatra, Phys. Lett. B **150**, 191 (1985); G. Gelmini and J. W. F. Valle, Phys. Lett. B **142**, 181 (1983); K. Choi and A. Santamaria, Phys. Lett. B **267**, 504 (1991); A. Joshipura and S. Rindani, Phys. Rev. D **48**, 300 (1992).

2.18 J. Frieman, H. Haber and K. Freese, Phys. Lett. B **200**, 115 (1988).

2.19 S. Pakvasa and K. Tennakone, Phys. Rev. Lett. **28**, 1415 (1972); J. N. Bahcall, N. Cabibbo and A. Yahil, Phys. Rev. Lett. **28**, 316 (1972).

2.20 V. Barger, J. Learned, P. Lipari, M. Lusignoli, S. Pakvasa and T. Weiler, Phys. Lett. B **462**, 109 (1999).

2.21 F. Wilczek, Phys. Rev. Lett. **49**, 1549 (1982).

3 Direct Measurements of Neutrino Mass

J. F. Wilkerson and R. G. H. Robertson

3.1 Introduction

In the seventy years since Pauli's brilliant conjecture that neutrinos must exist, our understanding of neutrino properties has proven to be remarkably limited. There are two key questions that have important consequences at the most basic level of our understanding of neutrinos. Do neutrinos have nonzero masses? And if so, what are their mass values? The answers to these questions are important both for our theoretical understanding of fundamental interactions of nuclei and particles and because the answers have significant astrophysical and cosmological implications.

In the current theoretical Standard Model framework of nuclear and particle physics, the neutrinos are presumed to consist of left-handed neutrinos and right-handed antineutrinos, all having zero rest mass. However, there is no compelling theoretical motivation underlying this assumption. In fact, all theoretical extensions to the Standard Model which have the goal of unifying the strong and electroweak interactions predict that neutrinos will have nonzero rest masses. There will be significant astrophysical and cosmological ramifications if neutrinos have mass. For example, from Big Bang cosmology we believe that the Universe is populated by a significant density (110 neutrinos/(flavor cm^3)) of relic neutrinos. On the basis of current astrophysical constraints, including recent measurements of the cosmic microwave background, it is possible that neutrinos could account for as much as one third of the dark matter that is observed to influence the dynamical behavior of the universe. From the results of atmospheric neutrino oscillation experiments, neutrinos contribute at least as much mass as do visible stars. Neutrinos also play a crucial role in stellar processes such as supernova explosions and the formation of heavy elements. (See Chap. 2 for details of the theoretical aspects of neutrinos and Chaps. 11 and 12 for the astrophysical and cosmological implications.)

Experimental techniques that probe the question of neutrino mass are classified into two categories, direct and indirect. The direct measurements are accomplished either through precise observations of decay kinematics in nuclear or particle decay processes, or by the utilization of time-of-flight measurements in the detection of supernova neutrinos incident on terrestrial neutrino detectors. Examples of decay measurements include studies of the shapes of β decay spectra, measurements of muon momentum in pion decay, and invariant-mass studies of multiparticle semileptonic decays of the τ. Limits on neutrino masses have also been derived from observation of neutrinos from Supernova 1987A, and a number of detectors are currently being used in a global supernova watch with the hope of making improved measurements by observing neutrinos from any supernova that might occur within our galaxy. A strength of direct techniques is that they make few, if any, theoretical assumptions about neutrino properties, since the measurements are based on purely kinematic observables. At present, the most precise direct measurements of neutrino mass yield only upper limits for the neutrino masses and hence do not answer the question of whether or not neutrinos have mass.

Indirect methods, such as searches for neutrino oscillations and measurements of neutrinoless double beta decay, require both that neutrinos have nonzero mass and, additionally, that lepton family violation or total-lepton-number nonconservation occurs. Oscillation measurements offer the most sensitive method of probing neutrino mass, but they measure only the squares of the mass differences. Thus, although they can tell us that neutrinos have mass, they can say nothing about the actual mass eigenvalues. There is recent compelling evidence from neutrino oscillation experiments that the answer to our first question, do neutrinos have mass, is yes. This is based on a clear signature of neutrino oscillations from atmospheric neutrino measurements [3.1]. There is also evidence consistent with oscillations from five independent solar-neutrino experiments [3.2]. The other process that can be used for an indirect method of probing for neutrino mass, neutrinoless double beta decay, has not yet been observed to occur. Hence these measurements only set conditional limits on neutrino masses, with the caveat that such limits are only meaningful if neutrinos are Majorana particles. Details about these indirect measurements of neutrino mass can be found in Chaps. 4–8.

3.2   Overview of Direct Measurements

During the past forty years there has been steady progress in probing neutrino masses through direct measurements of decay kinematics. Figure 3.1 presents a summary of these results. Note that in the figure the neutrinos are labeled according to their flavor eigenstates, and that, with the exception of a 1980 result reported by Lubimov's group of a 35 eV electron antineutrino mass [3.3], all points represent upper limits on their respective neutrino masses. The evidence from oscillation experiments that neutrinos are mixed complicates the question of what exactly one means by neutrino mass, since the neutrino flavor eigenstates (ν_e, ν_μ and ν_τ) will be admixtures of the actual mass eigenstates (ν_1, ν_2 and ν_3).

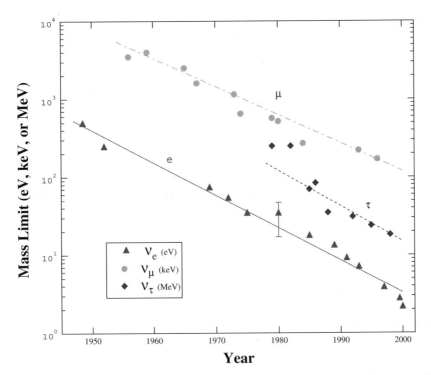

Fig. 3.1. History of direct neutrino mass measurements. *Points without error bars* represent upper limits on the mass values, while the *one point with error bar* represents a nonzero mass result, ruled out by subsequent measurements

Direct measurements take the form of the measurement of a transition rate, differential in energy, from an initial unstable state $|I\rangle$ to a multiparticle final state that includes a neutrino. Writing the transition rate in terms of the matrix element of an operator T,

$$\frac{\mathrm{d}N}{\mathrm{d}E} = |\langle l|\langle \nu_l|T|I\rangle|^2 = \left|\sum_i U_{li}^*\langle l|\langle \nu_i|T|I\rangle\right|^2 , \qquad (3.1)$$

one sees that, in kinematic regimes where the neutrino mass is irrelevant and the matrix elements are the same, the original flavor-basis result is returned since $\mathbf{U}$ is unitary. On the other hand, in a kinematic regime where only the k matrix element is nonzero (the other neutrinos being too heavy), then the rate is proportional to U_{lk}^2. In between, a more complicated situation occurs. The present evidence from oscillation experiments is that the mass differences are all very small on the scale of what is experimentally accessible to direct measurements, and the matrix elements are therefore effectively the same

Table 3.1. Limits on neutrino masses from direct measurements

Neutrino type	PDG evaluation (1998) [3.4]	Recent results	Group	Ref.
$\overline{M_{\nu_e}}$	< 15 eV (No C.L.)	< 2.2 eV (95% C.L.)	Mainz 2000	[3.5, 3.6]
$\overline{M_{\nu_\mu}}$	< 190 keV (90% C.L.)	< 170 keV (90% C.L.)	PSI 1996	[3.7]
$\overline{M_{\nu_\tau}}$	< 18.2 MeV (95% C.L.)	< 18.2 MeV (95% C.L.)	ALEPH 1998	[3.8]

over the full kinematic regime. The mass limits then are limits on a weighted average mass that can be defined as

$$\overline{M_{\nu_l}} = \sqrt{\sum_{i=1}^{k} |U_{li}|^2 m_{\nu_i}^2} \ . \tag{3.2}$$

Here $l = \{e, \mu, \text{or } \tau\}$, U_{li} are the amplitudes in the Maki–Nakagawa–Sakata lepton mixing matrix, and the sum over k includes all mass eigenstates that are kinematically allowed for a particular measurement. Note that if there are more than three neutrino families, for example if sterile neutrinos should exist, the above 3×3 matrix is simply replaced by the appropriate $N \times N$ matrix. The notation $\overline{M_{\nu_l}}$ will be used throughout this chapter, with minimal assumptions about the exact nature of the lepton-mixing-matrix values. The summaries below make no distinction between masses of neutrinos and antineutrinos, since we assume the validity of CPT invariance. But in discussing details of experimental measurements, the type of neutrino, ν or $\bar{\nu}$, will be explicitly indicated. The current limits on neutrino masses, as recommended by the Particle Data Group (PDG) [3.4], are given in Table 3.1. Also included in the table are the most precise limits that have been set by direct techniques to date. Differences between the PDG recommendations and the recent results will be discussed in later sections.

3.3   Beta Decay and Electron Capture Measurements (ν_e)

The concept of direct mass searches based on decay kinematics can be traced directly back to Pauli and Fermi. In 1932 Pauli proposed the neutrino as a solution to preserve conservation laws during beta decay [3.9]. On the basis of the experimental observations, he postulated that the neutrino must be a neutral particle with a small or zero rest mass. Shortly thereafter, Fermi, in developing his beta decay theory, showed that the neutrino must have a rest mass smaller than that of the electron [3.10]. He also realized that a measurement of the shape of the beta decay energy spectrum could provide a

Table 3.2. Limits on $\overline{M_{\nu_e}}$ from tritium beta decay measurements

Experiment	Measurement apparatus	Source	$\overline{M_{\nu_e}}^{2}$ (eV^2)	Limit (95% C.L.)	Year
Mainz University	Retarding	T$_2$ solid	$-1.6 \pm 2.5 \pm 2.1$	< 2.2	2000 [3.5, 3.6]
INR, Troitsk	Retarding	T$_2$ gas	$-1.9 \pm 3.4 \pm 2.2$	(< 2.5)	1999 [3.22]
Livermore	Toroidal magnet	T$_2$ gas	$-130 \pm 20 \pm 15$	$-$	1995 [3.23]
Zürich	Toroidal magnet	OTS-T	$-24 \pm 48 \pm 6$	< 11.7	1992 [3.18]
Los Alamos	Toroidal magnet	T$_2$ gas	$-147 \pm 68 \pm 41$	< 9.3	1991 [3.17]
INS, Tokyo	$\pi\sqrt{2}$ magnet	Cd ARC-T	$-65 \pm 85 \pm 65$	< 13.1	1991 [3.16]
ITEP, Moscow	Toroidal magnet	Valine-T	$+676 \pm 235$	$26\,^{+6}_{-5}$	1987 [3.15]

sensitive test of the mass of the electron antineutrino. In 1948 Curran, Angus and Cockroft [3.11, 3.12] made the first such shape measurement. Using a simple proportional counter, this group measured the electron energies for tritium beta decay in the endpoint energy region, where the electron carries away the maximum possible energy, and set a limit of $\overline{M_{\nu_e}} < 1$ keV [3.13]. Because of its low endpoint energy (18.6 keV), superallowed decay, simple nuclear properties and simple atomic structure, tritium has remained the nucleus of choice in electron antineutrino mass measurements via beta decay.

By the early 1970s the mass limit had been reduced to 55 eV by the seminal work of Bergkvist [3.14]. Then, in 1980, the physics community's attention was suddenly captured by the report of Lubimov and collaborators at the Institute for Theoretical and Experimental Physics (ITEP) in Moscow that the electron antineutrino had a mass of 35 eV [3.3]. This announcement spurred nearly two dozen groups around the world to embark on attempts to verify the ITEP measurement. After a decade of work, the revised ITEP report of $\overline{M_{\nu_e}} = 26$ eV [3.15] was clearly refuted by three independent groups [3.16–3.18]. During the past decade several groups have continued to make significant progress in pushing the sensitivity of tritium beta decay to the level of a few eV. Table 3.2 gives a summary of recent measurements.

In principle, electron capture decay provides an alternative and attractive means to measure neutrino mass. Capture channels with Q-values comparable to atomic electron binding energies can be hindered or blocked if the neutrino mass is appreciable. Photon spectroscopy techniques can be used to measure the probability distribution of shells left vacant in the primary capture. Unfortunately, rearrangement processes analogous to shake-up and shake-off make the interpretation of such spectra extremely complex, and consequently electron capture is no longer used in neutrino mass measurements.

Beta decays of nuclei and of the pion have also been used [3.4] to search for heavier-mass neutrinos than are currently considered probable. A flurry of excitement over an apparent 17 keV neutrino ended when it was traced to scattering effects, but it was a salutary demonstration of the sensitivity of mass determinations to small systematic effects [3.19].

Detailed reviews describing the methods and history of the earlier searches for neutrino mass can be found in the literature [3.20, 3.21].

3.3.1   Method of Measurement of the Beta Decay Endpoint

The technique for detecting neutrino mass by means of beta decay is essentially to search for a distortion in the shape of the beta spectrum in the endpoint energy region. The spectrum can be described using the Fermi form for the probability of emitting an electron with total energy E in beta decay:

$$\mathrm{d}N(E) = K|M|^2 F(Z,R,E)p_e E(E_0 - E)\sqrt{(E_0 - E)^2 - \overline{M_{\nu_e}}^2 c^4}\,\mathrm{d}E \;,$$

$$(3.3)$$

where

$$K = G_F^2 \frac{m^5 c^4}{2\pi^3 \hbar^7} \cos^2(\theta_c),$$

$$(3.4)$$

G_F is the weak coupling constant, M is the nuclear matrix element for the transition (for tritium $|M|^2 = 5.55$), p_e is the electron momentum, E_0 is the endpoint energy (the maximum electron energy for zero neutrino mass), m is the mass of the electron, and $\overline{M_{\nu_e}}$ is the mass of the electron antineutrino. The Fermi function, $F(Z,R,E)$, is a Coulomb correction term that results from the influence of the nuclear charge on the wave function of the emitted electron. The spectrum can be linearized by plotting $\sqrt{N(E)/Ep_e F(Z,R,E)}$ versus E. This form, known as the Kurie amplitude, is particularly useful for seeing the effect of the neutrino mass.

As one can see in Fig. 3.2, a plot of the beta spectrum expected from tritium, the spectrum shape is only sensitive to the neutrino mass in the energy region of the beta spectrum from a few m_ν below the endpoint to the endpoint energy of the decay. If the neutrino has a mass, than the number of decays in this region will be less than that expected for the case of a massless neutrino. Note that, in fitting the measured shape to a calculated shape, the functional form depends on m_ν^2 and is not defined for $m_\nu^2 < 0$. One immediate difficulty apparent in the figure is that there are very few decays in the region of interest (only 3×10^{-11} of the total rate is in the last 5 eV). A related problem is that one must carefully eliminate or minimize backgrounds. The other serious problem is that subtle effects such as instrumental resolution, energy loss of the electrons and atomic-physics excitations during the decay can cause shape distortions that are of similar size to the effect of nonzero neutrino mass but of opposite sign. Thus, precise determination of a value or limit for $\overline{M_{\nu_e}}$ requires complete and accurate understanding of all systematic effects that can alter the shape of the spectrum.

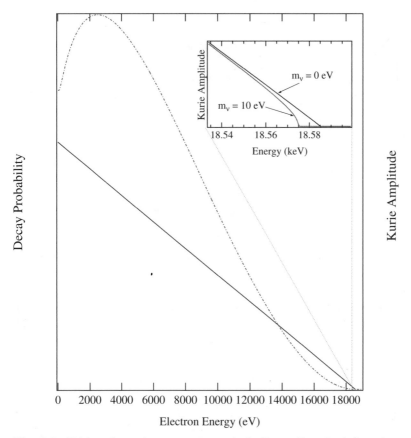

Fig. 3.2. Tritium beta decay spectrum, including a linearized form known as a Kurie plot. The *inset* shows a blowup in the region of interest and assumes a neutrino mass of 10 eV. Note that no atomic corrections have been applied

Two important modifications to the beta spectrum occur because of the presence of atomic electrons. The Fermi function must be corrected for electrons screening the nuclear charge, and the probabilities of exciting atomic states during the beta decay process must be accounted for and included in the calculations. The screening corrections are significant at low electron energies, those less than 2 keV, but are less important and are well understood at higher energies [3.20, 3.24]. Atomic final-state effects require extensive knowledge of the allowed states of the decaying daughter atom or molecule and dramatically influence the observed beta spectrum. The modified decay probability must be written as a sum over all possible atomic or molecular final states:

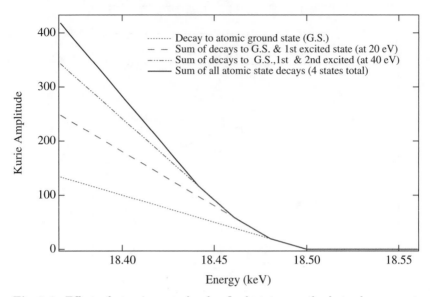

Fig. 3.3. Effect of atomic or molecular final states on the beta decay spectrum. There are four final states here, of equal probability, at excitation energies of 0, 20, 40, and 60 eV

$$\mathrm{d}N(E) = K|M|^2 \sum_i w_i F(Z, R, E) p_e E(E_0 - E - V_i)$$

$$\times \sqrt{(E_0 - E - V_i)^2 - \overline{M_{\nu_e}}^2 c^4}\ \mathrm{d}E\ , \tag{3.5}$$

for

$$E \le E_0 - V_i - \overline{M_{\nu_e}} c^2\ .$$

Here V_i is the excitation energy of the ith state in the daughter atom or molecule and w_i is the probability of a transition to the final state i. In Fig. 3.3 a hypothetical decay spectrum is shown that includes four atomic states, populated with equal probability and separated by 20 eV from each other. Notice that the effect on the spectrum shape is opposite to the effect that a neutrino mass would produce.

3.3.2 Tritium Beta Decay Experiments

Recent tritium beta decay experiments have utilized either magnetic spectrometers or gridless solenoidal retarding electrostatic spectrometers to analyze the momentum of the electrons. The magnetic spectrometers, which measure the differential beta spectrum, are relatively insensitive to backgrounds, but have a less favorable relationship of resolution to luminosity (see Sect. 3.3.6). Thus, most experiments using magnetic spectrometers have

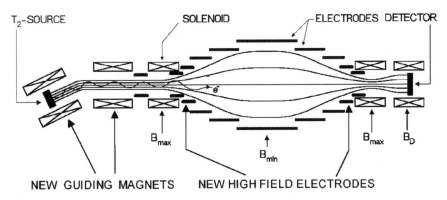

Fig. 3.4. Schematic drawing of the Mainz solenoidal retarding spectrometer [3.25]

been forced to accept a modest energy resolution on the order of 10–20 eV (FWHM). The solenoidal retarding electrostatic spectrometers measure an integral beta spectrum, integrating all energies above the acceptance energy of the spectrometer. They differ from pure electrostatic spectrometers in that the electrons are guided along the field lines of an inhomogeneous magnetic field as they traverse the spectrometer. A schematic drawing of the spectrometer used by the Mainz group is shown in Fig. 3.4. These spectrometers are able to achieve excellent energy resolution, on the order of a few eV, without loss of source acceptance. Initially these systems were limited by backgrounds, because the electrons of interest are decelerated to essentially zero energy at the center of the spectrometer. However, both the Mainz and the Troitsk groups have successfully managed to overcome these difficulties, and currently these groups have set the most precise limits on $\overline{M_{\nu_e}}$.

Extracting a neutrino mass from a measured spectrum requires the use of a calculated set of atomic or molecular final states and fitting the expected functional form with the following free parameters: neutrino mass squared, endpoint energy, amplitude and background. For the case of magnetic spectrometers, (3.5) is modified by multiplying it by additional linear and quadratic efficiency terms,

$$1 + \alpha_1(E_0 - E) + \alpha_2(E_0 - E)^2 \,, \tag{3.6}$$

which are either fitted or calculated. For both spectrometer types, the theoretical spectrum must be convoluted with the detector resolution function and with a function that accurately represents energy loss processes in the source.

3.3.3 Sensitivity to Shape Effects

Precisely characterizing all possible processes that modify the shape of the observed beta spectrum is the key to making an accurate measurement of neutrino mass. We know from the example of comparing the ITEP result with recent tritium measurements that two statistically significant results can be in total disagreement. It is also important to recognize that as one attempts to increase the sensitivity and measure smaller values of $\overline{M_{\nu_e}}$, the total uncertainty budget varies as the square of the neutrino mass sensitivity. This budget is the sum of both statistical and systematic uncertainties. Thus, while a 10 eV mass limit has a total uncertainty budget of 100 eV^2, a measurement with 2 eV sensitivity has a total uncertainty budget of 4 eV^2.

There are a four primary processes that contribute to modifying the shape of the observed beta spectrum: atomic final-state effects, energy loss, instrumental resolution and energy-dependent efficiencies of the analyzing system. All of these shape effects must be correctly accounted for before one can extract a neutrino mass from the beta spectrum. Because the elimination of most of these systematic shape effects is impossible, one must minimize and accurately account for them in a model independent manner where possible. Usually this means that one should attempt to perform a measurement of the particular shape-modifying process.

Atomic Final-state Effects. As discussed earlier, one must account for atomic-physics corrections that arise when the decaying tritium atom decays to an excited state in the ^{3}He$^+$ ion, or for the molecule when T$_2$ decays to the ^{3}He–T$^+$ molecular ion. Extracting a reliable value for the neutrino mass requires either a precise knowledge of the branching ratios and energies for all possible final states or that one is able to make high-resolution measurements in the limited energy region above the region where excited states contribute to the decay spectrum. At present all groups rely on theoretical calculations for estimating these effects. To quantify the magnitude of these effects, one can simply note that going from the simplest source of atomic tritium to the a source of molecular tritium produces a change of over 200 eV^2 in the value of $\overline{M_{\nu_e}}^2$. The only sources for which the decay probabilities can be accurately calculated are for atomic tritium and molecular tritium [3.26–3.28], where the uncertainties are estimated to be about 1 eV^2 for a fitting range covering 350 eV. (See [3.20] for an earlier discussion of final-state calculations for atomic and molecular tritium.)

For more complex sources, exact calculations are impossible, and certain simplifications and assumptions that introduce model-dependent uncertainties are required. The complicated amino-acid-based valine source used by the ITEP group offers a telling example of possible model dependences. The calculation of the variance of the final states, which is strongly correlated with the extracted m_ν^2 value, changed by over 600 eV^2, from 1500 to 892 eV^2, when the assumptions used in the model were changed [3.29, 3.30]. It is also important to note that several of the experiments using complex sources assumed final-state configurations different from the actual source configurations. For example, the INS group had a source consisting of $C_{20}H_{40-x}O_2Cd-(xT)$ on

Ru_2O_3 attached to an aluminum substrate. What the group assumed, however, was that the final states were those of valine, $C_5H_{10}NO_2$–T. Similarly, the Zürich group used a monolayer-type source of $C_{18}H_{30}SiO_3$–(6T) on SiO on Si, on a carbon substrate. What this group assumed was that the final states were those for propane, C_3H_7–T. Given the importance of accurate knowledge of final-state effects, the use of such complex sources clearly limits the ultimate sensitivity of such measurements.

Recognizing that to reliably probe the mass region below 10 eV one must be certain of the final-state contributions, the Los Alamos group, followed by the Livermore, Mainz and Troitsk groups, developed molecular tritium sources. The Livermore and Los Alamos groups used gaseous sources, while the Troitsk and Mainz groups, which are still acquiring data, use a gaseous and a frozen solid tritium source, respectively. More importantly, with their high resolution and acceptance, the Mainz and Troitsk groups have the capability of acquiring data in the 10–15 eV range below the endpoint energy, where excited states are not expected to contribute.

Shape Effects Due to Energy Loss. Energy loss of the electrons emitted from the source will modify the observed beta spectrum. Solid sources may have additional nonnegligible contributions from backscattering and surface contamination. The energy loss in the source is often quoted as the percentage of electrons that transverse the source without losing any energy. This "no loss fraction" varies between experiments, with modern experiments having values between 80 and 98%. The Los Alamos experiment offers an example of how accurately one must know the source thickness. In this experiment, which used a molecular T_2 gaseous source, the source thickness was calculated to have a "no loss fraction" of 91.4%. However, in order to verify this calculation the Los Alamos group made an in-situ measurement of the energy loss of the source using a $^{83}Kr^m$ line source introduced simultaneously with the tritium gas. The result was a measured "no loss fraction" of 93.5%. This small 2% difference between calculated and measured values results in a change in $\overline{M_{\nu_e}}$ of 25 eV^2. Thus, it is clear that in experiments probing at a level of 5 eV sensitivity one must know the source thickness to better than 0.5%. Another example is provided by the Mainz group, which uses a solid molecular tritium source. The initial results reported by the group [3.31, 3.32] yielded excessively negative m_ν^2 values. These results have now been attributed to unexpectedly high energy losses in the frozen substrate because of dewetting from the substrate at temperatures around 4 K with the subsequent formation of individual crystals [3.6].

Instrumental Shape Effects. There are instrumental effects introduced by the spectrometer systems used to analyze the momentum of the electrons. These effects include the finite energy resolution of the system and energy-dependent extraction (or acceptance) efficiencies. Very tiny low- or high-energy tails, usually from scattering in the analyzing system, can have dramatic effects. For example an unaccounted-for linear tail of 0.015% per eV that extended over a 500 eV region would introduce a 300 eV^2 error in the derived value. Initially some experiments relied on estimating the resolution function by Monte Carlo calculations. But, since the calculations depend totally on the input assumptions, it is quite possible and even likely that small scattering tails or resolution effects would not be included. The Los Alamos, Livermore and Mainz groups made direct measurements of their instrumental resolution using a ^{83}Kr^m conversion line source. The Troitsk group has utilized an electron gun to determine the resolution.

For magnetic spectrometers, one must also characterize the energy efficiency of the analyzing system. Although this parameter is allowed to be a free parameter in the fit, it is important to estimate the systematic uncertainty associated with the assumption made in picking the functional form. For example, both the Zürich and the Livermore groups make the assumption, from Monte Carlo calculations, that their energy efficiency has a linear functional dependence. In several instances they have made measurements over an extended range that show that the calculations are in reasonable agreement with the measurements. However, in the actual region of interest near the endpoint, the tritium spectrum itself is far more sensitive than any of the calibration measurements performed to date. The Los Alamos group took a different approach and assumed that they did not know the exact functional form of the energy efficiency to the accuracy required, so they performed two fits, one using a linear term and one using a quadratic term. The different fits cannot be distinguished on the basis of the chi-squared goodness of fit. The group then took the average between these two fits to obtain the mean best value, and the difference was used to estimate the systematic uncertainty. For the Los Alamos data the difference was 32 eV^2, the group's largest systematic uncertainty. The Mainz and Troitsk groups do not include such corrections in fitting their data, most likely because they cover a significantly smaller energy region below the endpoint. In principle, such effects could be present in their data.

Systematic Checks. There is one nice independent check of the systematics [3.33]. In fitting the beta spectrum shape, one also fits the endpoint energy value. One can then calculate from this value the ^{3}H–^{3}He atomic-mass difference, which has been measured in totally independent experiments, most recently by Van Dyck et al. using an ion cyclotron resonance (ICR) technique, which yielded a value of 18590.1(17) eV [3.34]. A correct tritium experiment with no systematic errors must yield the correct atomic-mass difference. However, the converse is not true; getting the right ^{3}H–^{3}He atomic-mass difference does not guarantee the correctness of the beta spectrum shape. The original agreement between the ITEP ^{3}H–^{3}He atomic-mass difference and the value measured from an earlier ICR measurement [3.35] was cited as supporting the ITEP neutrino mass value. However, on the basis of improved measurements, both the earlier ICR result and the ITEP result give incorrect values for the ^{3}H–^{3}He atomic-mass difference, while recent tritium beta decay experiments are consistent with the current ICR value. The recent Mainz and Troitsk results [3.6, 3.22] do not discuss this systematic check, but hopefully future experiments will include such a comparison.

The other systematic check that should be performed is to vary the fitting region used to extract a neutrino mass. If the extracted neutrino mass or other fit parameters vary with the region included in the fit, it is clearly a sign of unknown systematic problems. For example, the initial Mainz results showed such behavior, and the group later discovered and reported that this was an indication of anomalies in their source thickness and hence in the energy loss.

3.3.4   Discussion of Tritium Beta Decay Results

Until very recently, the ability to extract a reliable neutrino mass limit from any of the tritium beta decay experiments has been questionable. The reason is because the experiments least subject to systematic uncertainties, those using molecular tritium as a source, all yielded statistically significant negative best-fit values of m_ν^2, implying that the experiments were seeing excess counts near the endpoint. The key point is that if one observes an excess of counts that cannot be understood, then one cannot be certain that this unknown excess is not masking a positive m_ν^2 signal. This problem has become known as the "negative mass squared" problem and was reinforced by the fact that the experiments based on more complicated sources also reported negative m_ν^2 values.

An obvious question is: are the data sets from the four molecular-tritium-source experiments consistent with one another? Table 3.3 gives a summary of various parameters pertaining to the four experiments. The negative m_ν^2 values given for each experiment were obtained by fitting over the energy region stated in the next row of the table. For these fits and m_ν^2 values, no additional arbitrary final states or step functions were included. One sees that the values of $\overline{M_{\nu_e}}^2$ are not in good agreement. In fact, the values are not even directly comparable, because each experiment uses a different prescription for modifying the functional form to accommodate negative m_ν^2 values. For simplicity, we can define $\epsilon = E_0 - E - V_i$ and $m = \overline{M_{\nu_e}}$; then all experiments use the form

$$m^2 > 0 : \epsilon\sqrt{\epsilon^2 - m^2 c^4} \tag{3.7}$$

for positive m_ν^2 values, but each uses a different functional form for negative m_ν^2 values:

- Mainz:

$$m^2 < 0 : (\epsilon + \mu)\left[1 + \frac{\mu}{\epsilon + \mu}\exp\left(-\frac{\epsilon}{\mu - 1}\right)\right]\sqrt{\epsilon^2 - m^2 c^4}\,\Theta(\epsilon + \mu)$$
$$\mu = 0.76\sqrt{-m^2 c^4} \tag{3.8}$$

Table 3.3. Comparison of molecular-tritium beta decay measurements as of 1998

Result or correction	Mainz	Troitsk 96 data	Livermore	Los Alamos
$\overline{M_{\nu_e}}^2$ (eV^2)	$-48 \pm 12 \pm 19$	$-15.4 \pm 7.2 \pm 2.8$	$-130 \pm 20 \pm 15$	$-147 \pm 68 \pm 41$
Fit range	170 eV	200 eV	1100 eV	825 eV
Atomic final state	Calculated	Calculated	Calculated	Calculated
Energy loss	Measured, Kr	Measured, e-gun	Measured, Kr	Measured, Kr
Trapping	N/A	Calculated	Measured	Measured
Dead time	Calculated	Calculated	Negligible	Negligible
Energy efficiency	–	–	Calculated	Fitted
Account for excess with				
Arbitrary final state	90 eV, 4%	Not given	35 eV, 10%	60–70 eV, 5%
Line/spike (location)	0–13 eV	5–16 eV	23 eV	0 eV
Line/spike (amplitude)	0–0.9×10^{-10}	0.6×10^{-10}	30×10^{-10}	10×10^{-10}

- Troitsk:

$$m^2 < 0 : 2\epsilon^2 - \epsilon\sqrt{\epsilon^2 - k^2 c^4} \, , k^2 = -m^2 \tag{3.9}$$

(discontinuities in the derivatives with respect to ϵ and m are smoothed out using bidimensional splines)

- Livermore:

$$m^2 < 0 : \left| \epsilon^2 + \frac{k^2 \epsilon}{2|\epsilon|} \right| \Theta(\epsilon + k) \tag{3.10}$$

- Los Alamos:

$$m^2 < 0 : \left(\epsilon^2 + \frac{k^2}{2} \right) \Theta(\epsilon) \, . \tag{3.11}$$

The results are shown graphically in Fig. 3.5, for a hypothetical, exaggerated excess of $m_\nu^2 = -10^4$ eV^2. The conclusion is that, because of the different prescriptions used, it is not possible to make a direct comparison of the negative m_ν^2 fit values obtained by the different groups. It must be recognized that there is no single "correct" form for $m_\nu^2 < 0$. What is appropriate depends on the experiment. The guiding principle has usually been to obtain χ^2 plots that are parabolic through $m_\nu^2 = 0$. Uncertainties from such plots are normally distributed.

If data were fitted by all groups to a "standard" functional form that assumed zero neutrino mass and used an agreed set of final states, the resulting

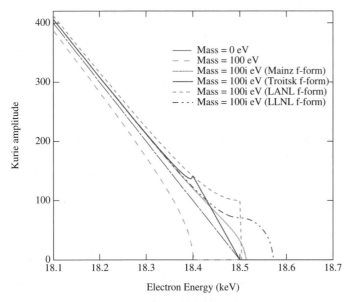

Fig. 3.5. Functional forms used in fitting excess counts

fit could then be presented as a residual plot, data minus fit, which would allow one to compare results.

There is another way to attempt to compare the measurements. In the bottom third of Table 3.3, parameters are shown for fitting the observed spectra with a fixed zero neutrino mass while also including additional fitting parameters: either an arbitrary additional final state or an additional spike or line in the spectrum. For the arbitrary final state, the excitation energy of the additional state and its strength relative to the sum of the other final states is given. For the spike (step function in an integral spectrum), the location of the spike below the endpoint is given along with the relative decay strength. Clearly the effects seen in the Livermore and Los Alamos data are not consistent with the much smaller effects seen in the Mainz and Troitsk data. It is apparent that with their high-resolution and high-luminosity spectrometers, the Mainz and Troitsk experiments have significantly higher sensitivity to neutrino mass than the Livermore and Los Alamos experiments. Furthermore, by measuring a limited region 10–15 eV below the endpoint, the Mainz and Troitsk experiments become relatively insensitive to any unknown problems with the molecular final-state calculations. The consistency between the Mainz and Troitsk experiments will now be discussed in detail.

3.3.5   Tritium Beta Decay Results from Troitsk and Mainz

Although the Mainz and Troitsk experiments seem to have the potential to probe the electron antineutrino mass to the level of a few eV, they have continued to be limited by systematic problems. Moreover, for the results reported up to 1999, neither the Mainz nor the Troitsk group was able to make measurements that were reproducible within the estimated uncertainties.

Review of the Troitsk Results. In the case of the Troitsk data, Lobashev et al. claim that there is a step function anomaly that varies in both its amplitude and its position below the endpoint [3.22]. These results, including the variations of the amplitude and the energy location of the step function, are shown in Fig. 3.6, binned according to month. This binning is used to illustrate Lobashev's hypothesis that the observed variation arises because of interactions as the Earth passes through a cloud of relict neutrinos with a density 10^{13} times higher than the mean universal neutrino density. Since the period is half a year, this requires that the cloud have a different inclination to the ecliptic than the earth does. Lobashev states that this explanation is extremely speculative. Recently Stephenson, Goldman and McKellar have examined the Troitsk results while considering possible non-Standard Model interactions [3.36]. They find that hypothetical interactions weaker than the Standard Model V–A interaction and having a different Lorentz structure could in principle produce such effects. However, there are several seriously disturbing aspects of the Troitsk data. Foremost is the revelation that data sets not exhibiting a step function are excluded from the published analysis [3.37]. Second, searching for correlations over a short number of periods is a notoriously difficult exercise, known to be prone to false correlations. Very recently, the Troitsk group reported additional data that are now in disagreement with their previous cyclical hypothesis [3.38]. A step function is still found, but it no longer seems to adhere to the half-year cycle. This is a positive development, and the group is examining other possible apparatus-related origins for these systematic effects. Given the anomalous behavior of the Troitsk data and the fact that data have been selectively excluded from the analysis, it is difficult to have confidence in the Troitsk group's current (2000) neutrino mass limit.

Review of the Mainz Results. The recently published Mainz data also exhibit inconsistencies from run to run, as seen in Fig. 3.7 [3.6]. Three of the data sets do not require a step function at the energy predicted by the Troitsk group, while the fit to one data set, Q4, is improved by including a step function at an energy that is consistent with the Troitsk results. One also observes that for data sets Q2–Q4 there are variations of the fits to m_ν^2 as a function of the range of data included in the fits. Again, this is a clear indication of unknown systematic effects. The data set that looks

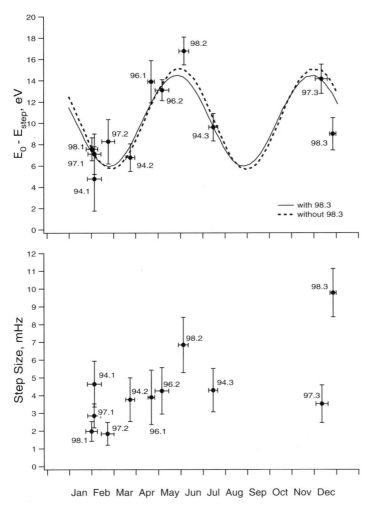

Fig. 3.6. Anomalous step functions reported for the Troitsk data [3.22]. Notice that both the amplitude and the energy location of the step function relative to the endpoint vary with time

consistent, both as one changes the range of data fit and by having good χ^2 values, is data set Q5. The group noted that these data were acquired under very similar conditions to those of Q3 and Q4, but that a voltage of ± 20 V with a frequency of 1 MHz was applied at one of the electrodes at the detector side of the spectrometer during measurement pauses, to destroy possible storage conditions for charged particles and to reduce the rate and fluctuations of the background. Potential trapping of particles in the type of spectrometer used here has been recognized as a possible problem. The

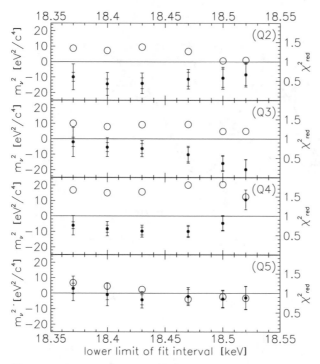

Fig. 3.7. Fits to m_ν^2 for four different Mainz data sets, Q2–Q5 [3.6]. The *solid dots* are the best-fit values of m_ν^2 while the *open circles* show the reduced values χ^2_{red}. For each of the data sets, fits to six different energy intervals are shown

group continued this procedure during three additional data collection runs in 1999, and found that these data sets were consistent with Q5 and did not show the systematic variation as a function of fit range. The group recently announced a new result [3.5] of

$$\overline{M_{\nu_e}}^2 = -1.6 \pm 2.5 \pm 2.1 \ \mathrm{eV}^2 \ , \tag{3.12}$$

which yields a limit of

$$\overline{M_{\nu_e}} < 2.2 \ \mathrm{eV} \ . \tag{3.13}$$

This result certainly looks reliable, and is based on consistent data that has passed several systematic checks.

3.3.6   Mass Sensitivity of Future Beta Decay Experiments

With the demonstration of neutrino mass and oscillations now all but certain, there is heightened interest in beta decay as a means for determining the masses of not just the "electron" neutrino, but of all active neutrinos.

Because the mixing angles may be large and the mass differences small, a determination of one mass becomes effectively a determination of all. Thus the historical difficulty that the electron neutrino was very likely not the right one to study, but was the one for which the sensitivity was greatest, may now be alleviated.

It is useful to ask what is the smallest mass detectable in beta decay, and what experimental approach is most likely to be fruitful.

In order to detect the effect of a neutrino mass m_ν at the endpoint of a beta spectrum of total kinetic energy E_0, an instrumental resolution of order m_ν/E_0 is needed. The spectral fraction per decay that falls in the last m_ν of the beta spectrum is approximately $A(m_\nu/E_0)^3$, where A is a constant of order unity.

For spectrometric experiments in which the source and the detector are physically separated, a limit on the source thickness is set by the cross section for inelastic interactions of outgoing electrons, such that one must have $\sigma n \leq 1$, where σ is the inelastic cross section (3.4×10^{-18} cm^2 for molecular T$_2$ [3.39]) and n is the superficial source density. The dimensions of the source (radius R_s) and the dimensions of the spectrometer (length or radius R) are related through the resolution needed, specifically,

$$\frac{\Delta E}{E} \simeq \left(\frac{R_s}{R}\right)^\alpha . \tag{3.14}$$

For magnetic spectrometers, $\alpha = 1$, and for magnetic–electrostatic retarding-field analyzers, α is a more favorable 2.

The rate in the last m_ν of the spectrum is of order

$$\frac{dN}{dt} = \left(\frac{R_s}{R}\right)^{3\alpha} \frac{\lambda \pi R_s^2}{\sigma} , \tag{3.15}$$

where λ, the mean decay rate, is approximately given in s^{-1} (for an allowed beta decay) by

$$\lambda = B\frac{E_0^4}{ft} . \tag{3.16}$$

In this expression, ft is the beta decay reduced matrix element (e.g. $\log_{10} ft = 3$ for tritium decay), E_0 is in eV and B is a constant, approximately 1.5×10^{-23}. Then,

$$\frac{dN}{dt} = m_\nu^{3+2/\alpha} E_0^{1-2/\alpha} \frac{B\pi R^2}{\sigma ft} \tag{3.17}$$

when background is absent and the detection efficiency is unity. For example, if one sets as a minimum practical limit one event per day in the last m_ν of the spectrum, then

$$m_{\nu(\text{ultimate})} = 1.7 \text{ eV}$$

for a magnetic spectrometer 10 m in length, and

$$m_{\nu(\text{ultimate})} = 0.3 \text{ eV}$$

for a retarding-field analyzer 3 m in radius. In practice these limits are eroded quickly by efficiency, background and the need to explore the rest of the spectrum both for extraction of the neutrino mass and for measurements related to systematic uncertainties.

Calorimetric experiments evade the source-thickness limit because energy loss is no longer an issue, but in practice they run into a different limit set by the presence in the data of the entire beta spectrum at a rate R_{tot}, which leads to pileup. The source must then be subdivided into N parts, each independently measured. The best prospect is ^{187}Re, for which $E_0 = 2.5$ keV. Initial experiments on ^{187}Re have been reported [3.40–3.42].

The experiments that at present most nearly approach the ideal conditions appear to be the T_2-based systems of Lobashev and Spivak [3.43] and Bonn et al. [3.25]. Results are already at the 2.2 eV statistical level [3.5, 3.6, 3.22].

Retarding-field analyzers have a high figure of merit at the kinematic endpoint, but do less well for structure away from the endpoint. Bandpass instruments based on the retarding-field principle have been proposed that make use of time of flight [3.25] or transverse drift across field lines in a gradient field [3.44].

While both the idealized statistical arguments and the experimental data indicate that the 1 eV level might be attained and surpassed, the anomalies observed at the endpoint must be quantitatively understood before a convincing value for or limit on the rest mass of ν_e can be given.

The structure of the two-electron molecular ion (HeT$^+$) produced in the decay of T_2 is presumably calculable to arbitrary accuracy given enough effort, but it would clearly be desirable to be able to work with the T atom, or even the T$^+$ ion, for which the atomic-structure calculations are different and simpler. Moreover, the rovibrational spectrum of the molecular ion broadens the spectrum by about 1 eV even if electronic excitations are rendered inconsequential by taking data in the last 20 eV of the spectrum only. Space-charge limits preclude a useful T$^+$ source, but it may not be out of the question to imagine an atomic-tritium source of worthwhile density. The Los Alamos source [3.17] was originally designed to use atomic tritium prepared by dissociation in an RF discharge. The gas was prevented from recombining immediately through the use of a highly polished aluminum source volume (tube), since it had been found that aluminum hindered surface recombination and volume recombination was negligible. The atomic spectrum was never taken, however, because the molecular contamination, even at less than 10%, had to be measured extremely precisely since its endpoint is about 8 eV higher than that of the atomic spectrum. The technology for vacuum ultraviolet absorption molecular spectroscopy was still primitive.

Initially a somewhat different approach had been considered, namely the use of the developing technology for preparing dense spin-polarized hydrogen at very low temperatures. The advantage of this method would be the essentially negligible contamination by molecular tritium; the technology was not sufficiently advanced in the 1980s to make this option attractive. Now it is possible to prepare hydrogen with atomic densities in excess of 10^{14} cm^{-3}, within striking distance of the 10^{15} to 10^{16} cm^{-3} needed [3.45–3.47]. To establish whether this could form the basis for a successful atomic-tritium experiment would require a significant research effort.

3.4   Particle Decay Measurements – ν_μ and ν_τ

The direct limits on both the muon and the tau neutrino masses are based on kinematic measurements using semileptonic, weak particle decays. The observables in these measurements are either the invariant mass or the decay particle momentum. The reliance on knowing particle masses and momenta dramatically limits these measurements' sensitivity to neutrino mass as compared with beta decay shape determinations (see Table 3.1). Theoretical bounds predicted by cosmology and nucleosynthesis, as well as estimated limits from SN1987A, set limits far lower then those placed by direct measurements, but such derived bounds have various model-dependent assumptions.

3.4.1 ν_μ Mass Measurements

The technique of performing a kinematically complete study of pion decay, $\pi^+ \rightarrow \mu^+ + \nu_\mu$, to probe the muon neutrino mass was first used by Barkas and collaborators [3.48] at the Lawrence Berkeley Laboratory 184 inch cyclotron in 1956. This simple two-body decay continues to be the most sensitive method of setting limits on $\overline{M_{\nu_\mu}}$. In this decay one must accurately measure the muon momentum produced from pion decay at rest and then combine this knowledge with values for the pion and muon masses to extract the muon neutrino mass:

$$\overline{M_{\nu_\mu}}^2 = m_\pi^2 + m_\mu^2 - 2m_\pi\sqrt{p_\mu^2 + m_\mu^2} \ . \tag{3.18}$$

From the above equation it is apparent that the neutrino mass value is determined by taking the difference of two large numbers; hence one expects any limit on $\overline{M_{\nu_\mu}}$ to be extremely sensitive to changes in these values. Historically, this has certainly proven to be a problem. For example, in 1992 Robertson [3.49] reevaluated the mass limits set by Abela et al. [3.50] by using an improved high-precision measurement of the muon momentum by Daum et al. [3.51]. The result was a value of $\overline{M_{\nu_\mu}}^2 = -0.127(25)$ MeV^2, which was more than 5σ away from zero. This clearly indicated a systematic problem in at least one of the two results that was used to derive the neutrino mass.

Current Limits on $\overline{M_{\nu_\mu}}$. Assamagan et al. [3.7] used the Paul Scherrer Institute accelerator to produce a surface muon beam to observe pion decay at rest. By measuring the outgoing muon momentum and using the improved value for the pion rest mass obtained by Jeckelmann, Goudsmit and Leisi [3.52], they found

$$\overline{M_{\nu_\mu}}^2 = -0.016 \pm 0.023 \text{ MeV}^2 \,,$$

which yields a mass limit of:

$$\overline{M_{\nu_\mu}} < 170 \text{ keV (90\% C.L.)} \,.$$

The uncertainty on $\overline{M_{\nu_\mu}}$ arises from the uncertainties on the two masses and the muon momentum:

$$\Delta(\overline{M_{\nu_\mu}}^2) = \sqrt{\left[\frac{\partial(\overline{M_{\nu_\mu}}^2)}{\partial m_\pi}\Delta m_\pi\right]^2 + \left[\frac{\partial(\overline{M_{\nu_\mu}}^2)}{\partial m_\mu}\Delta m_\mu\right]^2 + \left[\frac{\partial(\overline{M_{\nu_\mu}}^2)}{\partial p_\mu}\Delta p_\mu\right]^2} \,.$$

$$(3.19)$$

For the Assamagan et al. result, the partial-derivative terms in the uncertainty relationship are roughly equal in magnitude. The fractional uncertainty that Assamagan et al. determined for the muon momentum, 4.4×10^{-6}, is comparable to the Particle Data Group pion mass uncertainty of 2.6×10^{-6} [3.4]. (The uncertainty of the muon mass is nearly ten times smaller, at 3.2×10^{-7} [3.4], and can safely be ignored.) The difference between the Assamagan et al. limit and the Particle Data Group limit of

$$\overline{M_{\nu_\mu}} < 190 \text{ keV (90\% C.L.)}$$

comes about because the Particle Data Group uses the unified, classical analysis approach suggested by Feldman and Cousins [3.53] in calculating the upper limit.

An alternative experimental method to determine the muon neutrino mass that is far less sensitive to uncertainties associated with the pion mass was used in an experiment performed by Anderhub et al. [3.54]. This technique measures the muon momentum for pions that decay in flight while circulating in a magnetic figure-of-eight ring. The result from this experiment, which ran at the Paul Scherrer Institute, gives

$$\overline{M_{\nu_\mu}}^2 = -0.014 \pm 0.20 \text{ MeV}^2 \,,$$

which yields a limit of

$$\overline{M_{\nu_\mu}} < 500 \text{ keV (90\% C.L.)} \,.$$

While less sensitive than the decay-at-rest measurement, this experiment has less systematic uncertainty.

Future Measurements. A novel suggestion has been made to use the muon g-2 apparatus at Brookhaven to measure pion decay in flight [3.55]. Both the pion and the muon momenta would be measured by using the g-2 ring as a spectrometer. The expected sensitivity of such a device would be in the 10 keV range [3.56]. Such an improvement would have an impact on astrophysical bounds, which are valid for neutrino masses below 10 keV [3.55].

3.4.2 ν_τ Mass Measurements

Limits on the mass of the tau neutrino have been derived by looking at rare multi-particle semileptonic decays of the tau, with only one neutrino in the final state and where the effective mass of the detected particles is close to m_τ. The experiments ALEPH, ARGUS, CLEO, DELPHI and OPAL have all examined a variety of decay modes including anywhere from three to eight π daughter particles, although the decays to three or five charged πs are the preferred decay modes because these modes provide good resolution of the four visible momenta and because of their minimal contamination by lower-multiplicity modes. The general scheme is that one observes the decay of a $\tau^+\tau^-$ pair created in an e^+e^- storage ring or collider. Usually a simple tau decay mode with a single charged particle acts as a trigger while one searches for tau multi-particle decays, which are used to form an invariant-mass plot and measure the missing energy. An event with an amount of energy close to the tau mass can put restrictive limits on the tau neutrino mass. For a recent review see [3.57].

Current Limits on $\overline{M_{\nu_\tau}}$. The ALEPH collaboration at LEP has produced the strictest limit on the tau neutrino mass from the 1991–1995 LEP run by observing 2939 $\tau^- \to 2\pi^-\pi^+\nu_\tau$ and 55 $\tau^- \to 3\pi^-2\pi^+\nu_\tau$ final-state decays. The current limit is

$$\overline{M_{\nu_\tau}} < 18.2 \text{ MeV } (95\% \text{ C.L.}) \ .$$

Future Measurements. It is almost certain that the b-factory detectors BABAR and BELLE will be able to reduce the limit on $\overline{M_{\nu_\tau}}$. A perhaps optimistic estimate is that with an integrated luminosity of 300 fb^{-1} these detectors might be able to set a limit as low as 3 MeV [3.58].

3.5 Summary

The evidence for neutrino mass from the observations of atmospheric and solar neutrinos is compelling. Oscillation experiments cannot, however, provide a value for the mass of one species, only differences in the squares of the masses. The atmospheric data, which indicate $|m_2^2 - m_3^2| = 3.5 \times 10^{-3}$ eV^2, show that either m_2 or m_3 is at least 0.06 eV, and neutrinos therefore provide at least as much mass as do luminous stars. Cosmological arguments about the evolution of structure in the Universe indicate that $\Omega_\nu \leq 0.1$, and the Hubble constant is now known to be 70 km/s/Mpc to about 10%, from which it may be concluded that $\sum_i m_i \leq 5$ eV. These values define the mass scale that is now the object of intense experimental interest.

Supernovae are prolific sources of neutrinos at astronomical distances, and even from the limited number of events recorded during SN1987a it was possible to conclude that $m_{\overline{\nu_e}} < 23$ eV [3.59]. Unfortunately, it is unlikely that a future supernova, no matter what its location, will provide a mass measurement below about 30 eV for neutrinos that interact only via the neutral current [3.60]. If subsequent collapse to a black hole occurs, however, a limit as low as 4 eV might be obtainable [3.61].

Direct kinematic measurements on the μ and τ neutrinos are not at present able to approach the region of cosmological interest. Unstable neutrinos of MeV mass are permitted in some cosmological scenarios, but this window has effectively been closed with the observation that ν_μ and ν_τ are maximally mixed with a very small mass difference.

Experiments on tritium beta decay have undergone much refinement and now have statistical precisions of order 1–2 eV. Significant further improvements can be foreseen in both systematic and statistical accuracy, and it may be possible to reach, by a direct kinematic measurement, a mass of order 0.5 eV. The linkage between mass eigenstates provided by oscillation experiments means that a beta decay measurement at this level may constrain the sum of the masses (of cosmological interest) tightly. It remains, however, to establish the oscillation link between ν_1 and $\nu_{2,3}$. That link is perhaps to be found in the solar neutrino problem. Direct neutrino mass measurements, a pursuit spanning more than half a century, remain central to an understanding of the neutrino and its role in the evolution of the Universe.

References

3.1 Y. Fukuda et al., Phys. Lett. B **433**, 9 (1998); Y. Fukuda et al., Phys. Lett. B **436**, 33 (1998); Y. Fukuda et al., Phys. Rev. Lett. **81**, 1562 (1998).

3.2 P. Fisher, B. Kayser and K. S. McFarland, Annu. Rev. Nucl. Part. Sci. **49**, 481 (1999).

3.3 V. A. Lubimov et al., Phys. Lett. B **94**, 266 (1980).

3.4 Particle Data Group, Eur. Phys. J. C **3** 1, (1998).

3.5 C. Weinheimer et al., talk presented at the 19th International Conference on Neutrino Physics and Astrophysics, Sudbury, Canada, June 2000, to be published (2000).

3.6 C. Weinheimer et al., Phys. Lett. B **460**, 219 (1999).

3.7 K. Assamagan et al., Phys. Rev. D **53**, 6065 (1996).

3.8 R. Barate et al., Eur. Phys. J. C **2**, 395 (1998).

3.9 W. Pauli, in *Rapports du Septième Conseil de Physique Solvay*, Brussels, 1933 (Gauthier-Villars, Paris, 1934).

3.10 E. Fermi, Z. Phys. **88**, 11 (1934).

3.11 S. C. Curran, J. Angus and A. L. Cockroft, Nature **162**, 302 (1948).

3.12 S. C. Curran, J. Angus and A. L. Cockroft, Phil. Mag. **40**, 53 (1949).

3.13 S. C. Curran, J. Angus and A. L. Cockroft, Phys. Rev. **76**, 853 (1949).

3.14 K.-E. Bergkvist, Nucl. Phys. B **39**, 317 (1972); K.-E. Bergkvist, Nucl. Phys. B **39**, 371 (1972).

3.15 S. Boris et al., Phys. Rev. Lett. **58**, 2019 (1987).

3.16 H. Kawakami et al., Phys. Lett. B **256**, 105 (1991).

3.17 R. G. H. Robertson et al., Phys. Rev. Lett. **67**, 957 (1991).

3.18 E. Holzschuh, M. Fritschi and W. Kuendig, Phys. Lett. B **287**, 381 (1992).

3.19 A. Franklin, Rev. Mod. Phys. **67**, 457 (1995).

3.20 R. G. H. Robertson and D. A. Knapp, Annu. Rev. Nucl. Part. Sci. **38**, 185 (1988).

3.21 E. Holzschuh, Rep. Prog. Phys. **55**, 1035 (1992).

3.22 V. M. Lobashev et al., Phys. Lett. B **460**, 227 (1999).

3.23 W. Stoeffl and D. J. Decman, Phys. Rev. Lett. **75**, 3237 (1995).

3.24 J. J. Simpson, Phys. Rev. D **23**, 649 (1981).

3.25 J. Bonn et al., Nucl. Instrum. Meth. A **421**, 256 (1999).

3.26 S. Jonsell and H. Monkhorst, Phys. Rev. Lett. **76**, 4476 (1996).

3.27 P. Froelich and A. Saenz, Phys. Rev. Lett. **77**, 4724 (1996) .

3.28 S. Jonsell, A. Saenz and P. Froelich, Phys. Rev. C **60**, 034601 (1999).

3.29 I. G. Kaplan et al., Zh. Eksp. Teor. Fiz. **84**, 833 (1983) [Sov. Phys. JETP **57**, 483 (1983)].

3.30 I. G. Kaplan et al., Phys. Lett. B **161**, 389 (1985).

3.31 C. Weinheimer et al., Phys. Lett. B **300**, 210 (1993).

3.32 H. Backe et al., in *Proceedings of the 17th Conference on Neutrino Physics and Astrophysics, Neutrino 96*, Helsinki, June 1996 (World Scientific, Singapore, 1996), p. 259.

3.33 S. T. Staggs et al., Phys. Rev. C **39**, 1503 (1989).

3.34 R. S. Van Dyck, D. L. Farnham and P. B. Schwinberg, Phys. Rev. Lett. **70**, 2888 (1993).

3.35 E. Lippmaa et al., Phys. Rev. Lett. **54**, 285 (1985).

3.36 G. J. Stephenson Jr., T. Goldman and B. H. J. McKellar, Phys. Rev. D **62**, 093013 (2000).

3.37 V. M. Lobashev, private communication (1999).

3.38 V. M. Lobashev et al., talk presented at the 19th International Conference on Neutrino Physics and Astrophysics, Sudbury, Canada, June 2000, to be published (2000).

3.39 V. N. Aseev et al., Eur. Phys. J. D **10**, 39 (2000).

3.40 A. Alessandrello et al., Phys. Lett. B **457**, 253 (1999).

3.41 F. Gatti et al., Nature **397**, 137 (1999).

3.42 F. Gatti et al., talk presented at the 19th International Conference on Neutrino Physics and Astrophysics, Sudbury, Canada, June 2000, to be published (2000).

3.43 V. M. Lobashev and P. E. Spivak, Nucl. Instrum. Meth. Phys. Res. A **240**, 305 (1985).

3.44 V. M. Lobashev, private communication (1998).

3.45 I. F. Silvera, Bull. Am. Phys. Soc. **40**, 1003 (1995).

3.46 D. G. Fried et al., Phys. Rev. Lett. **81**, 3811 (1998).

3.47 T. J. Greytak et al., Physica B **280**, 1 (2000).

3.48 W. H. Barkas, W. Birnbaum and F. Smith, Phys. Rev. **101**, 778 (1956).

3.49 R. G. H. Robertson, in *Proceedings of the International Converence on High-Energy Physics, ICHEP92,* Dallas, August 1992 (AIP Conference Proceedings Vol. 272, ed. by J. R. Sanford, 1993), p. 140.

3.50 R. Abela et al., Phys. Lett. B **146**, 431 (1984).

3.51 M. Daum et al., Phys. Lett. B **265**, 425 (1991).

3.52 B. Jeckelmann, P. F. A. Goudsmit and H. J. Leisi, Phys. Lett. B **335**, 326 (1994).

3.53 G. J. Feldman and R. D. Cousins, Phys. Rev. D **57**, 3873 (1998).

3.54 H. G. Anderhub et al., Phys. Lett. B **114**, 76 (1982).

3.55 E. Adelberger et al., in *Particle and Nuclear Astrophysics and Cosmology in the Next Millennium: Proceedings of the 1994 Snowmass Summer Study,* Snowmass, Colorado, June–July 1994, ed. by E. W. Kolb and R. D. Peccei (World Scientific, Singapore, 1995), p. 195.

3.56 K. Jungmann et al, in *Physics and detectors for DAΦNE Frascati,* Nov. 1999, preprint nucl-ex/0002005, to be published (2000).

3.57 L. Passalacqua, Nucl. Phys. B, Proc. Suppl. **55C**, 435 (1997).

3.58 R. M. Roney, talk presented at the 19th International Conference on Neutrino Physics and Astrophysics, Sudbury, Canada, June 2000, to be published (2000).

3.59 T. J. Loredo and D. Q. Lamb, in *Fourteenth Texas Symposium on Relativistic Astrophysics,* ed. by E. J. Fenyves, Ann. N.Y. Acad. Sci. **571**, 601 (1989).

3.60 J. F. Beacom and P. Vogel, Phys. Rev. D **58**, 093012 (1998); Phys. Rev. D **58**, 053010 (1998).

3.61 J. F. Beacom, R. N. Boyd and A. Mezzacappa, preprint hep-ph/0006015, to be published (2000).

4 Neutrino Oscillations
and the Solar Neutrino Problem

W. C. Haxton

4.1 Introduction

Part of the interest in neutrino astrophysics has to do with the fascinating interplay between nuclear and particle physics issues – e.g. whether neutrinos are massive and undergo flavor oscillations, whether they have detectable electromagnetic moments, etc. – and astrophysical phenomena, such as the clustering of matter on large scales, the processes responsible for the synthesis of nuclei, the mechanism for core-collapse supernovae and the evolution of our sun. This summary addresses one of the oldest problems in neutrino astrophysics, the 30 year puzzle of the missing solar neutrinos. This puzzle grew out of attempts to test the standard theory of main-sequence stellar evolution, but has now led to speculations about physics beyond the Standard Model of electroweak interactions. I shall describe the work that defined the solar neutrino problem, the likelihood that its resolution is connected with massive neutrinos and the hopes we have for future experiments.

4.2 Open Questions in Neutrino Physics[1]

The existence of the neutrino was first suggested by Wolfgang Pauli in a private letter dated December, 1930. The motivation was to solve an apparent problem with energy conservation in nuclear β decay: the observable particles in the final state (the daughter nucleus and emitted electron) carried less energy than that released in the nuclear decay. Pauli suggested that an unobserved particle, the neutrino, accompanied the decay and accounted for the missing energy.

A number of important developments followed Pauli's suggestion. In 1934 Fermi [4.2] suggested a theory of β decay that was modeled after electromagnetism, except that there was no analogue of the electromagnetic field: the interaction occurred at a point. (Apart from the missing aspect of parity violation, this was the correct reduction of today's Standard Model to an

---

[1] A more extended,popular summary of the development of neutrino physics can be found in [4.1].

effective theory.) In 1934 Bethe and Critchfield described the role of β decay in thermonuclear reaction chains powering the stars,

$$(A, Z) \rightarrow (A, Z - 1) + e^+ + \nu_e ,$$

thus predicting that our sun produces an enormous neutrino flux. In 1956 Cowan and Reines [4.3] succeeded in measuring neutrinos emitted by a reactor through the reaction

$$\bar{\nu}_e + p \rightarrow n + e^+ ,$$

exploiting the positron and neutron coincidence. (The neutron was detected by a (n, γ) reaction on a Cd neutron poison.) In 1957 the weak force mediating neutrino interactions was found to violate parity maximally. Later experiments found that the ν_e was replicated twice more in nature – the ν_μ and ν_τ – each accompanying a distinct charged lepton;

$$\nu_e \leftrightarrow e^- , \quad \nu_\mu \leftrightarrow \mu^- , \quad \nu_\tau \leftrightarrow \tau^- .$$

Finally, all of this physics was embodied in the standard electroweak model, out of which came the prediction of a new neutral interaction mediating neutrino scattering.

Despite all of this progress, a remarkable number of questions remain. We now believe neutrinos are massive, but still have no measurement of an absolute neutrino mass (only mass differences). Many models attribute the puzzle of neutrino mass – why these neutrinos are so much lighter than other Standard Model particles – to scales well beyond the Standard Model, but we lack independent experimental tools for probing these scales. We do not know the particle–antiparticle conjugation properties of neutrinos: because they carry no Standard Model charges, both the Dirac (distinct antiparticle) and the Majorana (no distinction between particle and antiparticle) possibilities are open. An associated question is the existence of nonzero electromagnetic moments: magnetic, charge radius, anapole and electric dipole. No nonzero moment has been measured.

Finally, there are many questions about the role of neutrinos in astrophysics and cosmology. We suspect cosmic background neutrinos contribute to dark matter and may influence large-scale structure formation. However, direct experimental attempts to measure background neutrinos have failed by many orders of magnitude to reach the expected density. Type II supernovae convert approximately 99% of the energy released in the infall into neutrinos of all flavors. Yet only $\bar{\nu}_e$s were detected from SN1987A. Supernova modelers predict that neutrinos play an essential role in the explosion mechanism and in the associated nucleosynthesis, yet there is disagreement about the success of neutrino-driven explosions. Finally, there is great interest in mounting searches for very high-energy astrophysical neutrinos that might be associated with active galactic nuclei, gamma ray bursts, etc.

Given all of these open questions, 70 years after Pauli's original suggestion, it would be nice to have a few more answers. There is every indication that some answers will come with the resolution of the solar neutrino puzzle.

4.3   The Standard Solar Model [4.4, 4.5]

Solar models trace the evolution of the sun over the past 4.7 billion years of main-sequence burning, thereby predicting the present-day temperature and composition profiles of the solar core that govern neutrino production. Standard solar models (SSMs) share four basic assumptions:

- The sun evolves in hydrostatic equilibrium, maintaining a local balance between the gravitational force and the pressure gradient. To describe this condition in detail, one must specify the equation of state as a function of temperature, density and composition.
- Energy is transported by radiation and convection. While the solar envelope is convective, radiative transport dominates in the core region where thermonuclear reactions take place. The opacity depends sensitively on the solar composition, particularly the abundances of heavier elements.
- Thermonuclear reaction chains generate solar energy. The Standard Model predicts that over 98% of this energy is produced from the pp chain conversion of four protons into ^{4}He (Fig. 4.1),

$$4p \to {}^4\text{He} + 2e^+ + 2\nu_e , \tag{4.1}$$

with proton burning through the CNO cycle contributing the remaining 2%. The sun is a large but slow reactor: the core temperature, $T_c \sim 1.5 \times 10^7$ K, results in typical center-of-mass energies for reacting particles of ~ 10 keV, much less than the Coulomb barriers inhibiting charged-particle nuclear reactions. Thus reaction cross sections are small: in most cases laboratory measurements are only possible at higher energies, so that cross section data must be extrapolated to the solar energies of interest.
- The model is constrained to produce today's solar radius, mass and luminosity. An important assumption of the Standard Model is that the sun was highly convective, and therefore uniform in composition, when it first entered the main sequence. It is furthermore assumed that the surface abundances of metals (nuclei with $A > 5$) were undisturbed by the subsequent evolution, and thus provide a record of the initial solar metallicity. The remaining parameter is the initial ^{4}He/H ratio, which is adjusted until the model reproduces the present solar luminosity in today's sun. The resulting ^{4}He/H mass fraction ratio is typically 0.27 ± 0.01, which can be compared with the Big Bang value of 0.23 ± 0.01. Note that the sun was formed from previously processed material.

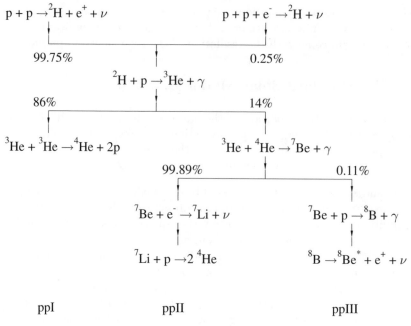

$$p + p \rightarrow {}^{2}\text{H} + e^{+} + \nu \qquad\qquad p + p + e^{-} \rightarrow {}^{2}\text{H} + \nu$$

99.75%   0.25%

$${}^{2}\text{H} + p \rightarrow {}^{3}\text{He} + \gamma$$

86%   14%

$${}^{3}\text{He} + {}^{3}\text{He} \rightarrow {}^{4}\text{He} + 2p \qquad\qquad {}^{3}\text{He} + {}^{4}\text{He} \rightarrow {}^{7}\text{Be} + \gamma$$

99.89%   0.11%

$${}^{7}\text{Be} + e^{-} \rightarrow {}^{7}\text{Li} + \nu \qquad\qquad {}^{7}\text{Be} + p \rightarrow {}^{8}\text{B} + \gamma$$

$${}^{7}\text{Li} + p \rightarrow 2\,{}^{4}\text{He} \qquad\qquad {}^{8}\text{B} \rightarrow {}^{8}\text{Be}^{*} + e^{+} + \nu$$

ppI   ppII   ppIII

Fig. 4.1. The solar pp chain

The model that emerges is an evolving sun. As the core's chemical composition changes, the opacity and core temperature rise, producing a 44% luminosity increase since the onset of the main sequence. The temperature rise governs the competition among the three cycles of the pp chain: the ppI cycle dominates below about 1.6×10^{7} K, the $ppII$ cycle between 1.7 and 2.3×10^{7} K, and the $ppIII$ above 2.4×10^{7} K. The central core temperature of today's SSM is about 1.55×10^{7} K.

The competition among the cycles determines the pattern of neutrino fluxes. Thus, one consequence of the thermal evolution of our sun is that the ^{8}B neutrino flux, the most temperature-dependent component, proves to be of relatively recent origin: the predicted flux increases exponentially with a doubling period of about 0.9 billion years.

A final aspect of SSM evolution is the formation of composition gradients on nuclear-burning timescales. Clearly, there is a gradual enrichment of the solar core in ^{4}He, the ashes of the pp chain. Another element, ^{3}He, can be considered a catalyst for the pp chain, being produced and then consumed, and thus eventually reaching some equilibrium abundance. The timescale for equilibrium to be established and the final equilibrium abundance are both sharply decreasing functions of temperature, and therefore increasing functions of the distance from the center of the core. Thus a steep ^{3}He density gradient is established over time.

Table 4.1. Solar neutrino sources and the flux predictions of the BP98 and Brun/Turck-Chieze/Morel (BTCM) SSMs in cm^{-2}s^{-1}

Source	E_ν^{max} (MeV)	BP98	BTCM98
$p + p \to {}^2\mathrm{H} + e^+ + \nu$	0.42	5.94×10^{10}	5.98×10^{10}
${}^{13}\mathrm{N} \to {}^{13}\mathrm{C} + e^+ + \nu$	1.20	6.05×10^8	4.66×10^8
${}^{15}\mathrm{O} \to {}^{15}\mathrm{N} + e^+ + \nu$	1.73	5.32×10^8	3.97×10^8
${}^{17}\mathrm{F} \to {}^{17}\mathrm{O} + e^+ + \nu$	1.74	6.33×10^6	
${}^8\mathrm{B} \to {}^8\mathrm{Be} + e^+ + \nu$	~ 15	5.15×10^6	4.82×10^6
${}^3\mathrm{He} + p \to {}^4\mathrm{He} + e^+ + \nu$	18.77	2.10×10^3	
${}^7\mathrm{Be} + e^- \to {}^7\mathrm{Li} + \nu$	0.86 (90%)	4.80×10^9	4.70×10^9
	0.38 (10%)		
$p + e^- + p \to {}^2\mathrm{H} + \nu$	1.44	1.39E8	1.41×10^8

The SSM has had some notable successes. From helioseismology (e.g. [4.6]), the sound speed profile $c(r)$ has been very accurately determined for the outer 90% of the sun, and is in excellent agreement with the SSM. Such studies verify important predictions of the SSM, such as the depth of the convective zone. However, the SSM is not a complete model, in that it does not explain all features of solar structure, such as the depletion of surface Li by two orders of magnitude. This is usually attributed to convective processes that operated at some epoch in our sun's history, dredging Li to a depth where burning takes place.

The principal neutrino-producing reactions of the pp chain and CNO cycle are summarized in Table 4.1. The first six reactions produce β decay neutrino spectra having allowed shapes, with endpoints given by E_ν^{max}. Deviations from an allowed spectrum occur for ${}^8\mathrm{B}$ neutrinos because the ${}^8\mathrm{Be}$ final state is a broad resonance. The last two reactions produce line sources of electron capture neutrinos, with widths ~ 2 keV characteristic of the temperature of the solar core. Measurements of the pp, ${}^7\mathrm{Be}$ and ${}^8\mathrm{B}$ neutrino fluxes will determine the relative contributions of the ppI, ppII and ppIII cycles to solar energy generation. As discussed above, and as later illustrations will show more clearly, this competition is governed in large classes of solar models by a single parameter, the central temperature T_c. The flux predictions of the 1998 calculations of Bahcall, Basu and Pinsonneault [4.4] (BP98) and of Brun, Turck-Chieze and Morel [4.5] are included in Table 4.1.

4.4 Solar Neutrino Experiments and their Implications

The first solar neutrino results were announced by Ray Davis Jr. and his Brookhaven collaborators in 1968, more than 30 years ago [4.7]. Located deep within the Homestake Gold Mine in Lead, South Dakota, the detector consists of a 100 000 gallon tank of C_2Cl_4. Solar neutrinos are captured by the reaction

$$^{37}Cl(\nu, e^-)^{37}Ar.$$

As the threshold for this reaction is 0.814 MeV, the important neutrino sources are the ^7Be and ^8B reactions. The ^7Be neutrinos excite just the Gamow–Teller (GT) transition to the ground state, the strength of which is known from the electron capture lifetime of ^{37}Ar. The ^8B neutrinos can excite all bound states in ^{37}Ar, including the dominant transition to the isobaric analogue state residing at an excitation energy of 4.99 MeV. The strength of excited-state GT transitions can be determined from the β decay $^{37}Ca(\beta^+)^{37}K$, which is the isospin mirror reaction to $^{37}Cl(\nu, e^-)^{37}Ar$. The net result is that, for SSM fluxes, 78% of the capture rate should be due to ^8B neutrinos, and 15% to ^7Be neutrinos. The measured capture rate [4.8], $2.56 \pm 0.16 \pm 0.16$ SNU (1 SNU = 10^{-36} captures/atom/s) is about one-third the SSM value.

Similar radiochemical experiments were begun in January, 1990 and May, 1991 by the SAGE and GALLEX collaborations, respectively, using a different target, ^{71}Ga. The special properties of this target include its low threshold and an unusually strong transition to the ground state of ^{71}Ge, leading to a large pp neutrino cross section (Fig. 4.2). The experimental capture rates are $66.6^{+6.8+3.8}_{-7.1-4.0}$ (SAGE) [4.9] and $77.5 \pm 6.2^{+4.3}_{-4.7}$ SNU (GALLEX) [4.10]. The SSM prediction is about 130 SNU [4.11]. Most importantly, since the pp flux is directly constrained by the solar luminosity in all steady-state models, there is a minimum theoretical value for the capture rate equal to 79 SNU, given Standard Model weak-interaction physics. Note there are substantial uncertainties in the ^{71}Ga cross section owing to capture of ^7Be neutrinos to two excited states of unknown strength. These uncertainties were greatly reduced by direct calibrations of both detectors using ^{51}Cr neutrino sources.

Experiments of a different kind, Kamiokande II/III and SuperKamiokande, exploit water Cerenkov detectors to view solar neutrinos event by event. Solar neutrinos scatter off electrons, with the recoiling electrons producing Cerenkov radiation that is then recorded in surrounding phototubes. Thresholds are determined by background rates; SuperKamiokande is currently operating with a trigger at approximately 6 MeV. The initial experiment, Kamiokande II/III, found a flux of ^8B neutrinos of $(2.80 \pm 0.19 \pm 0.33) \times 10^6/cm^2$ s after about a decade of measurement [4.12]. Its much larger successor SuperKamiokande, with a 22.5 kiloton fiducial volume, yielded the result $(2.45 \pm 0.04 \pm 0.07) \times 10^6/cm^2$ s after the first 825 days of measurements [4.13]. This is about 48% of the SSM flux. This result continues to improve in accuracy.

These results can be combined to limit the principal solar neutrino fluxes, under the assumption that no new particle physics distorts the spectral shape

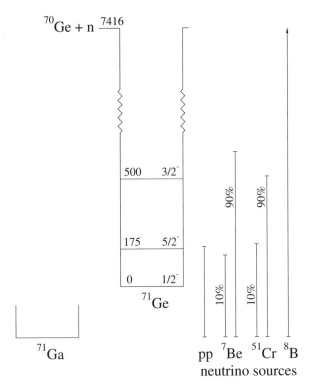

Fig. 4.2. Level scheme for ^{71}Ge showing the excited states that contribute to absorption of pp, ^{7}Be, ^{51}Cr and ^{8}B neutrinos

of the pp and ^{8}B neutrinos. One finds

$$\phi(pp) \sim 0.9\,\phi^{\mathrm{SSM}}(pp) \,,$$
$$\phi(^{7}\mathrm{Be}) \sim 0 \,,$$
$$\phi(^{8}\mathrm{B}) \sim 0.47\,\phi^{\mathrm{SSM}}(^{8}\mathrm{B}) \,. \tag{4.2}$$

A reduced ^{8}B neutrino flux can be produced by lowering the central temperature of the sun somewhat, as $\phi(^{8}\mathrm{B}) \sim T_{\mathrm{c}}^{18}$. However, such an adjustment, either by varying the parameters of the SSM or by adopting some nonstandard physics, tends to push the $\phi(^{7}\mathrm{Be})/\phi(^{8}\mathrm{B})$ ratio to higher values rather than the low one (4.2), since

$$\frac{\phi(^{7}\mathrm{Be})}{\phi(^{8}\mathrm{B})} \sim T_{\mathrm{c}}^{-10} \,. \tag{4.3}$$

Thus the observations seem difficult to reconcile with plausible solar-model variations: one observable ($\phi(^{8}\mathrm{B})$) requires a cooler core while a second, the ratio $\phi(^{7}\mathrm{Be})/\phi(^{8}\mathrm{B})$, requires a hotter one.

This physics was nicely illustrated by Castellani et al. [4.14]. These authors generated a series of nonstandard models by changing the S-factor for

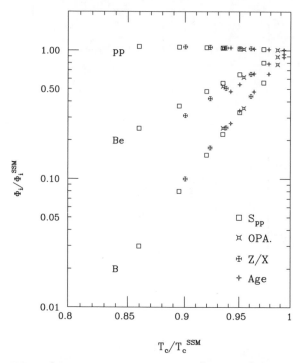

Fig. 4.3. The responses of the pp, ^{7}Be and ^{8}B neutrino fluxes to the indicated variations in solar-model input parameters, displayed as a function of the resulting central temperature T_c. From Castellani et al. [4.14]

the $p + p$ reaction, modifying the core metallicity, introducing weakly interacting massive particles as a new mechanism for energy transport, etc. The resulting core temperature T_c and neutrino fluxes were then determined, and the latter were plotted as a function of the former. The pattern that emerges is striking (Fig. 4.3): parameter variations producing the same value of T_c produce remarkably similar fluxes. Thus T_c provides an excellent one-parameter description of standard model perturbations. Figure 4.3 also illustrates the difficulty of producing a low ratio $\phi(^7\text{Be})/\phi(^8\text{B})$ when T_c is reduced. This result is consistent with our earlier argument and shows that even extreme changes in quantities such as the metallicity, opacity or solar age cannot produce the pattern of fluxes deduced from experiment (4.2).

Is it possible to change the solar model in a way that reduces the ^{7}Be/^{8}B neutrino flux ratio? It is appears the answer is no in models where the nuclear reactions burn in equilibrium, with plausible cross sections. However, Cumming and Haxton [4.15] pointed out that a remaining possibility was a nonequilibrium model in which the solar core was mixed on the timescale of ^{3}He evolution, about 10^7 years. Thus the pp chain is prevented from reaching equilibrium. This suggestion has some physical plausibility because it allows the sun to burn more efficiently, with a cooler core and enhanced ppI terminations. The SSM ^{3}He profile is known to be overstable, as was first discussed by Dilke and Gough [4.16]. Also, the possibility of a persistent convective core powered by the ^{3}He gradient has been discussed in the literature. (The SSM core is convective for about 10^8 years because of out-of-equilibrium burning of the CNO cycle.) A strong argument against a mixed core was offered by Bahcall et al. [4.17], who showed that homogenizing the core of the SSM led to very large changes in the helioseismology. While this is a sobering result, this test was not done in a self-consistent model. However, there is work in progress to test the helioseismology of a more realistic mixed-core model – one where the ^{4}He content of the core, the temperature gradient, the nuclear-reaction rates and the luminosity are handled consistently [4.18]. If the helioseismology remains unacceptable, this will rule out the only solar-model conjecture for producing a reduced ^{7}Be/^{8}B flux ratio, which is necessary to produce fluxes closer to those observed.

However, there is a popular argument showing that no SSM change can completely remove the discrepancy with experiment: if one assumes undistorted neutrino spectra, no combination of pp, ^{7}Be and ^{8}B neutrino fluxes fits the experimental results well [4.19]. In fact, in an unconstrained fit, the required ^{7}Be flux is unphysical, negative by about 2.5σ. This is clearly a strong hint that one should look elsewhere for a solution!

The remaining possibility is new neutrino physics. Suggested particle physics solutions of the solar neutrino problem include neutrino oscillations, neutrino decay, neutrino magnetic moments and weakly interacting massive particles. Among these, the Mikheyev–Smirnov–Wolfenstein (MSW) effect – neutrino oscillations enhanced by matter interactions – is widely regarded as perhaps the most plausible.

4.5 Neutrino Oscillations

One odd feature of particle physics is that neutrinos, which are not required by any symmetry to be massless, nevertheless must be much lighter than any of the other known fermions. For instance, the current limit on the $\bar\nu_e$ mass is $\lesssim 5$ eV. The Standard Model requires neutrinos to be massless, but the reasons are not fundamental. Dirac mass terms $m_{\rm D}$, analogous to the mass terms for other fermions, cannot be constructed, because the model contains no right-handed neutrino fields. Neutrinos can, however, have Majorana mass terms

$$\overline{\nu_{\rm L}^{\rm c}} m_{\rm L} \nu_{\rm L} \quad \text{and} \quad \overline{\nu_{\rm R}^{\rm c}} m_{\rm R} \nu_{\rm R} \,, \tag{4.4}$$

where the subscripts L and R denote left- and right-handed projections of the neutrino field ν, and the superscript c denotes charge conjugation. The first term above is constructed from left-handed fields, but can arise only as a nonrenormalizable effective interaction when one is constrained to generate

m_L with the doublet scalar field of the Standard Model. The second term is absent from the Standard Model because there are no right-handed neutrino fields.

None of these Standard Model arguments carries over to the more general, unified theories that theorists believe will supplant the Standard Model. In the enlarged multiplets of extended models it is natural to characterize the fermions of a single family, e.g. ν_e, e, u, d, by the same mass scale m_D. Small neutrino masses are then frequently explained as a result of the Majorana neutrino masses. In the seesaw mechanism,

$$M_\nu \sim \begin{pmatrix} 0 & m_D \\ m_D^T & m_R \end{pmatrix} . \tag{4.5}$$

Diagonalization of this matrix produces one light neutrino, $m_{\text{light}} \sim m_D$ (m_D/m_R), and one unobservably heavy, $m_{\text{heavy}} \sim m_R$. The factor (m_D/m_R) is the needed small parameter that accounts for the distinct scale of neutrino masses. The masses for the ν_e, ν_μ and ν_τ are then related to the squares of the corresponding quark masses m_u, m_c and m_t. Taking $m_R \sim 10^{16}$ GeV, a typical grand-unification scale for models built on groups like $SO(10)$, the seesaw mechanism gives the crude relation

$$m_{\nu_e} : m_{\nu_\mu} : m_{\nu_\tau} \leftrightarrow 2 \times 10^{-12} : 2 \times 10^{-7} : 3 \times 10^{-3} \text{eV} . \tag{4.6}$$

The fact that solar neutrino experiments can probe small neutrino masses, and thus provide insight into possible new mass scales m_R that are far beyond the reach of direct accelerator measurements, has been an important theme of the field.

Consider for simplicity just two neutrino flavors. The states of definite mass are the states that diagonalize the free Hamiltonian. Similarly, the weak-interaction eigenstates are the states of definite flavor, that is, the ν_e accompanies the positron in β decay, and the ν_μ accompanies the muon. There is every reason to assume that these two bases are not coincident, but instead are related by a nontrivial rotation,

$$|\nu_e\rangle = \cos\theta_v |\nu_1\rangle + \sin\theta_v |\nu_2\rangle ,$$
$$|\nu_\mu\rangle = -\sin\theta_v |\nu_1\rangle + \cos\theta_v |\nu_2\rangle , \tag{4.7}$$

where θ_v is the (vacuum) mixing angle.

Consider a ν_e produced at time $t = 0$ as a momentum eigenstate[2]

$$|\nu(t=0)\rangle = |\nu_e\rangle = \cos\theta_v |\nu_1\rangle + \sin\theta_v |\nu_2\rangle . \tag{4.8}$$

The resulting probability for measuring a ν_e downstream then depends on $\delta m^2 = m_2^2 - m_1^2$;

$$P_{\nu_e}(t) = |\langle \nu_e | \nu(t) \rangle|^2$$
$$= 1 - \sin^2 2\theta_v \sin^2 \left(\frac{\delta m^2 t}{4k} \right) \rightarrow 1 - \frac{1}{2} \sin^2 2\theta_v , \tag{4.9}$$

---

[2] There have been several treatments from the perspective of wave packets, rather than plane waves. For recent work, see [4.20].

where the limit on the right is appropriate for large t. (When one properly describes the neutrino state as a wave packet, the large-distance behavior follows from the eventual separation of the mass eigenstates.) If the the oscillation length

$$L_o = \frac{4\pi\hbar c E}{\delta m^2 c^4} \tag{4.10}$$

is comparable to or shorter than one astronomical unit, a reduction in the solar ν_e flux would be expected in terrestrial detectors.

The suggestion that the solar neutrino problem could be solved by neutrino oscillations was first made by Pontecorvo in 1958, who pointed out the analogy with $K_0 \leftrightarrow \bar{K}_0$ oscillations. From the point of view of particle physics, the sun is a marvelous neutrino source. The neutrinos travel a long distance and have low energies (~ 1 MeV), implying a sensitivity of

$$\delta m^2 \gtrsim 10^{-12} \text{ eV}^2 \ . \tag{4.11}$$

In the seesaw mechanism, $\delta m^2 \sim m_2^2$, so neutrino masses as low as $m_2 \sim 10^{-6}$ eV could be probed. In contrast, terrestrial oscillation experiments with accelerator or reactor neutrinos are typically limited to $\delta m^2 \gtrsim 0.1$ eV^2. (Planned long-baseline experiments, though, will soon push below 0.01 eV^2.)

From the expressions above one expects vacuum oscillations to affect all neutrino species equally, if the oscillation length is small compared with an astronomical unit. This is somewhat in conflict with the data, as we have argued that the ^{7}Be neutrino flux is quite suppressed. Furthermore, there is a weak theoretical prejudice that θ_v should be small, like the Cabibbo angle. The first objection, however, can be circumvented in the case of "just so" oscillations, where the oscillation length is comparable to one astronomical unit. In this case the oscillation probability becomes sharply energy-dependent, and one can choose δm^2 to preferentially suppress one component (e.g. the monochromatic ^{7}Be neutrinos). This scenario has been explored by several groups and remains an interesting possibility. However, the requirement of large mixing angles remains.

4.6   The Mikheyev–Smirnov–Wolfenstein Mechanism [4.21]

In order to include matter effects, we first consider vacuum oscillations for the more general case

$$|\nu(t=0)\rangle = a_e(t=0)|\nu_e\rangle + a_\mu(t=0)|\nu_\mu\rangle \ , \tag{4.12}$$

from which one easily calculates

$$i\frac{d}{dx}\begin{pmatrix} a_e \\ a_\mu \end{pmatrix} = \frac{1}{4E}\begin{pmatrix} -\delta m^2\cos 2\theta_v & \delta m^2\sin 2\theta_v \\ \delta m^2\sin 2\theta_v & \delta m^2\cos 2\theta_v \end{pmatrix}\begin{pmatrix} a_e \\ a_\mu \end{pmatrix} . \tag{4.13}$$

We have equated $x = t$, that is, set $c = 1$.

Mikheyev and Smirnov [4.21] showed in 1985 that the density dependence of the neutrino effective mass, a phenomenon first discussed by Wolfenstein in 1978, could greatly enhance oscillation probabilities: a ν_e is adiabatically transformed into a ν_μ as it traverses a critical density within the sun. It became clear that the sun was not only an excellent neutrino source, but also a natural regenerator for cleverly enhancing the effects of flavor mixing.

While the original work of Mikheyev and Smirnov was numerical, their phenomenon was soon understood analytically as a level-crossing problem. The vacuum oscillation evolution equation changes in the presence of matter to

$$
\mathrm{i}\frac{\mathrm{d}}{\mathrm{d}x}\begin{pmatrix} a_e \\ a_\mu \end{pmatrix}
= \frac{1}{4E}\begin{pmatrix} 2E\sqrt{2}G_\mathrm{F}\rho(x) - \delta m^2 \cos 2\theta_\mathrm{v} & \delta m^2 \sin 2\theta_\mathrm{v} \\ \delta m^2 \sin 2\theta_\mathrm{v} & -2E\sqrt{2}G_\mathrm{F}\rho(x) + \delta m^2 \cos 2\theta_\mathrm{v} \end{pmatrix}\begin{pmatrix} a_e \\ a_\mu \end{pmatrix}, \quad (4.14)
$$

where G_F is the weak coupling constant and $\rho(x)$ the solar electron density. The new contribution to the diagonal elements, $2E\sqrt{2}G_\mathrm{F}\rho(x)$, represents the effective contribution to M_ν^2 that arises from neutrino–electron scattering. The indices of refraction of electron and muon neutrinos differ because the former scatter by charged and neutral currents, while the latter have only neutral-current interactions. The difference in the forward-scattering amplitudes determines the density-dependent splitting of the diagonal elements of the new matter equation.

It is helpful to rewrite this equation in a basis consisting of the light and heavy local mass eigenstates (i.e. the states that diagonalize the right-hand side of the equation),

$$
|\nu_\mathrm{L}(x)\rangle = \cos\theta(x)|\nu_e\rangle - \sin\theta(x)|\nu_\mu\rangle ,
$$
$$
|\nu_\mathrm{H}(x)\rangle = \sin\theta(x)|\nu_e\rangle + \cos\theta(x)|\nu_\mu\rangle . \quad (4.15)
$$

The local mixing angle is defined by

$$
\sin 2\theta(x) = \frac{\sin 2\theta_\mathrm{v}}{\sqrt{X^2(x) + \sin^2 2\theta_\mathrm{v}}} ,
$$
$$
\cos 2\theta(x) = \frac{-X(x)}{\sqrt{X^2(x) + \sin^2 2\theta_\mathrm{v}}} , \quad (4.16)
$$

where $X(x) = 2\sqrt{2}G_\mathrm{F}\rho(x)E/\delta m^2 - \cos 2\theta_\mathrm{v}$. Thus $\theta(x)$ ranges from θ_v to $\pi/2$ as the density $\rho(x)$ goes from 0 to ∞.

If we define

$$
|\nu(x)\rangle = a_\mathrm{H}(x)|\nu_\mathrm{H}(x)\rangle + a_\mathrm{L}(x)|\nu_\mathrm{L}(x)\rangle , \quad (4.17)
$$

the neutrino propagation can be rewritten in terms of the local mass eigenstates

$$i\frac{d}{dx}\begin{pmatrix} a_H \\ a_L \end{pmatrix} = \begin{pmatrix} \lambda(x) & i\alpha(x) \\ -i\alpha(x) & -\lambda(x) \end{pmatrix}\begin{pmatrix} a_H \\ a_L \end{pmatrix}, \tag{4.18}$$

with the splitting of the local mass eigenstates determined by

$$2\lambda(x) = \frac{\delta m^2}{2E}\sqrt{X^2(x) + \sin^2 2\theta_v} \tag{4.19}$$

and with the mixing of these eigenstates governed by the density gradient,

$$\alpha(x) = \left(\frac{E}{\delta m^2}\right)\frac{\sqrt{2}\,G_F(d/dx)\rho(x)\sin 2\theta_v}{X^2(x) + \sin^2 2\theta_v}. \tag{4.20}$$

The results above are quite interesting: the local mass eigenstates diagonalize the matrix if the density is constant. In such a limit, the problem is no more complicated than our original vacuum oscillation case, although our mixing angle is changed because of the matter effects. But if the density is not constant, the mass eigenstates in fact evolve as the density changes. This is the crux of the MSW effect. Note that the splitting achieves its minimum value, $(\delta m^2/2E)\sin 2\theta_v$, at a critical density $\rho_c = \rho(x_c)$,

$$2\sqrt{2}EG_F\rho_c = \delta m^2 \cos 2\theta_v, \tag{4.21}$$

that defines the point where the diagonal elements of the original flavor matrix cross.

Our local-mass-eigenstate form of the propagation equation can be trivially integrated if the splitting of the diagonal elements is large compared with the off-diagonal elements,

$$\begin{aligned}\gamma(x) &= \left|\frac{\lambda(x)}{\alpha(x)}\right| \\ &= \frac{\sin^2 2\theta_v}{\cos 2\theta_v}\frac{\delta m^2}{2E}\frac{1}{|(1/\rho_c)(d\rho(x)/dx)|}\frac{[X(x)^2 + \sin^2 2\theta_v]^{3/2}}{\sin^3 2\theta_v} \gg 1,\end{aligned} \tag{4.22}$$

a condition that becomes particularly stringent near the crossing point, where

$$\gamma_c = \gamma(x_c) = \frac{\sin^2 2\theta_v}{\cos 2\theta_v}\frac{\delta m^2}{2E}\frac{1}{|(1/\rho_c)(d\rho(x)/dx)|_{x=x_c}} \gg 1. \tag{4.23}$$

The resulting adiabatic electron neutrino survival probability [4.22], valid when $\gamma_c \gg 1$, is

$$P_{\nu_e}^{\text{adiab}} = \frac{1}{2} + \frac{1}{2}\cos 2\theta_v \cos 2\theta_i, \tag{4.24}$$

where $\theta_i = \theta(x_i)$ is the local mixing angle at the density where the neutrino was produced.

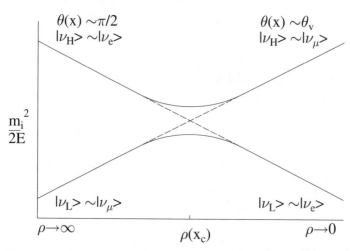

Fig. 4.4. Schematic illustration of the MSW crossing. The *dashed lines* correspond to the electron–electron and muon–muon diagonal elements of the M^2 matrix in the flavor basis. Their intersection defines the level-crossing density ρ_c. The *solid lines* are the trajectories of the light and heavy local mass eigenstates. If the electron neutrino is produced at high density and propagates adiabatically, it will follow the heavy-mass trajectory, emerging from the sun as a ν_μ

The physical picture behind this derivation is illustrated in Fig. 4.4. One makes the usual assumption that, in vacuum, the ν_e is almost identical to the light mass eigenstate, $\nu_L(0)$, i.e. $m_1 < m_2$ and $\cos\theta_v \sim 1$. But as the density increases, the matter effects make the ν_e heavier than the ν_μ, with $\nu_e \to \nu_H(x)$ as $\rho(x)$ becomes large. That is, the mixing angle at high density rotates to $\pi/2$. The special property of the sun is that it produces ν_es at high density that then propagate to the vacuum, where they are measured. The adiabatic approximation tells us that if initially $\nu_e \sim \nu_H(x)$, the neutrino will remain on the heavy mass trajectory provided the density changes slowly. That is, if the solar density gradient is sufficiently gentle, the neutrino will emerge from the sun as the heavy vacuum eigenstate, $\sim \nu_\mu$. This guarantees nearly complete conversion of ν_es into ν_μs, producing a flux that cannot be detected by the Homestake and SAGE/GALLEX detectors.

But this does not explain the curious pattern of partial flux suppressions coming from the various solar neutrino experiments. The key to this is the behavior when $\gamma_c \lesssim 1$. Our expression for $\gamma(x)$ shows that the critical region for nonadiabatic behavior occurs in a narrow region (for small θ_v) surrounding the crossing point, and that this behavior is controlled by the derivative of the density. This suggests an analytic strategy for handling nonadiabatic crossings: one can replace the true solar density by a simpler (integrable!) two-parameter form that is constrained to reproduce the true density and its derivative at the crossing point x_c. Two convenient choices are the linear

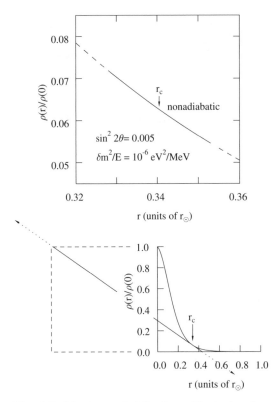

Fig. 4.5. The *top part* of the figure illustrates, for one choice of $\sin^2 2\theta$ and δm^2, that the region of nonadiabatic propagation (*solid line*) is usually confined to a narrow region around the crossing point r_c. In the *bottom part* of the figure, the *solid lines* represent the solar density and a linear approximation to that density that has the correct initial and final values, as well as the correct density and density derivative at r_c. Thus the linear profile is a very good approximation to the sun in the vicinity of the crossing point. The MSW equations can be solved analytically for this wedge. By extending the wedge to $\pm\infty$ (*dotted lines*) and assuming adiabatic propagation in these regions of unphysical density, one obtains the simple Landau–Zener result discussed in the text

($\rho(x) = a + bx$) and exponential ($\rho(x) = ae^{-bx}$) profiles. As the density derivative at x_c governs the nonadiabatic behavior, this procedure should provide an accurate description of the hopping probability between the local mass eigenstates when the neutrino traverses the crossing point. The initial and final points x_i and x_f for the artificial profile are then chosen so that $\rho(x_i)$ is the density where the neutrino was produced in the solar core and $\rho(x_f) = 0$ (the solar surface), as illustrated in Fig. 4.5. Since the adiabatic result ($P_{\nu_e}^{\text{adiab}}$) depends only on the local mixing angles at these points, this choice builds in that limit. But our original flavor-basis equation can then be integrated exactly for linear and exponential profiles, with the results given in terms of parabolic cylinder and Whittaker functions, respectively.

That result can be simplified further by observing that the nonadiabatic region is generally confined to a narrow region around x_c, away from the endpoints x_i and x_f. We can then extend the artificial profile to $x = \pm\infty$, as illustrated by the dashed lines in Fig. 4.5. As the neutrino propagates adiabatically in the unphysical region $x < x_i$, the exact solution in the physical region can be recovered by choosing the initial boundary conditions

$$a_L(-\infty) = -a_\mu(-\infty) = \cos\theta_i \exp\left[-i \int_{-\infty}^{x_i} \lambda(x)dx\right] ,$$

$$a_H(-\infty) = a_e(-\infty) = \sin\theta_i \exp\left[i \int_{-\infty}^{x_i} \lambda(x)dx\right] . \tag{4.25}$$

That is, $|\nu(-\infty)\rangle$ will then adiabatically evolve to $|\nu(x_i)\rangle = |\nu_e\rangle$ as x goes from $-\infty$ to x_i. The unphysical region $x > x_f$ can be handled similarly.

With some algebra, a simple generalization of the adiabatic result emerges that is valid for all $\delta m^2/E$ and θ_v:

$$P_{\nu_e} = \frac{1}{2} + \frac{1}{2}\cos 2\theta_v \cos 2\theta_i (1 - 2P_{hop}) , \tag{4.26}$$

where P_{hop} is the Landau–Zener probability of hopping from the heavy-mass trajectory to the light trajectory on traversing the crossing point. For the linear approximation to the density [4.23, 4.24],

$$P_{hop}^{lin} = e^{-\pi\gamma_c/2} . \tag{4.27}$$

As it must by our construction, P_{ν_e} reduces to $P_{\nu_e}^{adiab}$ for $\gamma_c \gg 1$. When the crossing becomes nonadiabatic (e.g. $\gamma_c \ll 1$), the hopping probability goes to 1, allowing the neutrino to exit the sun on the light-mass trajectory as a ν_e, i.e. no conversion occurs.

Thus there are two conditions for strong conversion of solar neutrinos: there must be a level crossing (that is, the solar core density must be sufficient to render $\nu_e \sim \nu_H(x_i)$ when it is first produced) and the crossing must be adiabatic. The first condition requires that $\delta m^2/E$ not be too large, and the second $\gamma_c \gtrsim 1$. The combination of these two constraints, illustrated in Fig. 4.6, defines a triangle of interesting parameters in the $(\delta m^2/E)$–$\sin^2 2\theta_v$ plane, as Mikheyev and Smirnov first found. A remarkable feature of this triangle is that strong $\nu_e \to \nu_\mu$ conversion can occur for very small mixing angles ($\sin^2 2\theta \sim 10^{-3}$), unlike in the vacuum case.

One can envision superimposing on Fig. 4.6 the spectrum of solar neutrinos, plotted as a function of $\delta m^2/E$ for some choice of δm^2. Since Davis sees *some* solar neutrinos, the solutions must correspond to the boundaries of the triangle in Fig. 4.6. The horizontal boundary indicates the maximum $\delta m^2/E$ for which the sun's central density is sufficient to cause a level crossing. If a spectrum straddles this boundary properly, we obtain a result consistent

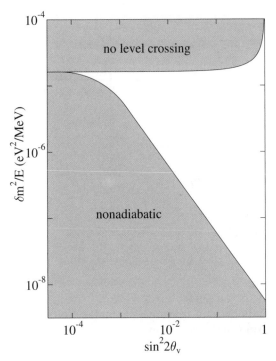

Fig. 4.6. MSW conversion for a neutrino produced at the sun's center. The *upper shaded region* indicates those values of $\delta m^2/E$ where the vacuum mass splitting is too great to be overcome by the solar density. Thus no level crossing occurs. The *lower shaded region* defines the region where the level crossing is nonadiabatic (γ_c less than unity). The *unshaded region* corresponds to adiabatic level crossings where strong $\nu_e \to \nu_\mu$ conversion will occur

with the Homestake experiment, in which low-energy neutrinos (large $1/E$) lie above the level-crossing boundary (and thus remain ν_es), but the high-energy neutrinos (small $1/E$) fall within the unshaded region, where strong conversion takes place. Thus such a solution would mimic nonstandard solar models in that only the ^{8}B neutrino flux would be strongly suppressed. The diagonal boundary separates the adiabatic and nonadiabatic regions. If the spectrum straddles this boundary, we obtain a second solution in which low-energy neutrinos lie within the conversion region, but the high-energy neutrinos (small $1/E$) lie below the conversion region and are characterized by $\gamma \ll 1$ at the crossing density. (Of course, the boundary is not a sharp one, but is characterized by the Landau–Zener exponential). Such a nonadiabatic solution is quite distinctive as the flux of pp neutrinos, which is strongly constrained in the standard solar model and in any steady-state nonstandard model by the solar luminosity, is now sharply reduced. Finally, one can imagine "hybrid" solutions where the spectrum straddles both the level-crossing

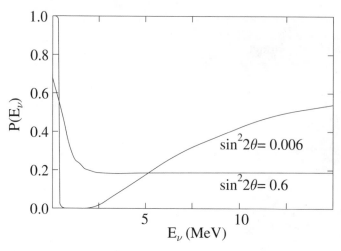

Fig. 4.7. MSW survival probabilities $P(E_\nu)$ for typical small–mixing-angle and large-mixing-angle solutions

(horizontal) boundary and the adiabaticity (diagonal) boundary for small θ, thereby reducing the ^{7}Be neutrino flux more than either the pp or the ^{8}B fluxes.

Several interesting solutions – i.e. oscillation scenarios that fit the Homestake, Kamiokande/SuperKamiokande and SAGE/GALLEX constraints well – can be found, depending on whether one considers oscillations into active or sterile states, oscillations enhanced by matter or vacuum oscillations, etc. These solutions can be distinguished only if new experiments are performed. For example, as various MSW solutions distort the neutrino spectrum in different ways, experiments with increased spectral sensitivity are important. This is illustrated in Fig. 4.7, where the survival probabilities $P_{\nu_e}^{\mathrm{MSW}}(E)$ for for two commonly discussed MSW oscillation solutions (large and small mixing angles) are contrasted.

An example of careful searches for solutions to the solar neutrino problem is the recent work of Bahcall, Krastev and Smirnov [4.25], the results of which are given in Figs. 4.8 and 4.9. Figure 4.8 shows three active oscillation solutions. Two of these, the small-mixing-angle (SMA) region (centered on $\delta m^2 \sim 5 \times 10^{-6}$ eV^2 and $\sin^2 2\theta_v \sim 7 \times 10^{-3}$) and the large-mixing-angle (LMA) region (centered on $\delta m^2 \sim 3 \times 10^{-5}$ and $\sin^2 2\theta_v \sim 0.8$), generate the spectral distortions shown in Fig. 4.7. The third solution region, LOW (low mass, low probability), corresponds to $\delta m^2 \sim 10^{-7}$ eV^2 and $\sin^2 2\theta_v \sim 0.9$. A fourth solution involving oscillations into sterile neutrinos is not shown, but roughly coincides with the SMA region. Similarly, Fig. 4.9 shows the vacuum oscillation regions allowed by the results of Bahcall et al., which correspond to much lower values of δm^2.

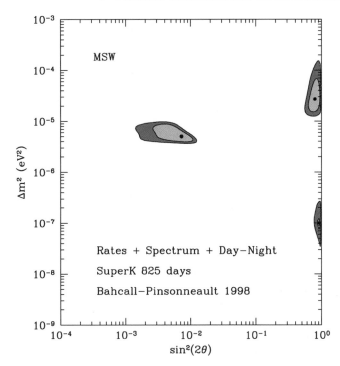

Fig. 4.8. The MSW SMA, LMA and LOW solution regions for oscillations into active states. These regions were determined from global fits to the rates for the chlorine, SAGE/GALLEX and SuperKamiokande experiments, as well as to the electron recoil energy spectrum and day–night effects measured by SuperKamiokande. From [4.25]

The MSW mechanism provides a natural explanation for the pattern of observed solar neutrino fluxes. While it requires profound new physics, both massive neutrinos and neutrino mixing are expected in extended models. The importance of nonzero neutrino masses is that they may provide a window on new physics far beyond the Standard Model. One illustration of this is the quadratic seesaw pattern mentioned earlier. As the SMA solution corresponds to $\delta m^2 \sim$ a few $\times 10^{-6}$ eV^2, it is consistent with a heavy neutrino mass $\sim$ a few $\times 10^{-3}$ eV. This is a typical ν_τ mass in models where $m_R \sim m_{\mathrm{GUT}}$. On the other hand, if this mass is associated with the second-generation neutrino (so that the seesaw Dirac mass is equated to the muon mass), this gives $m_R \sim 10^{12}$ GeV and predicts a heavy $\nu_\tau \sim 10$ eV. Such a mass is of great interest cosmologically as it would have consequences for supernova physics, the dark-matter problem and the formation of large-scale structure.

There are many interesting elaborations of the MSW effect that are not discussed here, but are treated in many papers: spin–flavor oscillations induced by the solar magnetic field [4.26, 4.27] (the mass difference between ν_e^{L}

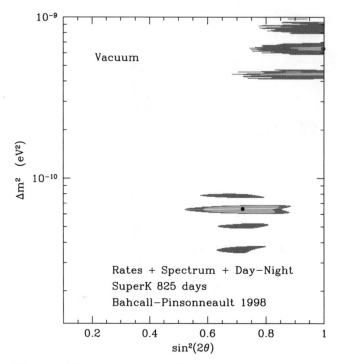

Fig. 4.9. Allowed vacuum oscillation regions as determined by Bahcall, Krastev and Smirnov [4.25]. The global fit was to the same data as were shown in Fig. 4.8

and a sterile ν_μ^R is compensated by matter effects); oscillations induced by density fluctuations [4.28–4.30]; "stochastic depolarization" effects in large, random magnetic fields [4.31]; and others. There are also interesting effects associated with the existence of three neutrinos: if the solar neutrino problem is due to MSW ν_e–ν_μ oscillations, one might expect a ν_e–ν_τ crossing at still higher densities. This has led to many interesting speculations about the role of the MSW mechanism in supernova explosions and in supernova nucleosynthesis, and about the possibility that ν_e–ν_τ oscillations governed by small mixing angles might be best probed using the supernova neutrino flux.

The solar neutrino problem, the strong evidence for neutrino oscillations from SuperKamiokande and other atmospheric neutrino experiments, and the claim by the LSND group for $\bar{\nu}_\mu \to \bar{\nu}_e$ oscillations at the LAMPF beamstop appear to establish that neutrinos have mass. Understanding the full pattern of neutrino masses and mixing angles and their implications for new physics will very likely have to await a great deal of additional work. Indeed, one current puzzle presented by the solar, atmospheric and LSND results is that they appear to require three independent δm^2s. That is, the values do not respect the relation

$$\delta m_{21}^2 + \delta m_{13}^2 + \delta m_{32}^2 = 0 \ . \tag{4.28}$$

Thus, either one or more of the experiments must be attributed to some phenomenon other than neutrino oscillations, or a fourth neutrino is required. That neutrino must be sterile to avoid constraints imposed by the known width of the Z_0.

4.7 Outlook

The argument that the solar neutrino problem is due to neutrino oscillations is, in a sense, circumstantial: this conclusion is derived from combining several experiments, no one of which requires new particle physics. There is no direct observation of new physics analogous to the zenith angle dependence of the SuperKamiokande atmospheric results. For this reason there is great interest in a new experiment now taking data in the Creighton nickel mine in Sudbury, Ontario, 6800 feet below the surface. The Sudbury Neutrino Observatory (SNO) has a central acrylic vessel filled with one kiloton of very pure (99.92%) heavy water, surrounded by a shield of 7.5 kilotons of ordinary water. SNO can detect electron neutrinos through the charged-current reaction

$$\nu_e + d \to p + p + e^- . \tag{4.29}$$

The Cerenkov light from the outgoing electron is then recorded in the array of 9800 phototubes that surround SNO's central vessel. The spectrum of electrons produced is quite hard, making reconstruction of the energy of the ν_e easier than in the case of neutrino–electron elastic scattering. Thus the experimenters may be able to detect distortions of the neutrino spectrum resulting from the MSW effect.

SNO will also study the neutral-current reaction

$$\nu_x(\bar{\nu}_x) + d \to \nu_x(\bar{\nu}_x) + p + n \tag{4.30}$$

by detecting the neutron produced, either through (n, γ) reactions on salt dissolved in the heavy water or in ^{3}He proportional counters. In this way the experimenters will obtain an integral measurement of the flux of active neutrinos, independent of flavor. Thus a neutral-current signal clearly larger than the corresponding ν_e signal would show that heavy-flavor neutrinos comprise a portion of the solar neutrino flux, providing definite proof of new physics. The SNO collaboration is expected to make its first announcement of results for the charged-current reaction early in 2001.

A great deal of effort [4.32] is also being focused on developing detectors sensitive to the lower-energy ^{7}Be and pp neutrinos. Indeed, the apparent absence of a ν_e ^{7}Be flux is one of the key inferences from existing data. Borexino [4.33], a detector consisting of 300 tons of liquid scintillator, will measure in real time scattered electrons from the reaction $\nu_x + e \to \nu_x + e$. The monoenergetic ^{7}Be neutrinos give rise to a Compton-like recoil spectrum with an edge at 660 keV, a feature that will be distinctive if backgrounds are suppressed sufficiently. The collaboration constructed and operated a five-ton prototype detector (the Counting Test Facility, or CTF) from 1995 to 1997 in order to assess background issues and other potential technical problems (deterioration of detector materials, radon penetration into the detector, etc.) The full-scale experiment, which will be conducted in Gran Sasso, could begin in 2001 or soon after.

The Homestake collaboration is developing a new radiochemical detector based on the reaction $^{127}I(\nu_e, e^-)^{127}Xe$. Like the reaction on chlorine, a noble gas is produced that can be extracted from very large volumes and readily counted. There are significant advantages to iodine because of its much larger cross section for solar-neutrino absorption, easier extraction and counting of the product Xe, and potential sensitivity to ^7Be neutrinos. The collaboration is also investigating hybrid detectors using both iodine and chlorine in which active detection of electrons produced in ^8B neutrino reactions is combined with rapid extraction and counting of the daughter atoms produced [4.34]. Such a determination of the ^8B neutrino contribution to the chlorine rate would allow the experimenters to deduce the charged-current ^7Be/CNO contribution too, clearly.

The GALLEX collaboration is conducting an improved and substantially larger experiment, the Gallium Neutrino Observatory (GNO) [4.35]. The improvements in size, to perhaps 100 tons, and in counting efficiency could result in a measurement accurate to about 5%. A result clearly below the minimum rate ($\sim$ 80 SNU) for a steady-state standard solar model, obtained by postulating solar burning via the ppI cycle only, would be very significant, demanding by itself either new neutrino physics or a transient sun. First results from GNO, a counting rate of $65.8^{+10.7}_{-10.2}$ SNU (1σ), were recently reported [4.35].

There are efforts under way that might produce detectors capable of recording pp neutrino events in real time. HERON [4.32] uses a target of superfluid ^4He, a detector medium that is exceptionally pure. The recoiling electrons produced in $\nu_x + e^-$ reactions lead to both prompt ultraviolet photons and a delayed roton/phonon evaporation signal at the free surface of the liquid. These signals are detected in an array of low-mass silicon or sapphire wafer calorimeters external to the liquid. The procedures have been demonstrated through early experiments with low-energy alpha particles; current work is focused on various improvements, such as more sensitive wafer detectors, necessary for the detection of neutrinos. A second development effort, HELLAZ [4.36], uses a gaseous helium detector and images the recoiling electron in a very large time projection chamber. The effort exploits a developed particle physics detection scheme, but requires improvements in scale and in operations at high pressure and low temperature.

These and a number of other detection schemes [4.37–4.40] promise to yield important results some years from now. In the immediate future, as the SNO neutral- and charged-current results become precise and as SuperKamiokande continues to amass data, the nature of the solar neutrino problem should become much clearer. The hope is that these results, in combination with Borexino and other next-generation experiments, with new atmospheric and (possibly) supernova neutrino measurements, and with precision tests of oscillations at accelerators and reactors, will allow us to completely characterize the neutrino mass matrix, providing a window on physics well beyond the Standard Model.

This work was supported in part by the US Department of Energy. I thank Plamen Krastev for providing Figs. 4.8 and 4.9.

References

4.1 W. C. Haxton and B. Holstein, Am. J. Phys. **68**, 15 (2000).

4.2 E. Fermi, Z. Phys. **88**, 161 (1934).

4.3 C. L. Cowan et al., Science **124**, 103 (1956).

4.4 J. N. Bahcall, S. Basu and M. H. Pinsonneault, Phys. Lett. B **433**, 1 (1998).

4.5 A. S. Brun, S. Turck-Chieze and P. Morel, Astrophys. J. **506**, 913 (1998); S. Turck-Chieze and I. Lopez, Astrophys. J. **408**, 347 (1993).

4.6 G. Fiorentini and B. Ricci, to appear in *Proceedings of Neutrino Telescopes '99* (Venice) (astro-ph/9905341).

4.7 R. Davis Jr., D. S. Harmer and K. C. Hoffman, Phys. Rev. Lett. **20**, 1205 (1968).

4.8 K. Lande, talk presented at Neutrino '98, Takayama, Japan, June 1998.

4.9 J. N. Abdurashitov et al., Phys. Lett. B **328**, 234 (1994); talk presented at Neutrino '98, Takayama, Japan, June 1998.

4.10 P. Anselmann et al., Phys. Lett. B **285**, 376 (1992); T. Kirsten, Rev. Mod. Phys. **71**, 1213 (1999).

4.11 J. N. Bahcall, *Neutrino Astrophysics* (Cambridge University Press, Cambridge, 1989).

4.12 Y. Suzuki, Nucl. Phys. B **38**, 54 (1995).

4.13 Y. Suzuki, talk presented at Lepton–Photon '99, Stanford, USA, August 1999.

4.14 V. Castellani, S. Degl'Innocenti, G. Fiorentini, M. Lissia and B. Ricci, Phys. Rev. D **50**, 4749 (1994).

4.15 A. Cumming and W. C. Haxton, Phys. Rev. Lett. **77**, 4286 (1996).

4.16 F. W. W. Dilke and D. O. Gough, Nature **240**, 262 (1972).

4.17 J. N. Bahcall, M. H. Pinsonneault, S. Basu and J. Christensen-Dalsgaard, Phys. Rev. Lett. **78**, 171 (1997).

4.18 R. Epstein and J. Guzik, private communication; V. Berezinsky, G. Fiorentini and M. Lissia, Phys. Rev. D **60**, 123002 (1999).

4.19 K. M. Heeger and R. G. H. Robertson, Prog. Part. Nucl. Phys. **40**, 135 (1998); Phys. Rev. Lett. **77**, 3720 (1996).

4.20 M. Nauenberg, Phys. Lett. B **447**, 23 (1999).

4.21 S. P. Mikheyev and A. Smirnov, Sov. J. Nucl. Phys. **42**, 913 (1985); L. Wolfenstein, Phys. Rev. D **17**, 2369 (1979).

4.22 H. Bethe, Phys. Rev. Lett. **56**, 1305 (1986).

4.23 W. C. Haxton, Phys. Rev. Lett. **57**, 1271 (1986).

4.24 S. J. Parke, Phys. Rev. Lett. **57**, 1275 (1986).

4.25 J. N. Bahcall, P. I. Krastev and A. Yu. Smirnov, Phys. Lett. B **477**, 401 (2000).

4.26 C. S. Lim and W. J. Marciano, Phys. Rev. D **37**, 1368 (1988).

4.27 E. Kh. Akhmedov, Sov. J. Nucl. Phys. **48**, 382 (1988).

4.28 A. Schaefer and S. E. Koonin, Phys. Lett. B **185**, 417 (1987).

4.29 P. I. Krastev and A. Yu. Smirnov, Phys. Lett. B **226**, 341 (1989).

4.30 W. C. Haxton and W. M. Zhang, Phys. Rev. D **43**, 2484 (1991).

4.31 F. N. Loreti, Y. Z. Qian, G. M. Fuller and A. B. Balantekin, Phys. Rev. D **52**, 2264 (1995); A. B. Balantekin, J. M. Fetter and F. N. Loreti, Phys. Rev. D **54**, 3941 (1995).

4.32 R. E. Lanou Jr., Nucl. Phys. B **77**, 55 (1999).

4.33 L. Oberauer, Nucl. Phys. B **77**, 48 (1999).

4.34 K. Lande, talk presented at the Carolina Symposium on Neutrino Physics, February 2000.

4.35 T. A. Kirsten, Nucl. Phys. B **77**, 26 (1999); M. Altmann et al., Phys. Lett. B **490**, 16 (2000).

4.36 C. Tao, in *Fourth International Solar Neutrino Conference*, ed. by W. Hampel (Max-Planck-Institut fur Kernphysik Press, Heidelberg, 1997), p. 238.

4.37 R. S. Raghavan, Phys. Rev. Lett. **78**, 3618 (1997).

4.38 M. Galeazzi et al., Phys. Lett. B **398**, 187 (1997).

4.39 H. Ejiri et al., submitted to Phys. Rev. Lett. (nucl-ex/9911008).

4.40 W. C. Haxton et al., nucl-th/0011014.

5 The Atmospheric Neutrino Anomaly: Muon Neutrino Disappearance

John G. Learned

5.1 Introduction

With the 1998 announcement of new evidence for muon neutrino disappearance observed by the SuperKamiokande experiment [5.30], the more than a decade old atmospheric neutrino anomaly moved from a possible indication of neutrino oscillations to an almost inescapable implication. In this chapter the evidence is reviewed, and indications are presented that the oscillations are probably between muon and tau neutrinos with maximal mixing. Implications and future directions are discussed.

The understanding of this phenomenon is now dominated by the data announced by the SuperKamiokande collaboration at Neutrino98, of which group the present author is a member. Much of this report dwells upon those results and updates to them, and so credit for this work is due to the whole collaboration, listed in the appendix to this chapter, who have labored hard to bring this experiment to fruition and who have been ably led by Prof. Yoji Totsuka of the University of Tokyo. That said, this report presents personal recollections and opinions of the author, particularly in matters of the previous history, interpretation of the present situation and future prospects for this line of research.

The phenomenon of neutrino oscillations is discussed in several other chapters of this volume (Chaps. 2 and 9 in particular), and to those the reader is directed for derivation of the expressions utilized in this chapter and for understanding of the origin and implications of neutrino oscillations generally. Model-building and implications in astrophysics and cosmology are likewise treated elsewhere, while in the following we focus narrowly upon the atmospheric neutrino anomaly, its experimental explication in terms of muon neutrino oscillations with tau neutrinos and the implications of those results.

5.1.1 Atmospheric Neutrinos

The neutrinos under discussion in this chapter arise from the decay of pions and other mesons and also muons, which are produced in the earth's atmosphere [5.1, 5.2]. The atmosphere is being constantly bombarded with cosmic rays, which consist mostly of protons but also include heavy nuclei and electrons and even neutral particles. The earth's magnetic field plus other

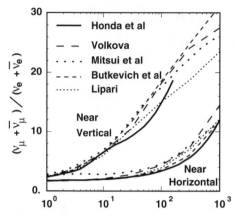

Fig. 5.1. The calculated ratio of the flux of atmospheric muon neutrinos to electron neutrinos versus neutrino energy. From Honda et al. [5.7]

magnetic fields cuts off the lower-energy particles from the sun and more distant sources, so that the mean incoming kinetic energy is around 1 GeV. Cosmic rays with lower energies do not cause effects which we can directly detect on earth or underground. Particles with energies in the multi-GeV range make showers in the roughly ten-interaction-length-thick (vertical column density) atmosphere. Cosmic-ray collisions with air nuclei produce pions and other particles in abundance, which themselves interact further or decay. This competition between interaction and decay leads to a steeper spectrum for the decay products. At energies below several GeV the muons produced in the decay of charged pions themselves decay:

$$\pi^+ \to \mu^+ \, \nu_\mu \, , \quad \pi^- \to \mu^- \, \bar{\nu}_\mu \, , \quad \mu^+ \to e^+ \, \nu_e \, \bar{\nu}_\mu \, , \quad \mu^- \to e^- \, \bar{\nu}_e \, \nu_\mu \, , \quad (5.1)$$

with decay lengths of $L_{\pi^\pm} = 0.056$ km $\times$ $E_\pi/$GeV and $L_\mu = 6.23$ km $\times$ $E_\mu/$GeV. Typical pion interaction lengths (roughly 150 g/cm^2) are on the order of a few km, depending upon altitude, angle and energy, while muons generally come to rest before decaying or being absorbed. Moreover (crucially and often ignored), the energy-sharing in the decays is such that the resulting neutrinos are also of nearly equal energy. These decay kinematics are of course well known, so the ratio of muon neutrinos to electron neutrinos can be calculated with rather good accuracy, about 5%, almost independently of the cosmic-ray spectrum [5.3, 5.5], as illustrated in Fig. 5.1.

Precise neutrino flux calculations (to a few percent) from man-made sources are difficult if not impossible, as indicated in other chapters in this book. The problem is even more difficult for the atmospheric neutrinos, since the absolute magnitude of the incoming cosmic-ray flux is not well known, being uncertain at present to perhaps 25% [5.3]. Calculations of the atmospheric neutrino flux in the few-GeV energy range require not only the input cosmic-ray flux, with appropriate modulation to account for solar-cycle variation and geomagnetic field, but also details of nucleon–nucleon and meson–nucleon interactions, not all of which have been well measured. The neutrino flux calculations also lead to *muon* flux predictions, and these can be (and have been) compared with data, though the appropriate data for low-energy muons at high altitude, as recorded in balloon measurements, are sparse and imprecise. Typically these calculations incorporate the approximation that the incoming cosmic rays, the secondaries and even the neutrinos all travel in the same direction. This is no doubt not a serious compromise in the few-GeV energy range, but has some effects in the energy range of a few hundred MeV. At the present time new calculations are in progress,[1] but the computer time required to do the full simulation is still a limiting factor and the job has yet to be done definitively.

Several features of the neutrino flux are worth mentioning. The muon neutrino flux can be approximated as a power law with a spectral index $\gamma \simeq -3.7$ [5.7] for energies between about 10 GeV and 100 TeV. The electron neutrino (and antineutrino) fluxes, which largely arise from muon decay, fall off more swiftly above several GeV, with strong angle dependence. As illustrated in Fig. 5.1, the ν_μ to ν_e flux ratio falls to a few percent at the higher energies, where the ν_es are mostly produced in kaon decay, $K^+ \to \pi^0 e^+ \nu_e$ (4.82% branching ratio). See Fig. 5.2 for atmospheric neutrino spectra of all three flavors, as expected under several assumptions about oscillations (two-flavor oscillations only).

There is a significant zenith angle variation in the atmospheric neutrino flux, more prominent at higher energies, called the "secant theta" effect. This is simply due to the fact that those pions and muons which are produced by incoming cosmic rays with trajectories nearly tangential to the earth have more flight time in a less dense atmosphere and hence more chance to decay. Thus there is a peak, becoming more prominent at higher energies, near the horizontal arrival direction in the atmospheric-neutrino angular distribution. This peak is symmetric about the horizon for any location except at the lowest neutrino energies, below around 400 MeV, where geomagnetic effects spoil the symmetry somewhat.

The atmospheric-neutrino energies practically accessible in underground experiments range from the few tens of MeV to the 1 TeV range, and the flight distances from roughly 20 km for down-coming neutrinos to 13 000 km for those traversing the earth from the far side. The neutrino cross section is sufficiently small that there should be negligible attenuation of these neutrinos: the attenuation is roughly $2.4 \times 10^{-5} E_\nu / \mathrm{GeV}$ for neutrinos traversing the earth's core, and thus negligible for any energies below about 100 TeV. In consequence of the large dynamic range in both energy and flight distance, the atmospheric neutrinos are potentially sensitive to oscillations over a range of mass-squared differences from about 10^{-4} to 10^{-1} eV^2, as is discussed below.

---

[1] New calculations by groups in Japan, Italy and the US are under way at the time of writing (June 2000). Preliminary updates may be found in [5.8].

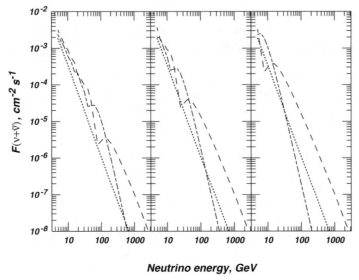

Neutrino energy, GeV

Fig. 5.2. Fluxes of neutrinos of all three flavors (*dots*, ν_e; *dashes*, ν_μ; *long-short dashes*, ν_τ) in the presence of $\nu_\mu \rightarrow \nu_\tau$ oscillations with maximal mixing and $\Delta m^2 = 10^{-2}$, $10^{-2.5}$, 10^{-3} eV^2, from *left* to *right*. Figure from Stanev [5.3]

5.1.2   Initial Indications

We shall not dwell upon the past history,[2] but shall note that the atmospheric neutrino anomaly has been around for some time, roughly since the mid-1980s. Indeed, the first notice of something peculiar in the atmospheric neutrino data stems from the 1960s when the seminal underground experiments in South Africa [5.9] and southern India [5.10] first detected the natural neutrinos and observed something of an absolute rate deficit, but not convincingly, as the flux predictions were rough and the statistics small.

A new round of instruments were built, beginning in the late 1970s, to search for nucleon decay as predicted by the (soon to be discarded) $SU(5)$ unification model. The problem with atmospheric neutrinos, a background to nucleon decay searches, became serious after the activation of the first large underground water Cerenkov detector, the IMB experiment, and by 1983 it was realized that the number of events containing muon decays was less than expected.[3] Soon this was confirmed by the second large water detector, the Kamioka experiment.[4] The Kamioka group extended the results with

---

[2] See Chap. 1 for an overview.

[3] The first published discussion of the muon decay deficit is apparently in [5.11], but the authors discount it as a serious discrepancy; the deficit is mentioned cryptically and no implications are elaborated in [5.12].

[4] [5.13] has no report of an anomaly; the first mention of an anomaly in the Kamioka data appears in [5.14].

good particle identification, giving a redundant measure of the relative muon deficit (as also did the IMB group). Some members of the IMB [5.15][5] and the Kamioka [5.16][6] groups began to suggest, at least in private, that oscillations were the cause of the deficit, but that conclusion was not widely taken seriously for nearly ten years. Indeed, the first published interpretations of the anomaly as due to oscillations were largely from outside the experimental groups [5.17]. To be fair, though, it seems to be the Kamioka group who first seriously believed that the anomaly was due to oscillations and not simply a detector or background problem.

The deficit is usually characterized as an R value, the ratio-of-ratios, the double ratio of muon to electron neutrinos, observed to expected. The effect was large; the observed R was about two-thirds of the expected value. On the initial evidence, the oscillations could have been from muon neutrinos to others (e.g. ν_τ or a new neutrino) or between the muon and electron neutrinos themselves. It was the ratio that was in deficit: one could not be sure whether there was an excess of electron neutrinos, a deficit of muon neutrinos or both. This led to suggestions of other possible "physics" causes, such as nucleon decay favoring electron modes (since the anomaly was not initially detected above the nucleon mass) or an excess of extraterrestrial electron neutrinos. See Table 5.1 in Sect. 5.2.10 for a summary of the situation. There were also suggestions of systematic problems, such as problems in muon identification, something wrong with flux calculations or neutrino interaction cross sections, backgrounds entering, or generic problems with the water Cerenkov detectors.

Over the intervening years between the emergence of this "atmospheric neutrino anomaly", as it became known, and the 1998 SuperK announcement, a great deal of effort went into study of these possible systematic causes of the anomaly. One troubling concern was that two European experiments, the NUSEX [5.18] and Frejus [5.19] detectors, did not observe any anomaly. Hence some people suspected a peculiarity of water as a target or with the employment of the Cerenkov radiation in vertex location. However, not only were the statistics of the European detectors relatively small, but also, as indicated by more recent work from a similar type of instrument in the US, the Soudan II detector [5.20], the presence of a surrounding veto counter is vital for the more compact type of slab detectors. Also, the MACRO experiment [5.23] has elucidated the nonnegligible production of low-energy (hundreds of MeV) pions by nearby cascades in rock; these particles enter cracks in nonhermetic detectors and appear to be neutrino interactions. In any case the Soudan II detector, now with significant exposure (4.6 kiloton years), finds an R value close to that of SuperK (and of IMB and Kamioka) [5.20].

---

[5] The IMB collaboration never claimed evidence for neutrino oscillations.

[6] The 1994 paper of [5.16] contains the first published suggestion of neutrino oscillations by the Kamioka collaboration.

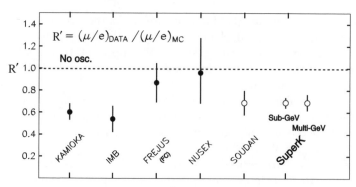

Fig. 5.3. The double ratio R of muon to electron neutrino events; data divided by expectations for various underground atmospheric neutrino detectors. From A. Mann [5.21]

Figure 5.3 shows these R values for the several experiments. Note that, under the assumption of no oscillations, all experiments should record $R = 1.0$, but if oscillations are taking place, then R should be reduced but not necessarily to the same value for all experiments, as it depends upon the energy range being studied. The European detectors had higher thresholds, which may partly explain their failure to detect the anomaly.

5.2   The SuperKamiokande Revolution

We now proceed to summarize the evidence for oscillations, which, while depending largely upon the SuperK experiment, has received important confirmation and consistency of results from Soudan II and MACRO (and consistency with previous, smaller experiments as well, except as discussed below). Before going on, it may be worthwhile to point out what permitted the big breakthrough with SuperK, which may not be obvious. The increase in size of detector, from near-kiloton fiducial volumes for Kamioka and Soudan and three kilotons for IMB to the twenty-two kilotons of SuperK, is not the whole story. As will be seen below, the most striking progress comes from the recording of muon events with good statistics in the energy region above 1 GeV. This is due to the detector linear dimensions as well as gross target volume: muon events with an energy more than 1 GeV and thus 5 m range were not likely to be fully contained in the Kamioka detector (or the IMB detector). SuperK, in contrast, has decent muon statistics up to almost 5 GeV, and this turns out to be crucial.

The most important data to be discussed below are those from the "fully contained" (FC) single-ring event sample, consisting of those events in which both the neutrino interaction vertex and resulting particle tracks remain entirely within the fiducial volume. For these events the relativistic-charged-particle energy and direction are well determined. We shall use the notation

FC for the single-ring events, which are about two-thirds of the total and arise mostly from quasi-elastic charged-current interactions in which the recoil nucleon is not seen. The multiring events have not yet been much used in analysis, owing to the ambiguous interpretation of overlapping rings from track segments in a Cerenkov detector (except, as discussed below, in the case of tests distinguishing the muon neutrino's oscillating partner). Moreover, the multipion final states are not modeled reliably in simulations as yet, and there are further complications of final-state nuclear scattering as well.

There are also "partially contained" (PC) events, in which a muon exits the fiducial volume from a contained vertex location. Such events are useful even though the total energy is not known, the energy observed being a lower limit. Of course, this is the case even with FC events, though to a lessor degree, because the observed particles are not of the same energy (or direction) as the incident neutrino, which is what one would like to know.

The particle types are identified by pattern recognition software, now well tested and verified by experiment with known particle beams at the accelerator [5.46]. Fortunately most of the contained events show single (Cerenkov-radiating) tracks in which the identification is quite clean (at the 98% level), as illustrated in Fig. 5.4. To be clear and cautious we usually refer to the reconstructed events as "muon-like" and "electron-like", though a safe approximation is that these represent muon and electron neutrino charged-current interactions.

The other two categories of events which we shall discuss are the through-going upwards-moving muons (UM), produced by neutrino interactions in the rock or outer detector and which come from directions below the horizon (as those from above the horizon can be confused with down-going muons from cosmic-ray interactions in the atmosphere near to overhead). Another category of event is the entering–stopping muon (SM). It is useful that these event categories probe approximately three different energy ranges of neutrinos: FC $\simeq$ 1 GeV; PC and SM $\simeq$ 10 GeV; UM $\simeq$ 100 GeV. This is illustrated in Fig. 5.5. It should be understood that, as far as we know, these neutrinos are all produced in the upper atmosphere by cosmic-ray interactions and are reasonably well described by models in terms of content, energy, and angular dependence (to a few percent) [5.3].

We shall not take up limited space here with the description of the SuperK detector, which is well documented elsewhere. The interested reader would do well to look at some of the theses from SuperK, which are available on the Web [5.26]. The short summary is that the SuperK detector consists of a large stainless steel cylinder (37 m high by 34 m diameter inside the inner detector) containing a structure holding 13 142 large (20 inch diameter) photo-multipliers. With extremely high photocathode coverage (40%), nearly an acre of photocathode and ten times more pixels than any earlier instrument, the instrument possesses a remarkable sensitivity of roughly eight

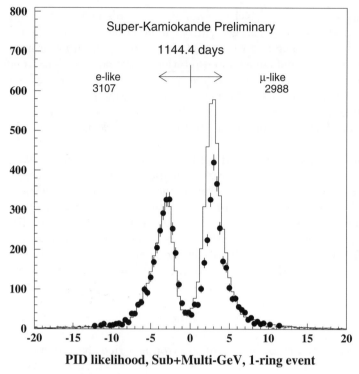

Fig. 5.4. Particle identification parameter distribution of the SuperK fully contained single-ring data and Monte Carlo simulation, illustrating the electron-like and muon-like separation [5.22]

photoelectrons per MeV of deposited (Cerenkov-radiating) energy. The latter permits detection of events down to < 5 MeV, so, for the present discussion, detection efficiency versus energy is not important, because the events we are discussing are all above ≃ 100 MeV. The inner volume is also well protected by a 2 m thick, fully enclosing veto Cerenkov counter, populated by 1800 recycled IMB (8 inch) photomultipliers with wavelength-shifting collars. Further, the inner "fiducial" volume is taken as 2 m inside the inner photomultiplier surface, resulting in the 22.5 kiloton volume used for most reported data.

The SuperK oscillations claim was first formally presented to the physics community in June 1998 at the Neutrino98 meeting in Takayama, Japan. The data were presented in several papers to the community [5.27–5.29], building upon past data from Kamioka [5.16] and IMB [5.15], and culminating in the claim of observation of oscillations of muon neutrinos, published in *Physical Review Letters* in August 1998 [5.30]. We now proceed to review the evidence, which has changed little except for new indications that the oscillating partner of the ν_μ is probably the ν_τ, and not a hypothetical sterile neutrino.

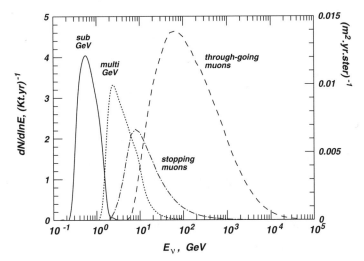

Fig. 5.5. Event rates as a function of neutrino energy for fully contained events ($E < 1.3$ GeV and $E > 1.3$ GeV), stopping muons (and similarly partially contained events) and through-going muons. From Engel et al. [5.4]

5.2.1 Up–Down Asymmetry

One way to look at the FC (and PC) data is in terms of a dimensionless up-to-down ratio, difference over sum (which has symmetrical errors, in contrast to just up/down) [5.32]. Downwards-going neutrinos have flown ~ 20–700 km, while upwards-going neutrinos have traveled ~ 700–10 000 km. The angle between the neutrino and the observed charged lepton is on average of the order of $40°/\sqrt{E_\nu/\text{GeV}}$, and the typical observed energy is half the neutrino energy. Thus the mixing of the hemispheres of origin of the events is important only for the lowest energies (below roughly 400 MeV). This asymmetry quantity is exhibited as a function of charged-particle momentum in Fig. 5.6, for both electrons and muons, with the PC data shown as well (for which we know only a minimum momentum), from an exposure of 70.4 kiloton years in SuperK. One sees that the electron data fit satisfactorily to no asymmetry, whilst the muon data show strong momentum dependence, starting from no asymmetry and dropping to about $-1/3$ above 1.3 GeV.

From this figure alone, without need for complex and often opaque Monte Carlo simulations, assuming the deviation from uniformity to be due to neutrino oscillations, one can deduce the following.

1. The atmospheric neutrino anomaly is largely due to disappearing muons, not excess electrons.
2. There is little or no coupling of the muon neutrino to the electron neutrino in this energy/distance range.
3. The oscillations of the muon neutrinos must be nearly maximal for the asymmetry to approach one-third.

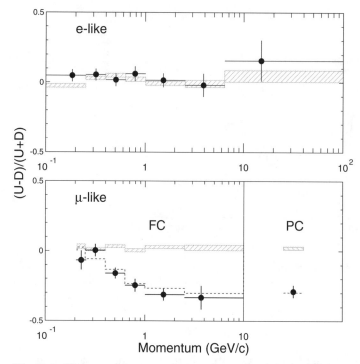

Fig. 5.6. The up-to-down asymmetry for muon (2486) and electron (2531) single-ring fully contained and partially contained (665) events in SuperK, from 1144.4 days of live time (analyzed by June 2000), as a function of observed charged-particle momentum. The muon data include a point for the partially contained events (PC) with a momentum of more than about 1 GeV/c. The *hatched region* indicates no-oscillation expectations, and the *dashed line* ν_μ–ν_τ oscillations with $\Delta m^2 = 3.2 \times 10^{-3}$ eV^2 and maximal mixing [5.22]

4. The scale of oscillations must be of the order of 1 GeV/200 km, within a factor of several times.

In fact, as can be seen from the dashed lines overlying the data points, the simulations produce an excellent fit to the muon neutrino oscillation hypothesis, while the no-oscillations hypothesis is strongly rejected. The deviation from the no-oscillaion hypothesis is so strong that statistical fluctuations as the cause of the deviation are completely improbable; one must look for systematic problems in order to escape the oscillations explanation.

One concern for some people has been the fact that the asymmetry is indeed maximal, which makes it appear that we are very lucky that the size of the earth and the cosmic-ray energies are "just so" to produce this dramatic effect. This appears to this author to fall in the category of lucky coincidences, such as the angular diameter of the moon and sun being the same as seen from earth. There is another oscillations-related peculiar coincidence, that the matter oscillation scale turns out to be close to one earth diameter, and this depends upon the Fermi constant and the electron column density of the earth. The phase space for "coincidences" is very large, and we humans are great recognizers of such patterns.

5.2.2 Neutrino Flux Dependence Upon the Terrestrial Magnetic Field

The effect of the earth's magnetic field on the atmospheric neutrino flux is a little complicated, but only important for very low energies. For example, for energies of a few GeV, the magnetic field provides some shielding from straight-downwards-going charged cosmic rays in regions near the magnetic equator. For higher energies and incoming trajectories near the horizon, the magnetic field still prevents some arrival paths. The effect is not up–down symmetric, and this spoils the symmetry otherwise expected from the neutrinos about the horizontal plane. However, the effects are mostly limited to neutrino energies below about 1 GeV, corresponding to cosmic-ray primaries below about 10 GeV. The picture is made a bit more complicated by the earth's magnetic field not being a nice, symmetrical dipole. Fortunately, there are good models of the magnetic field, and the people who have made flux calculations take these effects into account, though (in the past) largely through a simple cutoff momentum depending on location. More recent calculations trace particles backwards in the magnetic field and determine trajectories that escape to infinity ([5.24] and references therein). Lipari has, however, recently shown that the double-humped cosmic-ray spectra seen in the AMS experiment, with a space-borne magnetic spectrometer in low earth orbit, may be due to particles in trapped orbits [5.25]. Moreover, Lipari points out that there are hints in the AMS data of a North–South asymmetry, which could bias the neutrino flux calculations and even pull the derived value of Δm^2. However, it should be emphasized that the effect of such variation from simple expectations will only bias the lowest-energy data from SuperK (roughly below 400 MeV), and analysis has demonstrated that the results quoted herein are stable against raising the acceptance energy for the data sample.

The SuperK group has published a paper [5.29] examining the azimuthal variation of the SuperK data ($\pm30°$ about the horizon) for intermediate to higher energies (400–3000 MeV), in an energy region where the calculations are thought to be reliable. Indeed, the SuperK data do exhibit significant variation from uniformity while fitting the flux predictions very well, giving one confidence in the modeling [5.29].

5.2.3 Natural Parameters for Oscillations: L/E

In an ideal world, one would assuredly study these data as a function of distance divided by energy, L/E, since that is the parameter in which one

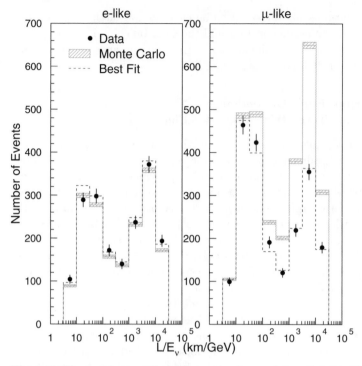

Fig. 5.7. The numbers of SuperK events observed, compared with the predicted numbers, as a function of the natural oscillations parameter L/E, distance divided by energy. The results are not normalized. The two peaks correspond to generally down-going (*left*) and up-going (*right*) particles. One sees that the muon deficit begins even in the upper hemisphere. The *shaded area* indicates no-oscillations expectations, and the *dashed line* the fit for ν_μ–ν_τ oscillations with maximal mixing and $\Delta m^2 = 0.0032$ eV^2 [5.22]

expects to see oscillatory behavior. For two-neutrino mixing with a mass-squared difference Δm^2 and a mixing angle θ, the probability of a muon neutrino of energy E_ν remaining a muon neutrino at distance L is given by [5.56]

$$P_{\mu\mu} = 1 - \sin^2 2\theta \sin^2 \left(1.27 \frac{\Delta m^2}{\text{eV}^2} \frac{L}{\text{km}} \frac{\text{GeV}}{E_\nu} \right). \tag{5.2}$$

However, since we observe only the secondary charged particle's energy and direction, badly smeared at the energies available (L/E_ν smeared by about a factor of two), plots in which one would wish for visible oscillations can at best show a smooth slide from the no-oscillations region to the oscillating regime. This is illustrated in Figs. 5.7 and 5.8, where the numbers of events, and the ratios of those numbers of events observed to those expected with no

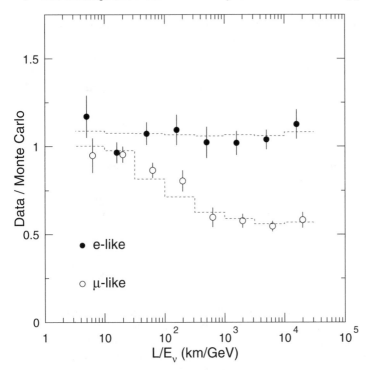

Fig. 5.8. The ratio of the number of events observed to the number predicted as a function of the natural oscillations parameter, distance divided by energy. The results are not normalized and overall there is a slight excess (about 8%, compared with a systematic uncertainty of 25%) compared with expectations. Electrons show no evidence for oscillations, while muons exhibit a strong drop with L/E. This is consistent with ν_μ–ν_τ oscillations with maximal mixing and $\Delta m^2 = 0.0032$ eV^2, as indicated by the *dashed lines* from the simulation [5.22]

oscillations, are plotted versus "L/E",[7] for muon and electron (type) events. The updated data are preliminary data from the SuperK 1144 day sample.

The plot is not "normalized", and we see something of an excess of electron-type events overall (+8%). This is a little worrisome, but acceptable since (as already noted) the absolute flux is uncertain to a larger extent. In contrast to the electron data, the muon points fall relative to the no-oscillations expectations with increasing L/E beyond about 50 km/GeV, reaching a plateau at about one-half of their initial value, consistent with maximal mixing. The results of including muon (to tau) neutrino oscillations in the Monte Carlo simulation are indicated by dotted lines and fit the data reasonably well.

---

[7] See [5.33] for a derivation of the correction used for translating the observed energy into L/E. The figure presented here is the SuperK official plot.

As noted, these data do not (and could not) show oscillations, owing to convolutions washing out the oscillatory behavior. It was this smooth fall, however, that caused the author and some colleagues to wonder if another model might fit the data, one in which one component of the muon neutrino decays rather than oscillates with distance. Two papers [5.34, 5.35] suggested neutrino decay to explain the atmospheric neutrino anomaly. I shall not discuss details here, but shall note that in order to construct a viable model we had to push on all available limits and invoke neutrino mass and mixing in any case. Consequently such models do not pass the economy test of Occam's razor, though, most annoyingly, they remain not ruled out as yet.

One may note that detecting multiple oscillation peaks is not ruled out in principle for detectors such as SuperK or Soudan II that employ atmospheric neutrinos. It is a matter of recording the final state of the muon neutrino charged-current events, including nuclear recoil, with sufficient accuracy and statistics. Detectors such as a liquid-argon device of the ICARUS type are claimed to have the resolution, if large enough. Soudan II has, apparently, good enough resolution to accumulate a "golden sample" in which the nuclear recoil is detected, permitting reconstruction of the incident neutrino energy and direction. Unfortunately, Soudan II does not have enough mass to achieve definitive statistics in a practical observing period [5.20]. Another possibility is that SuperK, with enough exposure and more highly developed analysis, would be able to accumulate an adequate sample of events in which the recoil proton is detected above the Cerenkov threshold. At the moment none of the above promises success.

Considering future experiments, the attempt to discern oscillations as a function of L/E is one area in which improvement may indeed be made. The MINOS [5.53] detector in Minnesota, with a neutrino beam from Fermilab, and the large detectors to be constructed in Gran Sasso, ICANOE [5.54] and OPERA [5.55], detecting a neutrino beam from CERN, give some hope of being able to yield oscillatory plots. A hypothetical detector, such as a megaton version of the Aqua-RICH instrument studied by Ypsilantis and colleagues, could have the resolution to see a multipeaked L/E plot [5.36, 5.37]. Nearer to technical development, the proposed MONOLITH experiment would consist of a 30 kiloton pile of magnetized iron and tracking-detector layers, and employ cosmic rays to detect the first dip in L/E in the up-going muon flux through the earth [5.38].

5.2.4 Energy and Angle Variation

The SuperK collaboration's preferred method of fitting the ensemble (single-ring) FC and PC data is to employ a χ^2 test on numbers of events binned by particle type, angle and energy, a total of 70 bins. The bin choices may seem a bit peculiar, but they have historical precedent (they are the same as employed for Kamiokande) and, though not optimal for the new data set,

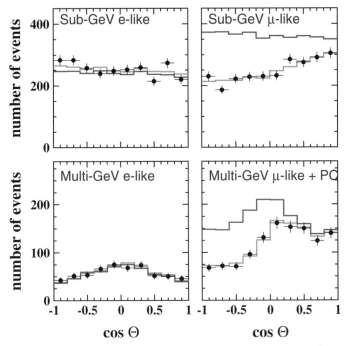

Fig. 5.9. Cosine-of-zenith-angle distributions of the contained and partially contained event data for two different energy ranges (above and below 1.3 GeV), for electron and muon single-ring events; 1144 live days of SuperK data (preliminary analysis) are indicated by *dots* with statistical error bars. The *black line* shows the no-oscillations simulation result, and the *gray line* that for oscillations between muon and tau neutrinos with the best-fit $\Delta m^2 = 0.0032$ eV^2 and maximal mixing [5.22]

this choice permits avoidance of any statistical (or confidence) penalty for choosing arbitrary bins. The fit employs a set of parameters to account for potential systematic biases. Details cannot be presented here, but it has been shown that the numerical results are quite insensitive to the selection of the parameters or their supposed "errors" (except for the overall normalization).[8] This method of systematic-error handling has been shown to be equivalent to employment of the correlation matrix of parameters [5.33].

Figure 5.9 illustrates the data plotted for two energy intervals (sub-GeV and multi-GeV, less or more than 1.3 GeV) for single-track events identified

---

[8] Nonetheless, the present author (personal opinion, not that of the collaboration) suspects that the process pulls the minima slightly towards lower values of Δm^2. The reasoning is that the parameters introduced unweight the effect of R on the fitting, which pulls upwards, while the shape pulls downwards. The author's bet remains that Δm^2 settles at around 5×10^{-3} eV^2, whereas the official fits give 2.5 to 3.5×10^{-3} ev^2.

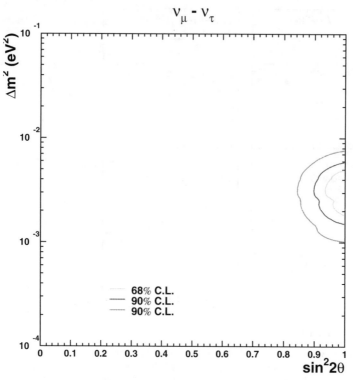

Fig. 5.10. Inclusion plot, showing the regions for three levels of statistical accept-ability in the plane of mixing angle and mass-squared difference for ν_μ–ν_τ oscilla-tions. This is from a preliminary analysis (June 2000) of the SuperK contained and partially contained event data from 1144 days [5.22]

as either electron-like or muon-like. The partially contained data is displayed with the multi-GeV muon data. The data are shown as a function of the cosine of the zenith angle, with +1 being down-going. One sees that the data fit very well the curves from the Monte Carlo simulation, at the values obtained from the grand-ensemble fit, $\Delta m^2 = 0.003$ eV^2 and $\sin^2(2\theta) = 1.0$. The limit on Δm^2 is 0.002 to 0.007 eV^2, and $\sin^2(2\theta) > 0.85$ at 90% confidence level.

The results of the fits are often presented in terms of an inclusion plot, showing an acceptable region(s) in the space of mixing angle ($\sin^2 2\theta$) and mass-squared difference (Δm^2), as presented in Fig. 5.10. The Δm^2 value at minimum χ^2 has moved a little upwards with the accumulated statistics, though not by much (good news for long-baseline experiments anyway), but remains uncertain to about a factor of two.

It is noteworthy that the earlier indications of and constraints upon the oscillation parameters from Kamiokande, IMB and Soudan gave somewhat larger values of Δm^2. All of these results depended upon fitting the R value, since no angular distribution was discerned (owing to limited statistics and lower mean energy due to containment). Later Kamiokande data did show angular variation in the PC data, but not statistically compellingly. For some reason not fully understood, the fits using R alone all seem to yield higher values of Δm^2. If one has some deficit in muons without angular determination, then one can fit that suppression with any Δm^2 above some threshold value by choosing an appropriate mixing angle. Thus the R constraints are open-ended upwards in Δm^2. Perhaps there is a systematic problem here due to the predicted neutrino spectra, or perhaps there is some physics yet to be elucidated. This is to suggest not that it seems possible for the preferred two-neutrino solution to move much, but that more complex, small effects at the $< 10\%$ level could be superposed on the present simple solution. Accelerator-based experiments should clarify this issue.

5.2.5   Muon Decay Events

It is not often emphasized, but the original indication of the anomaly, a deficit in stopped-muon decays ($\simeq 2.2$ μs after the initial neutrino event), remains with us, and constitutes a nice alternative sample, almost independent and with quite different systematics. It is not so clean a sample (there are muon decays from pions produced in electron CC and all-flavor NC events) and the statistics are lower, but the complete consistency of the muon decay fraction remains a reassuring complement to the energy and angle analysis employing track identification.

5.2.6   Through-Going and Entering–Stopping Muons

Another cross-check comes from the UM and SM samples, which are particularly attractive because the source energies are factors of 10 and 100 higher and the detector systematics rather different (for example, the target is mostly rock, not water). A drawback to these samples is that one is restricted to using muons arriving from below the horizon, owing to the overwhelming number of down-going cosmic-ray muons penetrating the mountain (at 50 000 times the rate in SuperK).

In going from the earlier instruments to SuperK, however, the gain is not so great (the 1200 m^2 of SuperK being about a factor of three more than the previously largest underground instrument, IMB, for example), since the rate of collection of through-going muons depends upon area, not volume. However, the much greater thickness of the detector (and the efficient tagging of entering and exiting events in the veto layer) yields many more entering–stopping (SM) events.

The angular distribution of the flux derived from 1260 UM events from below the horizon, each with more than 7 m track length in the detector, is shown in Fig. 5.11, where one sees that the angular distribution is nicely consistent with the assumption of oscillations and not with the no-oscillations

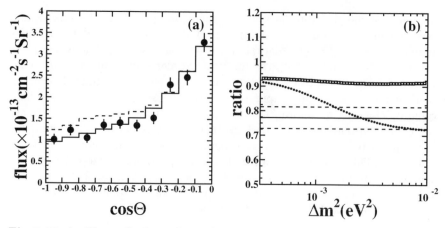

Fig. 5.11a,b. The preliminary flux calculated from 1260 up-going muons during 1138 live days of SuperK data, with 9 events background-subtracted in the bin nearest the horizon, $\cos\Theta = 0$. The *error bars* show statistical errors only. Expectations for the no-oscillations case (*solid line*) and the best fit for ν_μ–ν_τ oscillations with $\Delta m^2 = 0.0032$ eV^2 and maximal mixing (*dashed line*) are shown [5.22]

assumption. However, since much of the effect is close to the horizon, where oscillations for the energies in question are just setting in, one worries about contamination of the near-horizon events with in-scattered events from the much greater numbers of down-going muons. There is no room for detail here, but SuperK does perform a small background subtraction (9 events out of 247, or 3.6% in that one bin) for events within 3° of the horizon, but otherwise finds no evidence for significant contamination [5.31].

In SuperK the SM sample was predicted to be 33–42% of the UM sample, as indicated in Fig. 5.12, yet in fact SuperK sees only about 24%±2%. Fitting the data to the oscillation hypothesis, one can make the now usual inclusion plot, which shows that the UM and SM results are completely in accord with those from the FC and PC data (see Fig. 5.23). However, as the statistics are smaller and the physics leverage not as great, the muon result does not add much to the FC and PC constraints, though it does stiffen the lower bound on Δm^2. The joint fit to the UM and SM data alone yields values of χ^2/number of degrees of freedom of 35.4/15 and 13/13 for the cases of no oscillations and ν_μ–ν_τ oscillations.

One may note that earlier experiments, such as IMB, with a final sample of 647 events, no veto counter and less mature flux calculations, did not find any net deficit in the UM sample, nor any significant deviation from no-oscillations expectations. A similar case obtained with other, smaller data sets. Indeed, one may note that, on the strength of the angular distribution of the SuperK UM data alone, one would hardly be making discovery claims. All UM data from IMB and Kamioka were and are in accord with the present results, but did not demand the conclusion of oscillations.

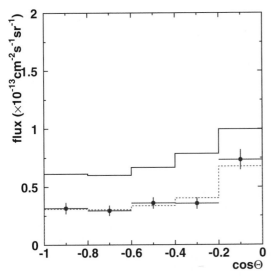

Fig. 5.12. The flux of stopping muons versus zenith angle in 1117 days of SuperK data. There are 311 events and a background of 21.4 with track length greater than 7 m, or about 1.6 GeV. The large overall deficit in the data (*dots* with statistical *error bars*) compared with the no-oscillations case (*solid line*) is less significant than it appears because of the 20% uncertainty in absolute flux [5.22]

There is a lengthy tale about an SM/UM analysis from the IMB experiment [5.15], which claimed an exclusion region very close to the now preferred solution. The IMB stopping-muon sample was small and it was not clean, owing to lack of a veto layer. More importantly, the interpretation seems to have been flawed owing to older flux models and Monte Carlo simulations. Work is in progress to reassess the old data with new flux calculations and an updated quark model [5.39]. Thus there remains a cloud upon the horizon, but one which may fade away upon reanalysis. It might also be worth recalling that in the 1980s the absolute rate of up-going muons, as measured in the IMB, Kamioka and other detectors, agreed with rates calculated employing the then available flux calculations. Also at that time, peculiar angular distributions which fitted no expectation were reported at conferences from the MACRO and Baksan detectors. These results all tended to give pause to claiming oscillations as the resolution of the atmospheric neutrino anomaly. These concerns were swept away by the clean and statistically convincing FC muon angular distributions from SuperK.

5.2.7 The Muon Neutrino's Oscillation Partner

Given that the muon neutrino is oscillating, is it oscillating with a tau neutrino or with a new, sterile neutrino which does not participate in either the charged-current (CC) or the neutral-current (NC) weak interaction? Fortunately there exist several means to explore this with SuperK data. The NC interactions should show an up–down asymmetry for sterile neutrinos but not for tau neutrinos (since the NC interactions for all ordinary neutrinos are the same). Another avenue for discrimination is that sterile neutrinos would have an additional oscillation effect due to "matter effects". The consequence would be a unique signature in the angular distribution of intermediate energy muons, as illustrated in Fig. 5.13.

Early SuperK efforts focused upon the attempt to collect a clean sample of π^0 events. As it turned out, this was frustrated because the rings (from the two decay γs) cannot be separated at energies above $\simeq 1$ GeV, and the net result is that there are not so many reconstructed events as to permit a good discrimination. In fact, the absolute rate is consistent with expectations, but the cross section is uncertain to about 20%, making the hint at tau coupling not significant. The K2K experiment should soon measure this cross section to perhaps 5%, however, making the π^0 rate a useful discriminant.

More recently, tests have been devised employing a multiring sample (MR), the PC event sample and the UM sample, all of which are independent of the single-ring FC sample, which yields the strongest oscillation parameter bounds.

The MR sample is cut by energy (> 1.5 GeV) and the requirement of the dominant ring being electron-like to enhance the NC content of the sample. A test parameter is constructed from the ratio of events from within $60°$ of the zenith and nadir. This is illustrated in Fig. 5.14, where one sees consistency of ν_τ and disfavoring of ν_{sterile}.

The PC sample can be cut on energy (requiring > 4 GeV) in order to achieve a higher neutrino source energy, and the up-going number compared with the down-going number of events. In this instance one is seeking matter effects, and the results are shown in Fig. 5.15, indicating again a preference for ν_τ over ν_{sterile}.

For the muons, the near-horizontal number can be compared with the number of nearly straight up-going events for another test of matter oscillations. The relevant results are presented in Fig. 5.16.

Finally, the three tests can be combined in a single χ^2 test for the case of $\nu_\mu \leftrightarrow \nu_\tau$ and for the two cases of ν_{sterile} heavier or lighter than ν_μ. The results are presented in Fig. 5.17, where one sees that the entire region in mixing-parameter space is eliminated for sterile neutrinos at more than the 99% confidence level, whilst the ν_τ case fits perfectly [5.41].

As to muon neutrinos coupling to electron neutrinos, SuperK can say only that the Δm^2 is out of range on the low side or that the sine of the mixing angle is less than about 0.1 if Δm^2 is large. As indicated in the earlier plots with up–down asymmetry, there is surely not much mixing in this energy range. One clever scenario [5.43] has the electron neutrino oscillation loss being just compensated by muon neutrinos splitting their oscillations between

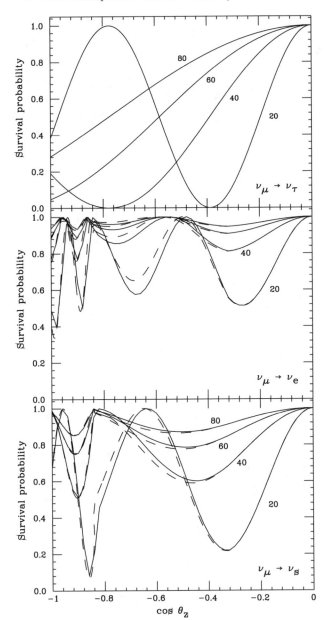

Fig. 5.13. Survival probability $P(\nu_\mu \to \nu_\mu)$ as a function of the zenith angle in the cases of maximal mixing of ν_μ with ν_τ (*upper panel*), ν_e (*middle panel*) and ν_s (*lower panel*). For $|\Delta m^2| = 5 \times 10^{-3}$ eV^2 the curves correspond to neutrino energies of 20, 40, 60 and 80 GeV. The *dashed curves* are calculated with the approximation of constant average densities in the mantle and in the core of the earth. From Lipari and Lusignoli [5.40]

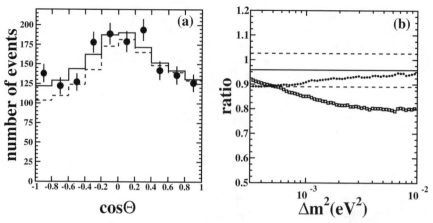

Fig. 5.14. (a) Zenith angle distributions of the multiring events satisfying cuts as described in the text; $\cos \Theta$ is $+1$ for down-going events. The *black dots* indicate the data and statistical errors. The *solid line* indicates the prediction for tau neutrinos, and the *dashed line* the prediction for sterile neutrinos with $(\Delta m^2, \sin^2 2\theta) = (3.3 \times 10^{-3} \mathrm{eV}^2, 1.0)$. These two predictions are normalized by a common factor so that the number of observed events and the predicted number of events for $\nu_\mu \leftrightarrow \nu_\tau$ are identical. (b) Expected up/down ratio as a function of Δm^2. The *horizontal lines* indicate data (*solid*) with statistical errors (*dashed*). The *black dots* indicate the prediction for tau neutrinos, and the *empty squares* the prediction for sterile neutrinos, both for the case of maximal mixing [5.41]

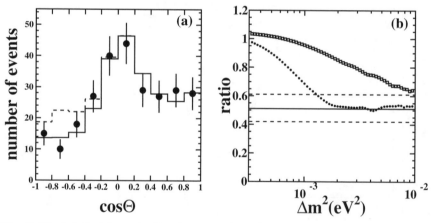

Fig. 5.15. (a) Zenith angle distributions of partially contained events satisfying cuts of $E_{\mathrm{vis}} > 5$ GeV; $\cos \Theta$ is $+1$ for down-going events. The *black dots* indicate the data and statistical errors. The *solid line* indicates the prediction for tau neutrinos, and the *dashed line* the prediction for sterile neutrinos with $(\Delta m^2, \sin^2 2\theta) = (3.3 \times 10^{-3} \mathrm{eV}^2, 1.0)$. (b) Expected up/down ratio as a function of Δm^2. The *horizontal lines* indicate data (*solid*) with statistical errors (*dashed*). The *black dots* indicate the prediction for tau neutrinos, and the *empty squares* the prediction for sterile neutrinos, both for the case of maximal mixing [5.41]

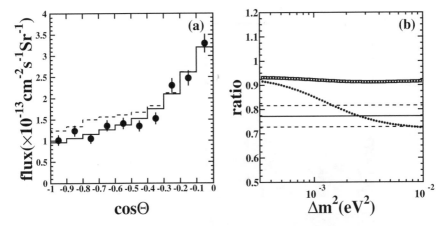

Fig. 5.16. (a) Zenith angle distributions of the upward-moving through-going muons; $\cos \Theta$ is -1 for vertical up-going events. The *black dots* indicate the data and statistical errors. The *solid line* indicates the prediction for tau neutrinos, and the *dashed line* the prediction for sterile neutrinos with $(\Delta m^2, \sin^2 2\theta) = (3.3 \times 10^{-3} \mathrm{eV}^2, 1)$. (b) Expected up/down ratio as a function of Δm^2. The *horizontal lines* indicate data (*solid*) with statistical errors (*dashed*). The *black dots* indicate the prediction for tau neutrinos, and the *empty squares* the prediction for sterile neutrinos, both for the case of maximal mixing [5.41]

electrons and taus. This seems to be ruled out by higher-energy SuperK data, however [5.33]. Further discussions of (many) other scenarios of oscillations can be found in Chap. 9.

5.2.8 Subdominant Oscillations

For all of the foregoing we have considered only two-flavor oscillations. A simple three-flavor analysis can be performed employing the assumption that solar-neutrino oscillations are driven by a much smaller mass difference than found for ν_μ and ν_τ; $m_1 < m_2 \ll m_3$. Then, using the standard notation for the three-neutrino MNS mixing matrix (see Chap. 9), the oscillation probabilities can be written as

$$P_{\nu_\mu \nu_\mu} = 1 - 4\sin^2 2\theta_{23} \cos^2 \theta_{13}(1 - \sin^2 \theta_{23} \cos^2 \theta_{13}) \sin^2(1.27\Delta m^2 L/E) \,, \quad (5.3)$$

$$P_{\nu_\mu \nu_\tau} = \cos^4 \theta_{13} \sin^2 2\theta_{23} \sin^2(1.27\Delta m^2 L/E) \,, \quad (5.4)$$

$$P_{\nu_\mu \nu_e} = \sin^2 2\theta_{13} \sin^2 \theta_{23} \sin^2(1.27\Delta m^2 L/E) \,, \quad (5.5)$$

$$P_{\nu_e \nu_\tau} = \sin^2 2\theta_{13} \cos^2 \theta_{23} \sin^2(1.27\Delta m^2 L/E) \,, \quad (5.6)$$

$$P_{\nu_e \nu_e} = 1 - \sin^2 2\theta_{13} \sin^2(1.27\Delta m^2 L/E) \,, \quad (5.7)$$

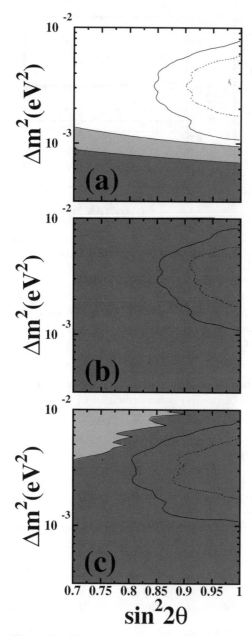

Fig. 5.17. Excluded regions for three alternative oscillation modes. (a) $\nu_\mu \leftrightarrow \nu_\tau$; the *light* and *dark shaded regions* are excluded at 90% and 99% C.L., respectively. (b) $\nu_\mu \leftrightarrow \nu_{\text{sterile}}$ with $\Delta m^2 > 0$; the whole region shown in this figure is excluded at 99% C.L.; (c) $\Delta m^2 < 0$; the whole region is excluded at 90% C.L., and the *dark shaded region* is excluded at more than 99% C.L. The thin *dotted* and *solid lines* indicate the 90% and 99% C.L. allowed regions, respectively, from the FC single-ring events [5.41]

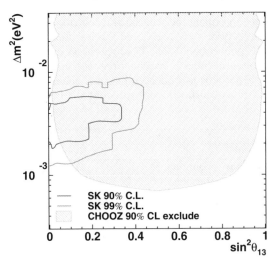

Fig. 5.18. Three-neutrino fits to the SuperK data showing 90% and 99% C.L. allowed regions. The *shaded region* is the 90% exclusion region from the Chooz experiment [5.44]. See text for qualifications

where $\Delta m^2 = m_3{}^2 - m_2{}^2$, and we neglect any consideration of CP or CPT violation.

The SuperK FC and PC 990 day data have been fitted with these equations [5.44], and the limits on $\sin^2 2\theta_{13}$ are less than 0.25 at 90% C.L., with $\Delta m^2 = 0.003$ eV^2. The best-fit value lies at $\sin^2 \theta_{13} = 0.03$, $\sin^2 \theta_{23} = 0.63$. At this value of Δm^2 the Chooz results allow either $\sin^2 \theta_{13} < 0.03$ or $\sin^2 \theta_{13} > 0.97$, so the latter is eliminated by the SuperK results [5.22], as illustrated in Fig. 5.18.

5.2.9   Nonstandard Oscillations

As discussed in Chap. 9 and elsewhere [5.45], there are models of neutrino oscillations which result from gravitational splitting, violations of Lorentz invariance and so on, which result in oscillation with a phase proportional to, say, $L \times E$ instead of L/E, or even with no energy dependence. The SuperK group has presented a fit to the data as a function of

$$P_{\nu_\mu \nu_\tau} = \sin^2 2\theta \sin^2(\beta L E^n) , \tag{5.8}$$

where n, β and $\sin^2 2\theta$ (0.7 to 1.3) are varied. As indicated in Fig. 5.19, the exotic solutions are strongly disfavored, while the normal function is perfectly acceptable, with $n = -1.06 \pm 0.14$.

5.2.10   Hypotheses to Explain the Anomaly

We conclude with a summary, presented in Table 5.1, of all hypotheses put

Table 5.1. List of hypotheses invoked to explain the atmospheric neutrino anomaly. Columns 2–4 contain criteria available prior to SuperK, and the last four contain data available after the 1998 SuperK publication [5.30]. The hypotheses divide into five systematics issues and eight potential physics explanations. As indicated in the text, the only remaining likely hypothesis is oscillation between muon and tau neutrinos. A "×" schematically indicates which evidence rules out the hypothesis in that row

Hypothesis	Evidence						
	Old			New			
	$R<1$ ($E<1$ GeV)	μ decay fraction	Volume fraction	$R<1$ ($E>1$ GeV)	A_e $\simeq 0$	A_μ <0	$R(L/E)$ $\simeq 0.5$
Atmosphereic flux calculation	×	√	√	×	√	×	×
Cross sections	×	√	√	×	√	×	√
Particle identification	√	×	×	√	√	×	√
Entering background	√	√	×	√	√	×	√
Detector asymmetry	√	√	×	√	×	√	√
Extraterrestrial ν_e	√	√	√	√	√	×	×
Proton decay	√	√	√	×	√	×	×
ν_μ decay	√	√	√	√	√	√	√
ν_μ absorption	√	√	√	√	√	√	×
ν_μ–ν_e	√	√	√	√	×	√	√
ν_μ–ν_s	√	√	√	√	√	×	√
Nonstandard oscillations	√	√	√	√	√	√	×
ν_μ–ν_τ	√	√	√	√	√	√	√

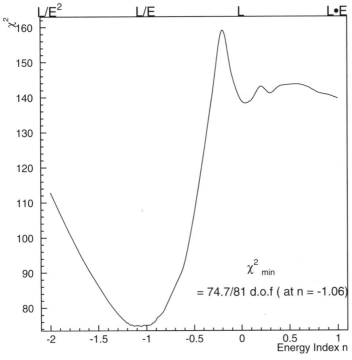

Fig. 5.19. The variation of χ^2 with the index n in the oscillating phase for muon neutrinos, as in const $\times LE^n$; $n = -1$ corresponds to ordinary oscillations. One sees that nonstandard solutions with values of n of 0 and 1 are strongly disfavored [5.22]

forth to explain the atmospheric neutrino anomaly. Space does not permit a full discussion here, but it is the case that the SuperK data now have eliminated almost all alternative hypotheses to explain the results.

The atmospheric-neutrino-flux calculations cannot be producing the anomaly since we see an effect, otherwise unexplained, in the muon zenith angle distribution. The cross sections cannot produce such a geographical effect, even if one could find some lepton-universality-breaking phenomenon. Particle identification has been heavily studied and verified at an accelerator with a 1000 ton Cerenkov detector tank [5.46]. Entering backgrounds, which would produce effects clustering near the outer walls, show no evidence for contribution to the anomaly in the SuperK data. Asymmetrical response of the detector is ruled out by the observed symmetry of the electron data, as well as by detailed calibration studies with isotropic sources.

Extraterrestrial neutrinos cannot be the culprits (unfortunately), since we see that it is a deficit of muon neutrinos causing the anomaly, not an excess of electron neutrinos. Proton decay is ruled out similarly, owing to geographical dependence of the deficit, and the extension of the anomaly to too high an energy (ruling out neutron–antineutron oscillations as the source as well). The decay of a muon neutrino component has been discussed and, while seemingly unlikely, is not totally ruled out as yet. Anomalous absorption of muon neutrinos by the earth, correlated with an exponential in the column density through the earth, is ruled out as well. This follows because the anomaly is not dominated by that small part of the solid angle going through the earth's core, whereas we see the muon deficit starting even above the horizon. Electron neutrino mixing is ruled out as the dominant effect, again owing to the muon angular distribution. Sterile neutrinos do not fit the data, as discussed above. As discussed in the previous section, nonstandard oscillations are also ruled out.

The only hypothesis which survives and which fits all the evidence is that muon neutrinos mix maximally with tau neutrinos with a Δm^2 in the range of 2–7×10^{-3} eV^2. It is noteworthy to this author that in all the tests made on the data sample, there appears to be great stability in the results against variations of all the parameters explored.

5.2.11   Results from Soudan II

The Soudan 2 detector, located in an old iron mine in Minnesota, USA, consists of a vertical-slab, fine-grained tracking calorimeter of 963 tons total mass. The cavern has a surrounding layer of proportional tubes, two or three layers thick on all sides. Data have been reported from 4.6 kiloton years of exposure [5.21], as illustrated in Fig. 5.20. The contained events plotted are selected for lepton energy > 700 MeV with no visible nuclear recoil, or for visible energy > 700 MeV, summed momentum > 450 MeV/c and lepton momentum > 250 MeV/c. The energy resolution is of order 20%, and the angular resolution of order 20–30°. In the figure, the predicted number of events has been normalized to the electron total. One can see that, with only of the order of 100 events of each type, the statistical significance is not great, but the depletion of up-going muon events is evident. The fits to the oscillation parameters are included later, in Fig. 5.23.

5.2.12   Results from MACRO

The MACRO detector, built primarily to seek monopoles, possesses a significant capability, with an effective mass of 5.3 kilotons, to detect through-going and stopping muons as well as contained and partially contained neutrino interactions [5.42]. The instrument, located in the deep underground Gran Sasso National Laboratory in Italy, consists of horizontal planes of a tracking instrument.

Figure 5.21 shows the results of an analysis of partially contained data, for which up and down cannot be distinguished, but which shows a clear deficit compared with expectations for no oscillations. The acceptance of such a planar instrument is small near the horizon, so most of the effect is from

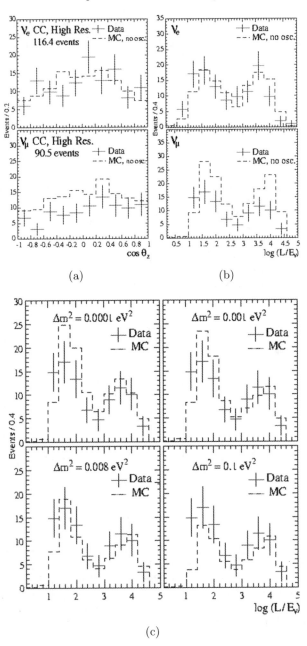

Fig. 5.20. (a) Distributions in $\cos\theta_z$ for the Soudan II ν_e and ν_μ flavor event samples. The data (*crosses*) are compared with the results of the no-oscillation Monte Carlo (MC) simulation (*dashed histogram*), where the MC results have been rate-normalized to the ν_e data. (b) Distributions of $\log(L/E_\nu)$ for ν_e and ν_μ charged-current events compared with the neutrino MC results with no oscillations; the MC results have been rate-normalized to the ν_e data. (c) Comparison of L/E_ν distribution for ν_μ data (*crosses*) and expectations from the hypothesis of neutrino oscillations for four Δm^2 values, with $\sin^2 2\theta = 1.0$. From Mann [5.21]

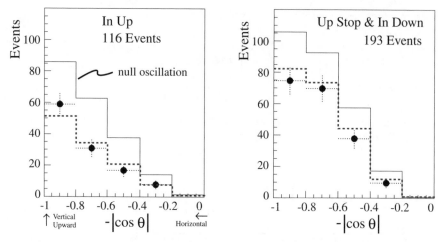

Fig. 5.21. Distributions in $\cos\theta$ for MACRO partially contained events. The data (*solid circles*) are seen to fall below the no-oscillation expectation in every bin of both samples. The *dashed line* shows expectations for maximal mixing and $\Delta m^2 = 0.0025$ eV^2. From Surdo et al. [5.42]

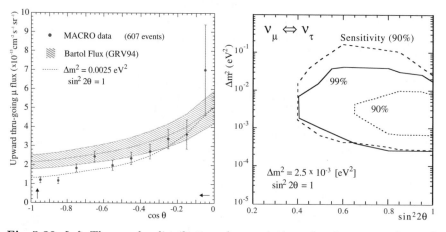

Fig. 5.22. *Left:* The angular distribution of upwards through-going muons observed in MACRO. The data distribution (*solid circles*) differs from the null oscillation expectation (*shaded band*) in shape and in rate. *Right:* the allowed regions for neutrino oscillations obtained by MACRO from the upwards through-going muons. The confidence-level and experimental-sensitivity boundaries were calculated using the Feldman–Cousins method. From [5.42]

the nearly vertical events. The In–Up sample has 116 events and exhibits a depletion of 0.57 ± 0.16 compared with expectations.

The MACRO detector also has a significant sample of upwards through-going muons, as illustrated in Fig. 5.22. Note that the muon energy threshold

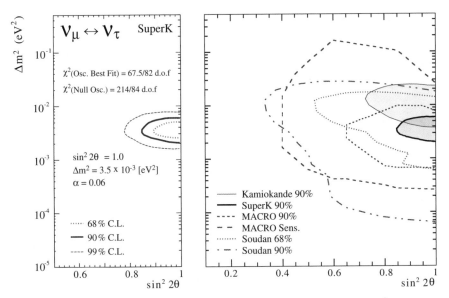

Fig. 5.23. *Left:* allowed regions obtained by SuperK on the basis of χ^2 fitting to FC and PC single-ring events, plus upward stopping muons, plus upward through-going muons. *Right:* allowed regions for oscillation parameters from Kamiokande (*thin-line boundary*), SuperK (*thick-line boundary*), MACRO (*dashed boundaries*) and Soudan 2 (*dotted* and *dot–dashed boundaries*). From Mann [5.21]

is low, in the 100 MeV range (varying with angle and entry location). The overall depletion (data/expectations) is $0.74\pm0.03\pm0.04\pm0.12$, where the last term reflects the uncertainty in the flux and cross section [5.42]. While the fit to the oscillations expectation appears not to be perfect and the minimum in χ^2 lies in the unphysical region, the confidence-level boundaries shown on the right in Fig. 5.22 indicate consistency with the SuperK results.

5.2.13   Combined Evidence

In the left part of Fig. 5.23 is the combined fit of the FC and PC data (848 live days) plus the UM (923 live days) plus the SM data (902 live days) from SuperK, with results as indicated, constraining the oscillation parameters to be roughly $0.002 < \Delta m^2/\text{eV}^2 < 0.006$ and $0.85 < \sin^2 2\theta < 1.0$, with the fit minimum in the physical region at maximal mixing and 0.0035 eV^2. The boundaries for MACRO and Soudan 2 are shown overlying the SuperK results in the right-hand panel. One can see that the results of the largest instruments now reporting atmospheric neutrino data, the water Cerenkov detectors and two dissimilar tracking calorimeters, all agree on muon neutrino disappearance, and are consistent with oscillation between muon and tau neutrinos. I cannot do better than to quote Tony Mann from his 1999

Lepton–Photon Conference plenary talk [5.21]: "I propose to you that congratulations are in order for the researchers of Kamiokande and of Super-K and more generally, for the non-accelerator underground physics community. For [Fig. 5.23b], Ladies and Gentlemen, is the portrait of a Discovery – the discovery of neutrino oscillations with two-state mixing."

5.2.14   Long-Baseline Results

The K2K experiment has been in operation for over one year at the time of writing. The 12 GeV KEK proton synchrotron has been equipped with a neutrino line, and a double detector has been built at about 100 m range, on the KEK campus. Neutrinos are aimed at the SuperK detector at a distance of 250 km, and events have been recorded [5.51]. Note that a 1 GeV neutrino would be near the first minimum after 250 km if $\Delta m^2 = 0.005$ eV^2.

At the time of writing, preliminary results are available, and it has been reported in conferences that the rate is low, consistent with the SuperK results, and inconsistent with no oscillations at the 2σ level [5.52]. For the period June 1999 to March 2000, the group received about 1.7×10^{19} protons on target (17% of proposal request), during the equivalent of about five months' running. The events were easily extracted from a GPS-synchronized 1.5 μs time window relative to the 12 GeV KEK proton synchrotron fast beam spill (2.2 s repetition period and about 5×10^{12} protons on target per spill). During this run 17 associated events were collected in the SuperK 22.5 kiloton fiducial volume, when 29.2 ± 3.1 were expected for no oscillations. The expected numbers are 19.3, 12.9 and 10.9 for 3, 5 and 7×10^{-3} eV^2. Of these SuperK events, 10 are single-ring, one of which was identified as electron-like (about equal to expectations).

Given the low energy of the neutrino beam (~ 1 GeV) and the frustratingly low data rate at SuperK, probably not much more can be expected from this experiment than a simple confirmation of the SuperK atmospheric-neutrino results. The full proposal run would collect 174 or so events with no oscillations or, assuming oscillations, about half that. Of those, about half should be single-ring muons, leaving perhaps 40–50 events from which to deduce the arriving neutrino spectrum. Given that the accelerator delivers a beam for about half a year each calendar year, one can see that it will take several years to achieve definitive results.

5.3   Implications

The ramifications of the explication of the atmospheric neutrino anomaly in terms of neutrino mixing and thus neutrino mass are great and span the known realms of fundamental physics from large to small. We have not discussed in this chapter the links to solar neutrinos (Chap. 4), nor the LSND results (Chap. 7). Certainly there is no conflict between the atmospheric muon neutrino results and the possible (nay, likely) solar oscillations. If, however, both the solar and the LSND results are correct, then we surely have some interesting physics to untangle, as it is generally admitted that no simple three-neutrino model can incorporate all three neutrino anomalies, and that new degrees of freedom would be required (see Chap. 9). From the evidence presented in this chapter alone, however, we can make some far-reaching, perhaps paradigm-shifting, conclusions.

5.3.1   Astrophysics and Cosmology

The implications of the oscillations results have been explored in other chapters in this book, so here we only outline those relevant to the muon neutrino oscillations. First, it appears that neutrinos with summed masses of the order of 0.1 eV will not make any major contribution to resolving the dark-matter quandary. Nonetheless, with a ratio of 2 billion to one for photons (and neutrinos) to nucleons from the Big Bang, even such a small neutrino mass may be greater in total than that of all the visible stars in the sky. Hence, while one must account for neutrino mass in further cosmological modeling, neutrinos are not likely to constitute the bulk of the "missing matter". However, if the neutrinos are nearly degenerate in mass and all have masses in a range near 1 eV (and hence we are observing only small splittings with the oscillations), then neutrino mass may dominate the Universe. While neutrinos are not favored by astrophysical modelers (fitting the spatial fluctuations in the cosmic microwave background for example), large neutrino masses are not ruled out ($\sum M_i \leq 6$ eV, $H_0 = 65$ km/s Mpc, $i = 1, 2, 3$) (Chap. 12). Nearly degenerate neutrino masses would not present a consistent picture with the quark and charged-lepton masses, which make large mass jumps between generations. But who knows? We do not have a viable grand unified theory with mass predictions (Chap. 10), so an open mind is appropriate.

The other major area of significance, perhaps of the deepest significance, has to do with baryogenesis, the origin of the predominance of matter by one part per billion over antimatter at the Big Bang time. There are claims that the old idea of accumulation of net baryon number via CP violation in the quark sector, while satisfying the Sakharov conditions [5.47], may not suffice for the early stages of expansion of the Universe [5.48]. If that is indeed the case, it may be that neutrinos provide the avenue for net baryon asymmetry generation, with its expression into hadrons becoming manifest relatively late in the game at the electroweak phase transition [5.49].[9] CP and even CPT violation in the neutrino sector, as yet almost unconstrained, could have dramatic implications.

Neutrino masses and possible sterile neutrinos have also been invoked to help resolve problems in understanding heavy-element synthesis in supernovae (Chap. 11).

---

[9] There are many papers dealing with this topic, as a search on "baryogenesis" in the Los Alamos preprint server or in SPIRES will reveal.

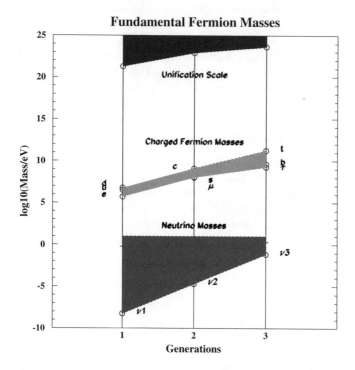

Fig. 5.24. The masses of the fundamental fermions. The *lower shaded region* indicates the rough range allowed by present oscillation results, with the upper boundary for nearly degenerate neutrino masses and the lower boundary for the approximate atmospheric and solar results, with an extrapolation to the first-generation mass. The unification scale marks the rough range implied by the seesaw mechanism

5.3.2   The Theoretical Situation: Why So Important?

Other chapters in this volume deal with the situation for particle theory, so we make only a few general remarks here, mostly from the experimentalists' phenomenological viewpoint. Figure 5.24 shows the masses of the fundamental fermions in three generations on a logarithmic scale in mass. Dramatically, one sees that if the neutrino masses are near the lower bounds (that is, at the presumed mass differences from present atmospheric and solar results), they lie 10–15 orders of magnitude below those of the other fundamental fermions (charged quarks and leptons). Graphically, one notes the spacing between the neutrino masses and the charged-fermion masses is just about the same as the distance (on the log scale) to the unification scale. This is a pictorial representation of the seesaw mechanism, as noted more than ten years ago [5.17]. This points up the task for grand-unification model building, and highlights the deep link between neutrino masses and nucleon decay (Chap. 10).

5.3.3   Future Muon Neutrino Experiments

The physics community seems to have rapidly accepted, with caution, the seeming inevitability of neutrino mass and oscillations (Chap. 2). Yet, most probably, the game has hardly begun and many a subtlety may await our exploration. However, if the LSND claims go quietly away (after the BooNE experiment runs), the mass and mixing picture could settle into the simple hierarchical pattern as explored in the bimaximal mixing scenario (or similar versions). This highlights the importance of experiments to follow up on the LSND results as one of the first items on the agenda in the current neutrino business (see Chap. 7).

Given present indications, it would seem that the K2K [5.51], and MINOS [5.53] experiments and the CERN–Gran Sasso experiments ICARUS [5.54] and OPERA [5.55] should confirm the SuperK results and make the oscillation parameters more precise. Of course, one would really like to see tau appearance, not just muon disappearance, to be sure we are not being misled. There has been some debate in the community as to what constitutes appearance. Because of the complexity of tau final-state identification, this author would prefer to see a real tau track recorded. In any case plans are in progress in the US, Japan and Europe for the obvious follow-up experiments to consolidate the oscillation scenario and refine the parameters, which experiments should take less than a decade. At this time it appears that only OPERA has the opportunity to record physical tau lepton tracks in emulsions, though the other experiments (including SuperK) may be able to detect tau kinematic signatures statistically.

More interesting for the long-range physics is filling in the MNS matrix (the lepton equivalent of the quark CKM matrix) for neutrinos [5.56]. This is not an easy business. The atmospheric neutrino measurements really only define, at best, three of the nine elements! Solar neutrinos get us another, and perhaps a constraint on two. Measuring the tau-related components directly seems pretty hopeless. Of course, if we could assume the matrix to be unitary and real we would be in good shape, as there are then only three independent parameters (plus the masses). But we do not know this, and if there exist CP violations we then have a total of three angles and one phase plus possibly another phase for the case of Majorana neutrinos, but this latter phase is only measurable in special circumstances such as neutrinoless double beta decay. If there are more (heavy or sterile) neutrinos, then things could be much more complicated (as the 3×3 submatrix will not be unitary). By analogy with the quarks (where the 3×3 CKM matrix with small mixing angles and one CP-violating phase suffices), perhaps we should not worry too much, except for lack of any guidance whatsoever from theory. CP violation is only very weakly constrained experimentally in the neutrino sector at present, so we could be in for big surprises, and, given the connection of neutrinos with cosmology and baryogenesis, one should indeed be suspicious, I believe.

As a whole, the particle physics community is just beginning to examine the newly illuminated possibilities in the neutrino sector. At the moment it appears as though muon colliders may provide our best route for next-generation explorations in this realm [5.57, 5.58]. The serendipitous realization that muon storage rings can provide neutrino beams more intense by about four orders of magnitude than previous artificial neutrino sources, along with the potential, via polarization of these beams, to make nearly pure beams of muon neutrinos or antineutrinos, with a relatively narrow energy spread, allows one to dream of experiments not heretofore thought possible. The prospects for this endeavor are just emerging at the time of writing and the dialog between physics possibilities and machine technical capabilities is in much ferment. The present excitement about this endeavor, probably at least ten years from realization, promises much evolution of thinking about critical tests and experiment planning. Aside from simple checking of previous results, precision measurement of oscillation parameters and filling out of the MNS matrix, the major goal in this author's mind is exploration for CP and CPT violation in the neutrino sector, not possible by any other means yet conceived. Of course, if θ_{13} turns out to be zero, which hopefully we shall know in a few years, then CP violation will not be possible.

Table 5.2 shows a rough comparison of the physics capabilities and reach of SuperK and of the various present and future long-baseline experiments. This table was generated by an informal working group at the 2000 Aspen "Neutrinos with Mass" workshop. Although the details may annoy proponents of some projects (and please others), the purpose was not to evaluate specific experiments, but to try and gauge the strengths of the various approaches as we move forward in the quest to complete the MNS matrix and explore for CP and CPT violations in the neutrino sector. We had inadequate information about OPERA and the proposed 50 GeV proton-driven neutrino beam in Japan (JHF to SuperK), so these two experiments may not be fairly represented.

The table assumes $\Delta m^2 = 0.003$ eV^2 and $\sin^2 2\theta \simeq 1.0$ with dominantly $\nu_\mu \leftrightarrow \nu_\tau$ mixing, large-mixing-angle MSW solar neutrino oscillations and no LSND/BooNE indication of oscillations. We also assumed a 20–50 GeV neutrino factory and a high-quality neutrino detector at a few thousand kilometers' distance.

There was considerable discussion of whether the new experiments will or will not be able to see at least one dip in the detected neutrino spectrum, unambiguously discriminating oscillations from disappearance (as in decay). My conclusion is that such a detection will be very difficult for all experiments with traditional neutrino beams if the Δm^2 is at the low end of the allowed region. ICARUS and MONOLITH claim to be able to detect oscillations in the cosmic ray beam, a consequence of their excellent resolution and the larger range of L/E values available from the atmospheric neutrinos than from a fixed distance long-baseline beam.

Table 5.2. Comparison of SuperK and long-baseline neutrino experiments in terms of physics capabilities and reach. See text for explanation

	SuperK (atm. ν)	K2K	MINOS	ICARUS	OPERA	JHF2K	ν factory
$d[\ln(\Delta m^2)]$	50%	20%	10%	10%	20%	10%	2%
$d(\sin^2 2\theta)$	5%	5%	5%	5%	?	4%	2%
See oscillations?	×	×	$\sqrt{}$(?)	×(atm.$\sqrt{}$)	?	×	$\sqrt{}$
τ appearance (kink)	×	×	×	×	$\sqrt{}$	×	$\sqrt{}$
τ appearance (kinematic)	$\sqrt{}$?	×	$2\sqrt{}$	$3\sqrt{}$	$4\sqrt{}$	×	$4\sqrt{}$
$\sin^2 2\theta_{13}$ limit	0.1	0.03?	0.03	0.015	(identify e?)	0.03	$1\text{–}3 \times 10^{-3}$
$d(\nu_s/n_\tau)$	20%	×	5%?	5%	?	×	1%
Eliminate decay models?	×(?NC/CC)	×	$\sqrt{}$(NC/CC)	×(atm.$\sqrt{}$)	?	×	$\sqrt{}$
Sign of Δm^2	×	×	×	×	×	×	$\sqrt{}$
$\nu_e \to \nu_\tau$	×	×	×	×	×	×	$\sqrt{}$
CP violation tests	×	×	×	×	×	×	$\sqrt{}$
CPT violation tests	×	×(?)	$\sqrt{}$	$\sqrt{}$	×	×(?)	$\sqrt{}$

I think what we all learned from this exercise is that there really is a great gulf between what we can accomplish with a neutrino factory and anything prior to that, even with more powerful traditional neutrino beams.

Measuring absolute neutrino mass remains a frustrating problem, which will not be resolved in the near future, it seems. While pushing to lower mass limits with tritium beta decay experiments will apparently not be able to reach below 0.1 eV, there is some hope from CMBR measurements (Chap. 12), there is a long shot via the "Weiler process" [5.59] and, optimistically, neutrinoless double-beta-decay experiments may eventually reach 0.01 or even 0.001 eV.

At high energies, explorations for cosmic neutrinos may be carried out in deep arrays in the ocean and under ice [5.61]. While the main goal of these attempts at high-energy neutrino astronomy will be aimed at astrophysics, with a high-energy-threshold detector capable of registering neutrinos in the PeV range, it may be that such instruments will be able to directly detect tau neutrinos (via the "double bang" signature [5.62]) and even determine the neutrino flavor mix, to the benefit of both particle physics and astrophysics.

A next-generation (megaton-scale) nucleon decay instrument to probe lifetimes to 10^{35} years would do wonders for advancing neutrino physics as well. Simply building a larger version of SuperK will not suffice, because of the need for greater resolution as well as size. The only candidate I see to go beyond SuperK is something like the AQUA-Rich style of imaging water Cerenkov detector [5.36, 5.37]. An attractive alternative, which need not be so massive to get to 10^{35} years in the kaon modes of nucleon decay, might be a 50–70 kiloton liquid-argon detector of the ICARUS style. Perhaps such a detector can be realized in concert with a long-baseline beam from a neutrino factory.

From the foregoing it should be apparent that we have entered a new era in elementary-particle physics, and that one can expect a long and interesting exploration into neutrino mass and mixing now that the door has been opened.

Acknowledgments

As noted earlier, the SuperKamiokande collaboration deserves the credit for most of the work reported herein, but any errors in interpretation are those of the author. Thanks to Sandip Pakvasa for many discussions and much help. Thanks also to Tony Mann, from whose excellent summary [5.21] I drew heavily. And, finally, thanks to the Aspen Center for Physics, and the year 2000 "Neutrinos with Mass" confreres for many lively discussions about neutrinos.

Appendix: Super-Kamiokande Collaboration, June 2000

Boston University M. Earl, A. Habig, E. Kearns, M. D. Messier, K. Scholberg, J. L. Stone, L. R. Sulak, C. W. Walter

Brookhaven National Laboratory M. Goldhaber

University of California, Irvine T. Barszczak, D. Casper, W. Gajewski, W. R. Kropp, S. Mine, L. R. Price, M. Smy, H. W. Sobel, M. R. Vagins

California State University, Dominguez Hills K.S. Ganezer, W.E. Keig

George Mason University R.W. Ellsworth

Gifu University S. Tasaka

University of Hawaii, Manoa A. Kibayashi, J. G. Learned, S. Matsuno, D. Takemori

Institute of Particle and Nuclear Studies, KEK Y. Hayato, T. Ishii, T. Kobayashi, K. Nakamura, Y. Oyama, A. Sakai, M. Sakuda, O. Sasaki

Kobe University S. Echigo, M. Kohama, A. T. Suzuki

Kyoto University T. Inagaki, K. Nishikawa

Los Alamos National Laboratory T. J. Haines

Louisiana State University E. Blaufuss, B. K. Kim, R. Sanford, R. Svoboda

University of Maryland M. L. Chen, J. A. Goodman, G. Guillian, G. W. Sullivan

State University of New York, Stony Brook J. Hill, C. K. Jung, K. Martens, M. Malek, C. Mauger, C. McGrew, E. Sharkey, B. Viren, C. Yanagisawa

Niigata University S. Inaba, M. Kirisawa, C. Mitsuda, K. Miyano, H. Okazawa, C. Saji, M. Takahashi, M. Takahata

Osaka University Y. Nagashima, K. Nitta, M. Takita, M. Yoshida

Seoul National University S. B. Kim

Tohoku University M. Etoh, Y. Gando, T. Hasegawa, K. Inoue, K. Ishihara, T. Maruyama, J. Shirai, A. Suzuki

Tokai University Y. Hatakeyama, Y. Ichikawa, M. Koike, K. Nishijima

Tokyo Institute for Technology H. Fujiyasu, H. Ishino, M. Morii, Y. Watanabe

University of Tokyo M. Koshiba

Institute for Cosmic Ray Research, University of Tokyo S. Fukuda, Y. Fukuda, M. Ishitsuka, Y. Itow, T. Kajita, J. Kameda, K. Kaneyuki, K. Kobayashi, Y. Kobayashi, Y. Koshio, M. Miura, S. Moriyama, M. Nakahata, S. Nakayama, Y. Obayashi, A. Okada, K. Okumura, N. Sakurai, M. Shiozawa, Y. Suzuki, H. Takeuchi, Y. Takeuchi, T. Toshito, Y. Totsuka (spokesman), S. Yamada

Warsaw University U. Golebiewska, D. Kielczewska

University of Washington, Seattle S.C. Boyd, A.L. Stachyra, R.J. Wilkes, K.K. Young

References

5.1 T. K. Gaisser, *Cosmic Rays and Particle Physics* (Cambridge University Press, Cambridge, UK, 1990).

5.2 V. S. Berezinsky, S. V. Bulanov, V. A. Dogiel, V. L. Ginzburg (ed.) and V. S. Ptuskin, *Astrophysics of Cosmic Rays* (North-Holland, Amsterdam, 1990).

5.3 T. Stanev, Phys. Rev. Lett. **83**, 5427 (1999); astro-ph/9907018.

5.4 R. Engel, T. K. Gaisser and T. Stanev, Phys. Lett. B **472** (2000); hep-ph/9911394.

5.5 L. V. Volkova, Yad. Fiz. **31**, 1531 (1980) [Sov. J. Nucl. Phys. **31**, 784 (1980)]; *DUMAND 1978 Workshop Proceedings*, ed. by A. Roberts, Vol. 1, p. 75 (1978).

5.6 T. K. Gaisser, M. Honda, K. Kasahara, H. Lee, S. Midorikawa, V. Naumov and T. Stanev, Phys. Rev. D **54**, 5578 (1996); T. Gaisser, Nucl. Phys. Proc. Suppl. **77**, 133 (1999); M. Honda, Nucl. Phys. Proc. Suppl. **77**, 140 (1999).

5.7 M. Honda, T. Kajita, K. Kasahara and S. Midorikawa, Phys. Rev. D **52**, 4985 (1995), hep-ph/9503439.

5.8 26th International Cosmic Ray Conference, Salt Lake City, Utah, August 1999; G. Battistoni et al., Astropart. Phys. **12**, 315 (2000), hep-ph/9907408.

5.9 F. Reines et al., Phys. Rev. Lett. **15**, 429 (1965); F. Reines et al., Phys. Rev. D **4**, 80 (1971).

5.10 H. Achar et al., Phys. Lett. **18**, 196 (1965); H. R. Krishnaswamy et al., Proc. Phys. Soc. Lond. A **323**, 489 (1971); H. Adarkar et al., Phys. Lett. B **267**, 138 (1991).

5.11 B. Cortez, PhD thesis, University of Michigan, 1983; W. Foster, PhD thesis, University of Michigan, 1983.

5.12 T.J. Haines et al., IMB Collaboration, Phys. Rev. Lett. **57**, 1986 (1986).

5.13 T. Kajita, PhD thesis, University of Tokyo, 1986; M. Nakhata et al., J. Phys. Soc. Japan **55**, 3788 (1986).

5.14 K. Hirata et al., Phys. Lett. B **205**, 416 (1988).

5.15 R. Clark et al. (IMB collaboration), Phys. Rev. Lett., **79**, 345 (1997); R. Becker-Szendy et al., Phys. Rev. D **46** (1992) 3720; D. Casper et al., Phys. Rev. Lett. **66** (1991) 2561.

5.16 Y. Oyama et al. (Kamiokande collaboration), hep-ex/9706008 (1997); K. S. Hirata et al., Phys. Lett. B **205**, 416 (1988); K. S. Hirata et al., Phys. Lett. B **280**, 146 (1992); Y. Fukuda et al., Phys. Lett. B **335**, 237 (1994).

5.17 J. G. Learned, S. Pakvasa and T. J. Weiler, Phys. Lett. B **207**, 79 (1988); V. Barger and K. Whisnant, Phys. Lett. B **209**, 365 (1988); K. Hidaka, M. Honda and S. Midorikawa, Phys. Rev. Lett. **61**, 1537 (1988).

5.18 G. Battistoni et al. (NUSEX collaboration), Phys. Lett. B **118**, 461 (1982); Nucl. Instrum. Meth. A **219**, 300, (1984); Nucl. Instrum. Meth. A **245**, 277 (1986).

5.19 C. Berger et al. (Frejus collaboration), Phys. Lett. B **245**, 305 (1990); Phys. Lett. B **269**, 227 (1991).

5.20 A. Mann et al. (Soudan-2 collaboration), in *Proceedings of the 8th International Symposium on Neutrino Telescopes*, Venice, Feb. 1999, ed. M. Baldo-Cedin (University of Padua, 1999), Vol. 1, p. 203, hep-ex/9912060.

5.21 A. Mann, Plenary talk at the 19th International Symposium on Lepton and Photon Interactions at High Energies, Stanford, Aug. 1999, hep-ex/9912007.

5.22 H. Sobel (SuperKamiokande collaboration), in *Proceedings of the NU2000 Conference*, Sudbury, Canada, June 2000, in press.

5.23 M. Ambrosio et al. (MACRO collaboration), Phys. Lett. B **478**, 5 (2000); Astropart. Phys. **9**, 105 (1998), hep-ex/9807032.

5.24 P. Lipari, T. K. Gaisser and T. Stanev, Phys. Rev. D **58**, 73003 (1998), astro-ph/9803093.

5.25 P. Lipari, *The East West Effect for Atmospheric Neutrinos,* hep-ph/0003013; *The Geometry of Atmospheric Neutrino Production,* hep-ph/0002282 (2000).

5.26 http://www-sk.icrr.u-tokyo.ac.jp/doc/sk/pub/.

5.27 Y. Fukuda et al. (SuperKamiokande collaboration), Phys. Lett. B **433**, 9 (1998), hep-ex/9803006.

5.28 Y. Fukuda et al. (SuperKamiokande collaboration), Phys. Lett. B **436**, 33 (1998), hep-ex/9805006.

5.29 T. Futagami et al. (SuperKamiokande collaboration), Phys. Rev. Lett. **82**, 5194 (1999), astro-ph/9901139.

5.30 Y. Fukuda et al. (SuperKamiokande collaboration), Phys. Rev. Lett. **81**, 1562 (1998), hep-ex/9807003.

5.31 Y. Fukuda et al. (SuperKamiokande collaboration), Phys. Rev. Lett. **82**, 2644 (1999), hep-ex/9812014.

5.32 J. W. Flanagan, J. G. Learned and S. Pakvasa, Phys. Rev. D **57**, 2649 (1998), hep-ph/9709438; J. W. Flanagan, PhD dissertation, UH 1997, available on the SuperKamiokande web page [5.26].

5.33 M. Messier, PhD thesis, Boston University, 1999.

5.34 V. Barger, J. G. Learned, S. Pakvasa and T.J. Weiler, Phys. Rev. Lett. **82**, 2640 (1999), astro-ph/9810121.

5.35 V. Barger, J. G. Learned, P. Lipari, M. Lusignoli, S. Pakvasa and T. J. Weiler, Phys. Lett. B **462**, 109 (1999), hep-ph/9907421.

5.36 P. Antonioli et al., "The AQUA-RICH atmospheric neutrino experiment", in *Proceedings of the RICH98 Workshop*, Israel, Nucl. Instrum. Meth., in press, preprint CERN-LAA/99-03, 5/5/99; T. Ypsilantis, Nucl. Instrum. Meth. A **433**, 104 (1999).

5.37 J. G. Learned, "The neutrino eye: a megaton low energy neutrino and nucleon decay detector", in *Proceedings of the International Workshop on Simulations and Analysis Methods for Large Neutrino Telescopes,* Zeuthen, Germany, July 1998, ed. by C. Spiering, DESY-PROC-1999-01.

5.38 K. Hoepfner et al. (MONOLITH collaboration), Nul. Phys. Proc. Suppl. **87**, 192 (2000); http://www.desy.de/~hoepfner/Neutrino/Monolith/.

5.39 D. Casper and R. Svoboda, private communication, 1999..

5.40 P. Lipari and M. Lusignoli, Phys. Rev. D **58**, 073005 (1998); hep-ph/9803440.

5.41 S. Fukuda et al., Phys. Rev. Lett. **85**, 3999 (2000).

5.42 A. Surdo et al. (MACRO collaboration), hep-ex/9905028; M. Spurio, hep-ex/9908066; S. Ahlen et al. (MACRO collaboration), Phys. Lett. B **357**, 481 (1995); M. Ambrosio et al., Phys. Lett. B **434**, 451 (1998); F. Ronga, hep-ex/9905025.

5.43 P. F. Harrison, D. H. Perkins and W. G. Scott, Phys. Lett. B **458**, 79 (1999), hep-ph/9904297; Phys. Lett. B **349**, 137 (1995).

5.44 Y. Obayashi (SuperKamiokande collaboration), in *Proceedings of the CIPANP2000 Meeting,* ed. by Z. Parsa and W. Marciano, paper 5.2.1 (in press 2000).

5.45 S. Pakvasa, "Exotic explanations for neutrino anomalies", in *Proceedings of the 8th International Symposium on Neutrino Telescopes,* Venice, Feb. 1999, ed. by M. Baldo-Ceolin (University of Padua, 1999), I, 283; hep-ph/9905426.

5.46 S. Kasuga et al., Phys. Lett. B **374**, 238 (1996); J. Breault, PhD thesis, University of California, Irvine, 1998.

5.47 A. D. Sakharov, JETP Lett. **5**, 24 (1967).

5.48 V. A. Kuzmin, JETP Lett. **12**, 228 (1970).

5.49 E. Kh. Akhmedov, V. A. Rubakov and A. Yu. Smirnov, Phys. Rev. Lett. **81**, 1359 (1998).

5.50 E. Church et al. (BooNE collaboration), nucl-ex/9706011 (1997).

5.51 Y. Oyama et al. (K2K collaboration), hep-ex/0004015; hep-ex/9803014; http://neutrino.kek.jp/.

5.52 K. Nakamura (K2K collaboration), in *Proceedings of the NU2000 Conference,* Sudbury, Canada, June 2000, in press.

5.53 E. Ables et al., FERMILAB-PROPOSAL-P-875, Feb. 1995; FERMILAB-PROPOSAL-P-875-ADD, NUMI-L-79, Apr. 1995.

5.54 F. Cavanna et al. (ICANOE collaboration), LNGS-P21-99-ADD-2; CERN-SPSC-99-40; CERN-SPSC-P-314-ADD-2, Nov. 1999.

5.55 K. Kodama et al. (OPERA collaboration), CERN-SPSC-98-25; CERN-SPSC-M-612; LNGS-LOI-8-97-A, Oct. 1998; hep-ex/9812015.

5.56 Z. Maki, M. Nakagawa and S. Sakata, Prog. Theor. Phys. **28**, 870 (1962).

5.57 K. T. McDonald, *Expression of Interest for R&D towards Neutrino Factory Based on a Storage Ring and a Muon Collider,* submitted to the US National Science Foundation, 7 Nov. 1999.

5.58 C. Albright et al., *Physics at a Neutrino Factory,* FERMILAB-FN-692, May 2000.

5.59 T. J. Weiler, Astropart. Phys. **11**, 303 (2000), hep-ph/9710431); D. Fargion, B. Meleand and A. Salis, astro-ph/9710029 (1997); T. Weiler, in Proceedings of Beyond the Desert 99, ed. by H. Klapdor-Kleingrothaus, hep-ph/9910316 (1999).

5.60 V. Barger, S. Pakvasa, T. J. Weiler and K. Whisnant, MADPH-00-1177; UH-511-964-00; VAND-TH-00-5; AMES-HET-00-04, May 2000; hep-ph/0005197.

5.61 J.G. Learned and K. Mannheim, Annu. Rev. Nucl. Sci., in press (2000).

5.62 J.G. Learned and S. Pakvasa, Astropart. Phys. **3**, 267 (1995); hep-ph/9408296.

6 Studies of Neutrino Oscillations at Reactors

Felix Boehm

6.1 Introduction

Neutrinos from reactors have played an important and decisive role in the early history of neutrino oscillations (e.g. [6.1]). After considerable controversy in the early 1980s, results from the reactors at ILL [6.2] in 1981, Goesgen [6.3] in 1986 and Bugey [6.4] in 1995 have shown no evidence for neutrino oscillations involving reactor $\bar{\nu}_e$. More recently, the Chooz [6.5] and Palo Verde [6.6, 6.7] experiments have confirmed these findings with greater sensitivity. The purpose of this chapter is to highlight the developments involving reactor neutrinos and to outline the current status and future studies.

We begin with a brief reminder of the parameters that play a role in neutrino oscillation physics [6.1]. Assuming, for simplicity, that there are only two neutrino flavors, then the two parameters describing oscillations are the mixing amplitude $\sin^2 2\theta$ and the mass parameter Δm^2. They are related to the probability of creating a weak-interaction state with flavor l' from a state l ($l \neq l'$) in an "appearance experiment" through the expression

$$P_{ll'} = \sin^2 2\theta \; \sin^2 \frac{1.27\Delta m^2(\text{eV})^2\text{L(m)}}{E_\nu(\text{MeV})} \; , \tag{6.1}$$

L being the distance between the neutrino source and detector ("baseline") and E_ν the neutrino energy. The probability that a state l disappears through oscillation is given by $P_{ll} = 1 - P_{ll'}$. While reactor $\bar{\nu}_e$ may oscillate into $\bar{\nu}_\mu$ or $\bar{\nu}_\tau$, these neutrinos cannot be observed via charged-current reactions as the energy of the $\bar{\nu}_e$ at a reactor is insufficient to create a μ or a τ. A reactor experiment thus explores only the disappearance of the $\bar{\nu}_e$.

As the oscillatory function depends on the ratio L/E_ν, it can be seen that low-energy reactor neutrinos are well suited to exploring the region of small Δm^2 at relatively modest baselines. For example, to explore the parameter Δm^2 down to 10^{-3} eV^2 a reactor experiment with E_ν around 5 MeV requires a baseline of $L = 1$ km, while an accelerator experiment with $E_\nu = 5$ GeV would require $L = 1\,000$ km.

It follows from (6.1) that oscillations manifest themselves through modifications of the energy spectrum of neutrinos arriving in the detector as well as by a change in the total neutrino yield. Both of these aspects can be explored in an experiment.

Reactor experiments, with their sensitivity to small Δm^2, have been directed towards exploring the physics of the atmospheric neutrino ratio [6.8], a topic described in Chap. 5 of this book. If extended to even larger baselines, these experiments are capable of shedding light on the large-mixing-angle solar-neutrino solution [6.9], as discussed in Chap. 4.

Most reactor neutrino detectors are based on the interaction with the proton

$$\bar{\nu}_e + p = e^+ + n \, , \tag{6.2}$$

with a threshold of 1.8 MeV. This inverse neutron decay has the largest cross section among neutrino–nuclear reactions. The presence of the time-correlated e^+, n signature provides a powerful way to retrieve the neutrino signal from the abundant neutron and low-energy radioactive backgrounds. The small anisotropy of the reaction products arising from the kinematics of the detection process can be used for "pointing" and thus for background suppression.

6.2   The Reactor Neutrino Spectrum

The neutrino sources for these experiments are large commercial power reactors, each producing about 3 GW of thermal power accompanied by neutrino emission at a rate of about 8×10^{20} $\bar{\nu}_e$/s. As a rule, these reactors run uninterruptedly at full power, except for a refueling cycle of about one month per year, which provide opportunities for studying the backgrounds of the detector system.

The $\bar{\nu}_e$ spectrum from a fission reactor and its relation to the reactor's power and status in the burn cycle are well understood today. Pioneering work on deriving this spectrum, taking into account the contributions of the fissioning isotopes ^{235}U, ^{239}Pu, ^{238}U, ^{241}Pu and ^{252}Cf and their evolution during the burn cycle, was reported by Vogel [6.10] in 1981. This extensive modeling work has been supplemented by experimental studies of the electron spectra of the fissioning isotopes ^{235}U, ^{239}Pu and ^{241}Pu with an on-line beta spectrometer at ILL Grenoble by Schreckenbach and others [6.11]. The combined uncertainty in the predicted reactor neutrino spectrum is about 3%.

Figure 6.1 shows the time evolution of the reactor power associated with the various fissioning fuel components, taken from [6.3]. This information was folded into the calculated neutrino spectrum [6.10].

High-statistics neutrino experiments involving the total neutrino yields were carried out at the Bugey reactor [6.12] at a short distance from the reactor (where possible oscillation effects are negligible). A 2 000 l water target was installed at a distance of 15 m from one of the 2 800 MW reactors at the Bugey site. As the detector responded to neutrons only, it provided an integral cross section for the reaction $\nu_e + p = e^+ + n$ for neutrinos with energies

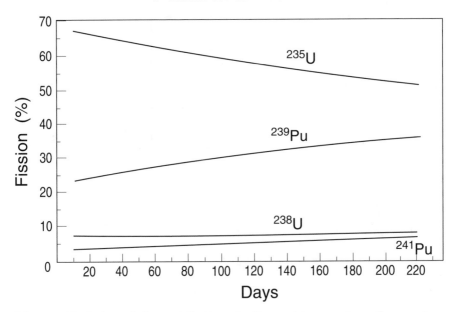

Fig. 6.1. Evolution of the contributions to the neutrino spectrum from various reactor fuel components (from [6.3])

above the reaction threshold of 1.8 MeV. The event rate was $3\,021/d$ with a background rate (reactor off) of $2\,600/d$. The cross section, obtained with an absolute accuracy of 1.4%, was compared with the cross section calculated on the basis of V–A theory. These results confirm that the reactor neutrino spectrum and its relation to reactor power and fuel composition are well understood. The results are listed in Table 6.1 together with previous results from Goesgen [6.3] and Krasnoyarsk [6.16]. There is an excellent agreement between the measured and calculated neutrino rates for all these experiments.

The parameterization by Vogel and Engel [6.13] serves as a convenient starting point for the present analyses of the measured reactor spectra. As an example, Fig. 6.2 shows the neutrino spectrum from ^{235}U fission together with the neutrino-proton reaction cross section and reaction yield.

Table 6.1. Integral cross sections for reactor neutrinos on protons

	Goesgen (1986) [6.3]	Krasnoyarsk (1990) [6.16]	Bugey (1994) [6.12]
$\pm\sigma_{\exp}$	3%	2.8%	1.4%
$\sigma_{\exp}/\sigma_{\text{V–A}}$	0.992 ± 0.04	0.985 ± 0.04	0.987 ± 0.03

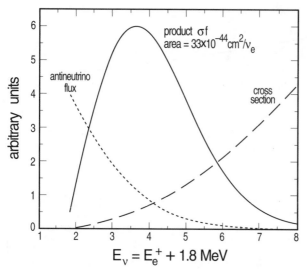

Fig. 6.2. Energy spectrum, cross section and yield of neutrinos from ^{235}U fission in a reactor [6.14]

6.3   Oscillation Experiments

6.3.1   The ILL Grenoble and Goesgen Experiments

Motivated by theoretical developments concerning neutrino mass and mixing in the 1970s (e.g. [6.15]), a group installed an early reactor experiment at the research reactor of the Institut Laue–Langevin (ILL) in 1977 with the aim of shedding light on oscillations involving $\bar{\nu}_e$. The neutrino detector in this ILL experiment [6.2] consisted of 30 individual cells with liquid scintillator which tracked the positron, sandwiched between ^{3}He proportional chambers to detect the neutron. The distance between the reactor and detector was 8.7 m. This "disappearance experiment" was searching for a possible reduction of the $\bar{\nu}_e$ flux as well as for a modification of the energy spectrum observed in the detector. It was found that the measured neutrino spectrum agreed with that calculated [6.10] and thus revealed no evidence for oscillations down to $\Delta m^2 = 0.15$ eV^2 for $\sin^2 2\theta \geq 0.25$.

To enhance the sensitivity of this experiment and to gain information on oscillations with smaller Δm^2, the ILL detector was modified and transferred to the more powerful Goesgen reactor in Switzerland. Three experiments were carried out between 1981 and 1985 with the detector at distances of 37.8 m, 45.9 m and 64.7 m from the reactor core.

The Goesgen detector [6.3] consisted of an array of liquid scintillation counters and ^{3}He multiwire proportional chambers, surrounded by an active scintillation veto counter and various shielding, as illustrated in Fig. 6.3.

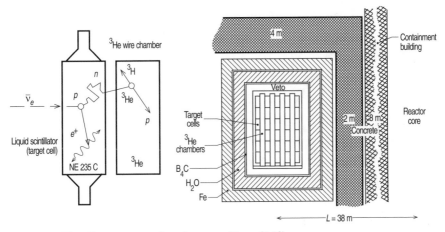

Fig. 6.3. The Goesgen neutrino detector (from [6.3])

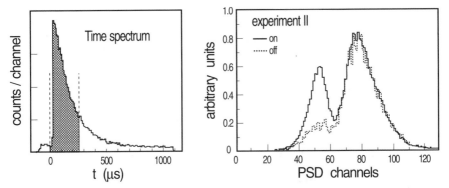

Fig. 6.4. *Left:* distribution of time intervals between positron and neutron in the Goesgen detector. *Right:* pulse shape spectra for reactor on (*solid curve*) and reactor off (*dotted curve*). The peak to the left represents the neutrino signal (positrons). The peak to the right is caused by high-energy cosmic-ray-induced neutrons (from [6.3])

The signature of an event was given by a positron pulse in the liquid scintillator followed by a neutron-induced reaction in the ^{3}He counter. The time correlation and time window chosen are shown in Fig. 6.4.

Pulse shape discrimination was instrumental in reducing background events associated with fast neutrons from cosmic rays, as illustrated in Fig. 6.4. No reactor-associated backgrounds were seen, as could be verified by comparing backgrounds with the reactor on and off. The observed correlated positron spectra, corrected for detector response and background, as a function of energy and position, are shown in Fig. 6.5.

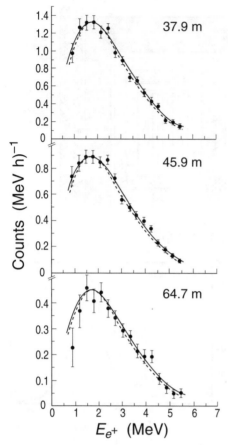

Fig. 6.5. Positron spectra at three positions of the detector (from [6.3]). The *solid curves* are the predicted positron spectra for no oscillations derived from fitting the reactor neutrino spectra to the data, while the *dashed curves* are obtained from the calculated neutrino spectra

In order to compare the spectra taken at various positions and at different times, the relevant reactor spectrum for each experiment had to be known. Small differences in reactor fuel composition were taken into account, although these differences were minimized by conducting each experiment over a full fuel cycle. Corrections for the difference in fuel composition varied by less than 5%, with a negligibly small uncertainty.

In the data analysis, the experimental positron spectra were compared with calculated spectra in two different ways. First, an analysis (analysis A) independent of the source neutrino spectrum was conducted. The neutrino spectrum was parameterized and a χ^2, calculated for the difference between the experimental yield and the expected yield, was minimized for a fixed set of parameters Δm^2 and $\sin^2 2\theta$. A maximum-likelihood test was used to

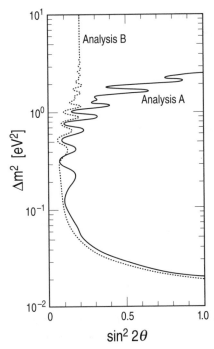

Fig. 6.6. Regions in the parameter space excluded at 90% C.L. by the Goesgen experiment [6.3]. Analysis A is based on the ratios of the neutrino spectra at three distances. Analysis B is based on the calculated neutrino spectrum

obtain the exclusion plot shown in Fig. 6.6. In a second analysis (analysis B) the measured spectra were compared with calculated spectra [6.10, 6.11] (also shown in Fig. 6.4) with the results also shown in Fig. 6.6.

It is important to appreciate the sources of the limitations on the accuracy of the mixing-angle sensitivity arising from the uncertainties in the absolute normalization of the neutrino spectrum (3%), the detector efficiency calibration using a calibrated neutron source (4%), the reaction cross sections (2%) and the reactor power (2%), compounding to an uncertainty of about 6%.

6.3.2 The Bugey Experiments

A high-statistics search for neutrino oscillations at the 2 800 MW Bugey reactor with the detectors at 15, 40 and 95 m, was reported by the Bugey group [6.4] in 1995. In this experiment, three identical 600 l segmented detectors were used. Each detector consisted of 98 prismatic cells viewed by a 3 inch photomultiplier on each side. The cells were filled with a ^{6}Li-loaded (0.15%) scintillator. Two principal advantages were quoted for the ^{6}Li loading: the neutron capture time in the scintillator is reduced to 30 μs (from about 170 μs on protons), and the resulting ^{3}H and $α$ particles can be distinguished from the reaction positrons by pulse shape discrimination. The relative and absolute normalization errors in these experiments were 2% and 5%, respectively. The 95 m experiment provided the most stringent results for the mass parameter Δm^2, of 10^{-2} eV^2.

6.3.3   The Experiments at Rovno and Krasnoyarsk

We mention briefly the experimental work at the Russian reactors at Rovno [6.17] and Krasnoyarsk [6.18]. At the Rovno power reactor a measurement of the total neutrino-induced neutron yield was carried out. The detector, at 18 m from the reactor, consisted of a water target into which a large number of ^{3}He proportional counters were embedded. The observed neutrino yield agreed with the calculated yield to within about 3%. A similar experiment was carried out at the three-reactor station at Krasnoyarsk. There, the ^{3}He neutron counters were embedded in polyethylene and stationed at 57 m from reactors 1 and 2, and 231 m from reactor 3. While only total yields were obtained, by comparing rates from reactors at different distances, information on oscillations could be derived. These detectors, however, could not provide information on the shape of the neutrino spectrum. Also, these detectors possessed much higher inherent backgrounds than the detectors discussed above, which identify the reaction neutron by an e^+n correlated signal.

6.3.4   The Long-Baseline Experiments at Palo Verde and Chooz

Results from atmospheric neutrino experiments, such as those from Kamiokande [6.8], have triggered reactor neutrino studies aimed at exploring the parameter region of Δm^2 between 10^{-2} and 10^{-3} eV^2. Two experiments, both at a distance L around 1 km, have been conducted recently, one at the French reactor station (two reactors) at Chooz [6.5] and the other at the Palo Verde [6.6] site in Arizona, USA (three reactors). Both experiments now have results and we describe them below. While these experiments were in progress, the new SuperKamiokande [6.19] results, which appeared in 1998, favored the ν_μ–ν_τ channel over the ν_μ–ν_e in some regions of the parameter plane.

Another experiment in the Kamioka mine in Japan, referred to as Kam-LAND [6.20], at a much larger distance from a number of power reactors, is still in the proposal stage.

To illustrate the effect of oscillations on the positron spectrum, Fig. 6.7 shows the expected spectrum for Chooz and Palo Verde for the case of no oscillations, as well as for the set of oscillation parameters favored by the Kamiokande results. Clearly, the effects from Kamiokande-type oscillations on the spectrum should be quite pronounced.

The Chooz and Palo Verde experiments are based on the reaction $\bar{\nu}_e + p = e^+ + n$ and rely on an (e^+, n) correlated signature. Both experiments make use of a Gd-loaded liquid scintillator. Gd loading reduces the capture time,

Expected PV no-osc. spectra

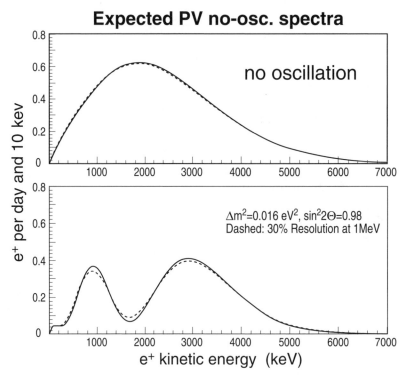

Fig. 6.7. Expected positron spectra for the Chooz and Palo Verde experiments for "no oscillations" and for oscillations given by the Kamiokande parameters

owing to its large thermal-neutron capture cross section, and also gives rise to a high-energy gamma cascade of up to 8 MeV. Both features are valuable; the short capture time helps reduce random coincidences, and the large gamma ray energy allows reduction of backgrounds as the energy threshold can be set above that of radioactive decay products. In both experiments, the amount of Gd dissolved in the scintillator is about 0.1% by weight. At a distance of ca. 1 km from the reactor, the detector response is about 5 events per day per ton of scintillator. The Chooz experiment takes advantage of an existing deep tunnel, reducing the cosmic-ray muon background substantially. The Palo Verde experiment, being in a shallow underground laboratory, has to cope with a considerably larger muon rate and thus has to rely on powerful background rejection. Because of this, the two detectors are designed quite differently. The Chooz detector consists of a homogeneous central volume of Gd scintillator, while the Palo Verde detector is made from finely segmented detector cells.

The Palo Verde Experiment. The Palo Verde experiment [6.6] is situated near the Palo Verde nuclear-power plant in Arizona (three reactors, 11 GW

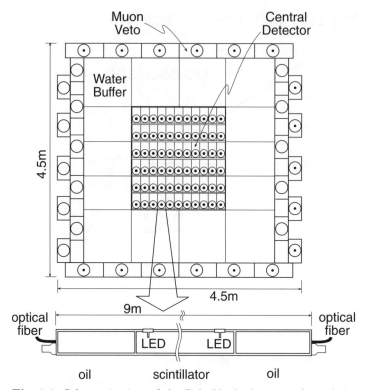

Fig. 6.8. Schematic view of the Palo Verde detector. One of the cells, with photomultiplier tubes, oil buffers, calibration LEDs and optical-fiber flashers is shown lengthwise at the *bottom*

thermal power). The detector is installed in an underground cave with 32 mwe (meter water equivalent) overhead at a distance of $L = 890$ m from reactors 1 and 3, and 750 m from reactor 2. Each reactor is shut down for refueling for a period of ca. 40 days every year, providing the opportunity for establishing the detector background.

The detector, shown schematically in Fig. 6.8, has a fiducial volume of 12 tons. Its liquid scintillator, whose composition is 60% mineral oil, 36% pseudocumene, 4% alcohol and 0.1% Gd, was developed in collaboration with Bicron [6.21]. It has an effective light attenuation length of 10 m for 440 nm light. The detector consists of 66 cells, each 9 m long, of which 7.4 m is active and 0.8 m on each end serves as an oil buffer. There is a 5 inch, low-radioactivity photomultiplier attached to each end, allowing both the anode and the last dynode to be read out. A blue LED installed at 0.9 m from each photomultiplier, in conjunction with optical fibers, allows each individual cell to be monitored. A passive water shield, 1 m thick, surrounds the block of active cells to help shield against radioactivity as well as muon-induced neutrons. An active-veto counter consisting of 32 12 m long MACRO cells is

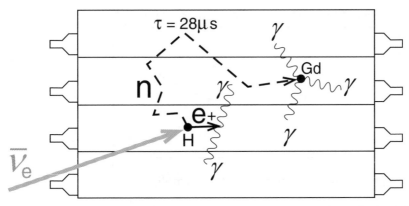

Fig. 6.9. Illustration of the neutrino reaction in the matrix of the Gd-loaded scintillator

placed on all four long sides, while a removable end-veto counter protects the ends of the cell matrix.

A diagram of the detector response, showing the $\bar{\nu}_e$ reaction and the gamma rays from Gd capture, is given in Fig. 6.9.

A neutrino signal consists of a fast (30 ns) $e^+\gamma\gamma$ trigger within a block of 3×5 cells, with the first hit having $E \geq 500$ keV, and the second hit $E \geq 30$ keV. This second hit includes the Compton response from the 511 keV annihilation gammas. This fast triple coincidence is followed by a slow (200 μs) signal associated with the 8 MeV gamma cascade following neutron capture in Gd within a 5×7 scintillator cell matrix.

Energy calibrations could be carried out with the help of small sources that were introduced through a set of Teflon tubes installed alongside a group of detector cells. The response from these sources at various positions made it possible to monitor the attenuation length of the scintillator which exhibited only a negligible decline over the period measured. Figure 6.10 shows the light yield along the scintillator cell. The linearity of the photomultiplier tube was obtained with the help of a fiber-optics flasher, while single photoelectron peaks were monitored with a blue LED.

Inasmuch as the experiment aims at extracting absolute $\bar{\nu}_e$-induced reaction rates, knowledge of the detection efficiency is essential. The positron efficiency was established with the help of the positron emitter ^{22}Na. (A calibrated ^{68}Ge source [6.22] dissolved in a special cell will also be implemented.) To obtain the neutron efficiency, a calibrated AmBe source was used in a tagged mode, i.e. in coincidence with the 4.4 MeV gamma from ^{12}C*. From these calibrations, combined with Monte Carlo simulations, an average (over the detector) efficiency was obtained. For the 1999 run this efficiency was found to be 0.112, yielding a neutrino event rate in the detector of 225 ± 8 per day.

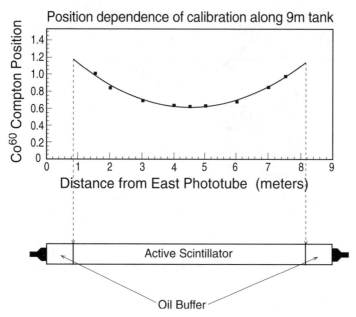

Fig. 6.10. Light yield along scintillator cell. The attenuation length of the Gd scintillator is 10 m

From the 1998/99 data, the observed rates for a 147 d run with full reactor power (three reactors on) and a 54.7 d run with reduced power (two reactors on) yielded the positron spectrum shown in Fig. 6.11. The time structure of the correlated signal is depicted in Fig. 6.12. The measured decay time of 35 µs agrees well with that modeled with a Monte Carlo simulation.

To test the oscillation scenario, a χ-squared analysis in the $(\Delta m^2$-$\sin^2 2\theta)$ plane was performed, taking into account the small variations in $\bar{\nu}_e$ flux from the burn-up-dependent fission rate of the reactor. The 90% C.L. acceptance region was defined according to a procedure suggested by Feldman and Cousins [6.23]. The data agreed well with the no-oscillation hypothesis. In an independent analysis, which does not rely on the "on" minus "off" scheme, the intrinsic symmetry of the dominant neutron background with respect to the time sequence of the e^+ and n signals was implemented to cancel a major part of the neutron-induced background. A small neutron background that remained after subtraction of the signal with reversed time sequence was obtained from a Monte Carlo simulation of muon spallation [6.24]. Experimental data points corresponding to energies ≥ 10 MeV (beyond the positron energy spectrum) served to normalize the calculated neutron spectrum. This analysis, which was based on subtraction of the neutron background, showed no evidence for $\bar{\nu}_e$-$\bar{\nu}_X$ oscillations, and thus agreed with the the results of the more traditional "on" minus "off" analysis. The region in the parameter space excluded at 90% C.L. is depicted by the curve "Palo Verde" in Fig. 6.16.

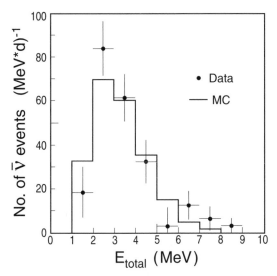

Fig. 6.11. Correlated positron spectrum derived from a three-reactor run and a two-reactor run: observed spectrum ("Data") and expected spectrum for no oscillations ("MC")

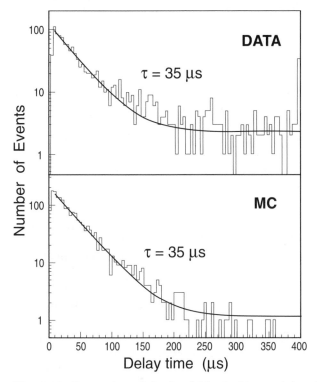

Fig. 6.12. Decay time of the fourfold coincidence giving the neutron capture time in our Gd scintillator

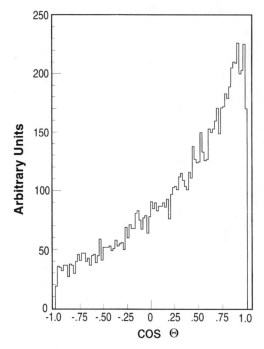

Fig. 6.13. Angular distribution of scattered (moderated) neutrons with regard to the neutrino direction

The segmentation of the Palo Verde detector makes it possible to study the $\bar{\nu}_e$–n angular correlation of reaction (6.2). This, in turn, establishes an independent background determination. From kinematics we find that the neutron moves preferentially in the direction of the incoming neutrino, with an angular distribution limited by

$$\cos(\theta_{\nu,n})_{\max} = [2\Delta/E_\nu - (\Delta^2 - m^2)/E_\nu^2]^{1/2} \,, \tag{6.3}$$

where $\Delta = M_n - M_p$.

From the Monte Carlo simulation it was found that the neutron scattering preserves the angular distribution, resulting in a shift of the mean coordinate of the neutron capture center $\langle x \rangle = 1.7$ cm [6.25]. The angular spread after scattering is very pronounced, as can be seen in Fig. 6.13. It should be noted that this effect was first studied by Zacek [6.26] in connection with the segmented Goesgen detector, where the forward/backward ratio was found to be as large as a factor of 2.

Preliminary results [6.27] give an asymmetry, expressed as events in the half-plane away from the reactor (forward) minus events in the half-plane towards the reactor (backward), of 109 ± 44, in agreement with a Monte Carlo simulation.

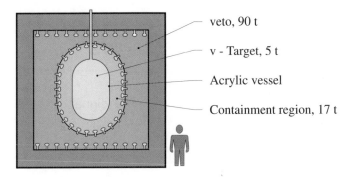

Fig. 6.14. Schematic arrangement of the Chooz detector

The Chooz Experiment. An experiment with a similar aim, but with a substantially different detector, was carried out at Chooz by a French–Italian–Russian–US collaboration. This experiment and its results [6.5] are reviewed below.

The Chooz detector is composed of three regions, a central region containing 5 tons of Gd-loaded liquid scintillator surrounded by an acrylic vessel, a containment region with 17 tons of ordinary liquid scintillator, and an outer veto region with 90 tons of scintillator. Figure 6.14 shows schematically the arrangement of the Chooz detector.

The inner two regions are viewed by a set of photomultipliers. An independent set of photomultipliers detects the light from the veto region. While the positron response is obtained from a signal in the inner region, the neutron response comprises signals from the inner region as well as from the containment region, resulting in a well-contained and well-resolved Gd capture sum peak at 8 MeV. As mentioned earlier, the Chooz detector is installed in a tunnel, thus reducing the correlated background to less than 10% of the signal.

The data were obtained at various power levels of the two Chooz reactors as these reactors were slowly brought into service. A total of 2991 neutrino events was accumulated in 8209 live hours with the reactor on, and 287 events in 3420 live hours with the reactor off. Normalized to the full power of the two reactors (8.5 GW th), the event rate corresponds to 27.4 ± 0.7 neutrino interactions per day, where the error includes contributions from the reaction cross section, the reactor power, the number of protons in the target and the detector efficiency. In comparison, the background rate was 1.0 ± 0.1 per day. The ratio of measured to expected neutrino signal was $1.01 \pm 2.8\%$ (statistical) $\pm 2.7\%$ (systematic). The total efficiency of the detector was found to be $69.8 \pm 1.1\%$.

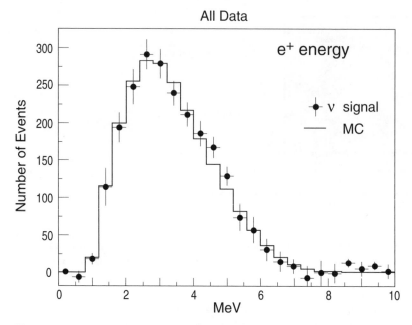

Fig. 6.15. Positron energy spectra from the Chooz experiment

The positron energy spectrum for the reactor-on and the reactor-off conditions is shown in Fig. 6.15, together with a plot of the ratio of the measured to the calculated spectrum.

The neutron capture event, characterized by an 8 MeV gamma peak, was localized to within $\sigma_x = 17.4$ cm. The energy resolution for the 8 MeV peak was $\sigma_e = 0.5$ MeV, or about 1 MeV FWHM. Calibrations for energy, neutron efficiency and timing were carried out with sources of ^{60}Co, ^{252}Cf and AmBe, respectively. The lifetime for neutron capture in the Gd scintillator was found to be 30.5 µs. The combined systematic error was 2.8%.

Figure 6.16 depicts the Chooz excluded area. Clearly, the Kamiokande region is excluded with a high confidence level, implying the absence of $\nu_e \leftrightarrow \nu_\mu$ oscillations. The mixing-angle limit for large Δm^2 from this analysis is $\sin^2 2\theta < 0.1$ at 90% C.L., again based on the widely accepted method of Feldman and Cousins [6.23]. The 90% limit for Δm^2 for maximum mixing from this experiment is 0.7×10^{-3} eV^2.

The Chooz collaboration has also compared the spectrum from reactor 2, which is at $L = 998$ m, with that of reactor 1 at $L = 1115$ m. The relative spectra from the two reactors at different distances provided information on oscillations independently of the absolute yields, as described above in the context of the "analysis A" of the Goesgen experiment. For the Chooz experiment, that analysis leads to an exclusion plot consistent with, but less stringent than, that of the analysis involving absolute neutrino yields.

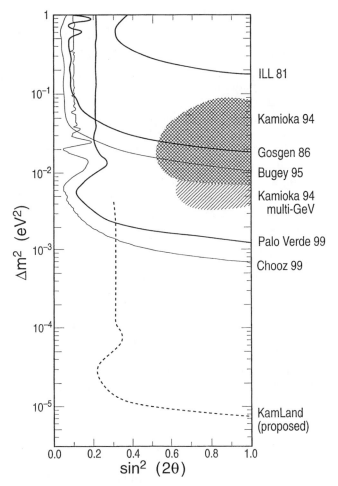

Fig. 6.16. An overview showing the evolution in time of the 90% excluded regions for the experiments reviewed in this chapter. The Kamiokande allowed regions [6.8] and the region allowed by the recent SuperKamiokande [6.19] results, if analyzed in the Δm^2 versus $\Theta_{1,3}$ plane [6.7, 6.28], are also shown by the *hatched areas*

The Chooz analysis also includes a discussion [6.29] of the neutron angular distribution, as mentioned in the section above on the Palo Verde experiment.

The parameter space that could be excluded by each of the aforementioned experiments is summarized in Fig. 6.16. The curves are labeled by the experiment and date, providing a historical account of the impressive gain in sensitivity over time.

6.3.5   The KamLAND Experiment

The KamLAND [6.20] experiment will be the ultimate long-baseline reactor experiment, destined to explore $\bar{\nu}_e$ disappearance at very small Δm^2. It will be sensitive to exploring the large-mixing-angle solar MSW solution. The experiment will also address the small-mixing-angle solar MSW solution at low neutrino energy by observing ν_e–electron scattering, as well as, by invoking seasonal variations, the vacuum oscillation solution.

The neutrinos originate from 16 nuclear power plants (130 GW thermal power) at distances between 80 and 800 km from the KamLAND detector, with 90% of the neutrino flux produced at sites at distances between 80 and 214 km. The detector will be a 1 kiloton liquid scintillator to be installed in the former Kamiokande cavity at a depth of 1 000 mwe. The spherical 1 kiloton scintillator and the surrounding 2.5 m mineral oil shielding are contained in an 18 m diameter stainless steel sphere that also supports the 2000 17 inch and 20 inch photomultipliers providing a 30% light coverage. The detector light yield is projected to be 100 photoelectrons per MeV. A water volume surrounding the sphere serves as a Cerenkov veto counter. A schematic view of the detector is shown in Fig. 6.17. The expected event rate associated with all power plants (51 reactors) is projected to be 750/y. The background rates due to radioactivity and neutrons from muon spallation were obtained from simulations and are predicted to be about 37/y. Under these conditions, a contour plot can be constructed for a three-year exposure which covers the large-mixing-angle solar MSW solution, as shown in Fig. 6.16, with a maximum sensitivity to Δm^2 of 4×10^{-6} eV^2. The KamLAND detector should be operational in 2001.

6.4   The $\bar{\nu}_e$–d Experiment at Bugey

Reactor neutrinos interacting with deuterons result in two reaction channels, a charged-current (CC) reaction

$$\bar{\nu}_e + d \rightarrow e^+ + n + n\,, \tag{6.4}$$

and a neutral-current (NC) reaction

$$\bar{\nu}_e + d \rightarrow \bar{\nu}_e + p + n\,. \tag{6.5}$$

As the CC reaction is sensitive to oscillations while the NC reaction is not, a measurement of the ratio of the reaction yields may serve as a test for oscillations. This scheme was first suggested by Reines et al. [6.30] and was implemented at the Savannah River reactor in 1980. However, owing to incomplete understanding of the neutron efficiency for single- and double-neutron events, the 1980 results turned out to be unreliable. The experiment was repeated recently by Riley et al. [6.31] using essentially the same 1980 apparatus installed in the Bugey reactor in France. The detector consisted

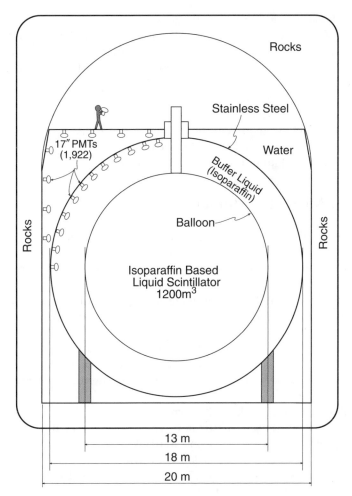

Fig. 6.17. Schematic view of the KamLAND detector (from [6.20])

of a central cylindrical volume containing 276 kg of deuterium into which ten ^{3}He proportional counters were immersed, serving as neutron detectors. A liquid-scintillator veto, as well as Pb and Cd shielding, surrounded the detector. In this experiment, only neutrons were counted. In addition to the reaction neutrons, however, there was a sizable number of background neutrons created by cosmic-ray muons that could not be tagged with the veto.

The event rate for one-neutron events (NC) was 37.7 ± 2.0 per day, with a background rate of 57.0 ± 1.5 per day. The rate for the two-neutron events (CC) was 2.45 ± 0.48, with a background of 3.26 ± 0.36 per day.

After correcting for the efficiencies for detecting one neutron and two neutrons of 0.29 ± 0.01 and 0.084 ± 0.006, respectively, as determined with the help of a ^{252}Cf source and model calculations, the ratio of the reaction yields for CC and NC events, divided by the calculated ratio, was found to be 0.96 ± 0.23, consistent with 1, and thus in agreement with the prediction. It appears that this result differs substantially from the 1980 work [6.30] with the same detector.

On account of the relatively large statistical errors and the close distance (18.5 m) to the reactor, the experiment's sensitivity to oscillations was only modest, allowing it to exclude an area in the parameter space with large mixing angles. The work also confirmed the calculated cross sections describing the neutrino–deuteron breakup in the CC and NC channels.

6.5   Neutrino Magnetic Moment

If neutrinos have mass, they may have a magnetic moment. An experimental effort to look for a neutrino magnetic moment, therefore, is of great interest. Some indications of a magnetic moment have come from a suggested correlation of the signal in the ^{37}Cl experiment with solar activity, suggesting a value of 10^{-11}–$10^{-10}\mu_B$. In addition, considerations of a possible resonant spin flavor precession (RSFP) and also of neutrino interactions in supernovae have been mentioned.

The neutrino magnetic moment contributes to the $\nu_e e$ scattering [6.32], as shown in Fig. 6.18. This contribution is most pronounced at low electron recoil energy. At about 300 keV the magnetic-moment scattering is roughly equal to the weak scattering.

Previous results by Reines et al. [6.33] from scattering reactor neutrinos on electrons in a 16 kg plastic scintillator have given $\mu_\nu = 2$–$4 \times 10^{-10}\mu_B$. More recently, Gurevitch et al. [6.34], using a 103 kg C_6F_6 target, found $\mu_\nu < 2.4 \times 10^{-10}\mu_B$, and Derbin et al. [6.35], with a 75 kg Si target, found $\mu_\nu < 1.8 \times 10^{-10}\mu_B$. A recent analysis [6.37] of the SuperKamiokande data [6.36] results in a limit of $\mu_\nu < 1.6 \times 10^{-10}\mu_B$.

An effort is now under way to obtain a value of, or stringent limit on μ_ν is by the MUNU experiment, a Grenoble–Munster–Neuchatel–Padova–Zurich collaboration [6.38], which has built a 1000 liter CF_4 TPC (time projection chamber) at 5 atm (18.5 kg), surrounded by an anti-Compton shield. This detector is now installed at the Bugey reactor in France. The expected event rate in the interval of 0.5 to 1 MeV recoil energy is 5.1 per day, with an expected background of 4.5 per day. Implementing the angular correlation of the scattered electrons with respect to the incoming reactor neutrinos is expected to enhance the signal-to-noise ratio significantly. A schematic view of the MUNU TPC is shown in Fig. 6.19.

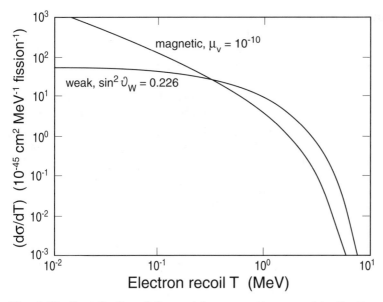

Fig. 6.18. Contribution of the neutrino magnetic moment to the $\bar{\nu}_e e \rightarrow \bar{\nu}_e e$ scattering, averaged over the reactor $\bar{\nu}_e$ spectrum. The purely weak cross section is also shown. (From Vogel and Engel [6.13])

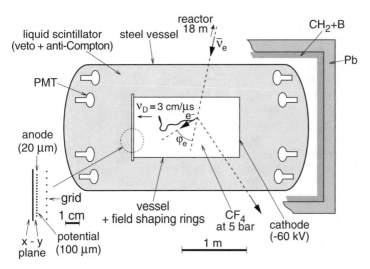

Fig. 6.19. Layout of the MUNU detector for the measurement of the neutrino magnetic moment

6.6   Conclusion

Reactor neutrinos, with their low energies, are well suited to explore small Δm^2 in the $\bar{\nu}_e$ disappearance channel. From the results of the Chooz and Palo Verde experiments, it can be concluded that the atmospheric ν_μ deficiency cannot be attributed to $\bar{\nu}_\mu \leftrightarrow \bar{\nu}_e$ oscillations. The Chooz experiment has ruled out this channel with a large confidence level, and the first data set from Palo Verde excludes it at 90% C.L. To improve on the mixing-angle sensitivity in these experiments so as to shed light on a possible three-flavor solution will be a challenging task. The KamLAND experiment, with a very large baseline, is now on the drawing board. Searches for a neutrino magnetic moment in the MUNU experiment are in progress.

References

6.1 F. Boehm and P. Vogel, *Physics of Massive Neutrinos*, 2nd edn. (Cambridge University Press, Cambridge, 1992).

6.2 H. Kwon et al., Phys. Rev. D **24**, 1097 (1981).

6.3 G. Zacek et al., Phys. Rev. D **34**, 2621 (1986).

6.4 B. Achkar et al., Nucl. Phys. B **434**, 503 (1995).

6.5 M. Apollonio et al., Phys. Lett. B **420**, 397 (1998); M. Apollonio et al., Phys. Lett. B **466**, 415 (1999).

6.6 F. Boehm et al., Phys. Rev. Lett., **84**, 3764 (2000); hep-ex/0003022.

6.7 F. Boehm et al., Phys. Rev. D **62**, 072002 (2000).

6.8 Y. Fukuda et al., Phys. Lett. B **335**, 237 (1994).

6.9 J. N. Bahcall, P. I. Krastev and A. Y. Smirnov, Phys. Rev. D **58**, 096016 (1998).

6.10 P. Vogel et al., Phys. Rev. C **24**, 1543 (1981).

6.11 K. Schreckenbach et al., Phys. Lett. B **218**, 365 (1989).

6.12 Y. Declais et al., Phys. Lett. B **338**, 383 (1994).

6.13 P. Vogel and J. Engel, Phys. Rev. D **39**, 3378 (1989).

6.14 P. Vogel, private communication.

6.15 S. M. Bilenky and P. Pontecorvo, Phys. Rep. **41**, 225 (1978); H. Fritzsch and P. Minkowski, Phys. Lett. B **62**, 72 (1976).

6.16 A. A. Kuvshinnikov et al., JETP Lett. **54**, 255 (1991).

6.17 A. I. Alfonin et al., JETP **67**, 213 (1998).

6.18 G. S. Vidyakin et al., JETP Lett. **59**, 390 (1994).

6.19 Y. Fukuda et al., Phys. Rev. Lett. **81**, 1562 (1998).

6.20 P. Alvisatos et al., *KamLAND, a Liquid Scintillator Anti-Neutrino Detector at Kamioka*, Stanford-HEP-98-03, Tohoku-RCNS-98-15, July 1998 (unpublished); A. Piepke, in *Neutrino 2000, International Conference on Neutrino Physics and Astrophysics*, Sudbury, Canada, June 2000.

6.21 A. Piepke at al., Nucl. Instrum. Meth. A **432**, 392 (1999).

6.22 A. Piepke and B. Cook, Nucl. Instrum. Meth. A **385**, 85 (1996).

6.23 G. J. Feldman and R. D. Cousins, Phys. Rev. D **57**, 3873 (1998).

6.24 Y-F. Wang, L. Miller and G. Gratta, Phys. Rev. D **62**, 013012 (2000); hep-ex/0002050.

6.25 P. Vogel and J. F. Beacom, Phys. Rev. D **60**, 053003 (1999); K. B. Lee, private communication.

6.26 G. Zacek, Thesis, Technical University of Munich (1984).

6.27 F. Boehm, in *Eighth International Workshop on Neutrino Telescopes*, Venice, Feb. 1999, ed. by M. Baldo-Ceolin (Edizioni Papergraf, 1999), p. 311.

6.28 K. Okumara, PhD thesis, University of Tokyo, unpublished; SuperKamiokande Collaboration, preliminary results.

6.29 M. Apollonio et al., Phys. Rev. D **61**, 012001 (2000).

6.30 F. Reines et al., Phys. Rev. Lett. **45**, 1307 (1980).

6.31 S. P. Riley et al., Phys. Rev. C **59**, 1780 (1999).

6.32 P. Vogel and J. Engel, Phys. Rev. D **39**, 3378 (1089).

6.33 F. Reines et al., Phys. Rev. Lett. **37**, 315 (1976).

6.34 I. I. Gurevitch et al., JETP Lett. **49**, 740 (1989).

6.35 A. I. Derbin et al., JETP Lett. **57**, 768 (1993).

6.36 Y. Fukida et al., Phys. Rev. Lett. **82**, 2430 (1999).

6.37 J. F. Beacom and P. Vogel, Phys. Rev. Lett. **83**, 5222 (1999).

6.38 C. Amsler et al., Nucl. Instrum. Meth. A **396**, 115 (1997).

7 Studies of Neutrino Oscillations at Accelerators

David O. Caldwell

7.1 Introduction

Until relatively recently a review of the search for neutrino oscillations at accelerators typically would include a statement such as "From first principles, there is no preferred region in the Δm^2–$\sin^2 2\theta$ parameter space, and therefore the whole has to be investigated experimentally." Using a two-flavor oscillation, $\sin^2 2\theta$ is the mixing angle giving the amplitude of the oscillation, and $\Delta m^2 = |m_2^2 - m_1^2|$ determines the oscillation length,

$$L = \frac{4\pi E \hbar}{\Delta m^2 c^3} = 2.48 \left(\frac{E}{\mathrm{MeV}} \right) \left(\frac{\mathrm{eV}^2}{\Delta m^2} \right) \text{ meters .}$$

As the preceding chapters have shown, the situation is now quite different, owing particularly to nonaccelerator experiments. At this time, a prudent person embarking on an accelerator experiment would consider it likely that ν_e oscillates to ν_μ, ν_τ or ν_s (a sterile neutrino) with $\Delta m^2_{ei} \lesssim 10^{-5}$ eV^2 and that ν_μ oscillates not to ν_e and almost surely not to ν_s, but rather to ν_τ with $\Delta m^2_{\mu\tau} \sim 10^{-3}$ eV^2. Since the high-intensity neutrino beams at accelerators are produced by protons on a fixed target, giving pions and kaons which yield neutrinos of dominantly the muon flavor, this does not leave many alternatives. Primarily ν_μ or $\bar{\nu}_\mu$ beams are used, although weaker ν_e beams have been employed, predominantly for appearance experiments. The detection mechanism is via charged-current weak interactions, $\nu_i + N \rightarrow i + X$, where $i = e$, μ, τ.

Checking the atmospheric neutrino result with long-baseline ν_μ beams has been discussed in Chap. 5 and will not be mentioned further here. Rather, the emphasis will be on intermediate-baseline experiments, particularly because the only accelerator experiment with evidence for oscillations is that of LSND [7.1, 7.2] which observes $\bar{\nu}_\mu \rightarrow \bar{\nu}_e$. The negative attempts to see $\nu_\mu \rightarrow \nu_\tau$ in the intermediate-baseline (i.e. larger $\Delta m^2_{\mu\tau}$) experiments of CHORUS [7.3] and NOMAD [7.4, 7.5] will also be discussed, with NOMAD's use of the weakness of the ν_e beam to check partially the LSND acceptance region now being of particular interest. A few other experiments will also be described, especially KARMEN [7.6], which so far provides the most stringent constraints on the LSND result.

The LSND result is of particular importance because it implies $\Delta m_{e\mu}^2 \sim$ eV^2, much larger than the already distinct mass differences required for the solar and atmospheric oscillations. This has at least two consequences: (1) since the three active neutrinos required by the Z^0 width can have only two mass differences, a fourth neutrino, ν_s, must exist which does not have the usual weak interaction; (2) neutrinos have enough mass to be a significant contributor to the missing mass of the Universe.

7.2   Motivations for the Experiments

Like most oscillation experiments, the LSND experiment was done to explore the parameter space available, in this case for the energy and intensity of the LAMPF accelerator, using both the ν_μ beam from $\pi^+ \to \mu^+ + \nu_\mu$ decay in flight and the $\bar{\nu}_\mu$ beam from $\mu^+ \to e^+ + \bar{\nu}_\mu + \nu_e$ decay at rest. There was some hint of an effect from a previous LAMPF experiment, E645 [7.7], however. Given the target and shielding requirements, the detector was placed as close to the source as possible, and this may have been a very fortuitous choice of distance. With a lower-intensity accelerator, it was necessary to put the KARMEN detector closer to the source, at about half the LSND distance, so the two experiments using the same $\bar{\nu}_\mu$ energy sample the oscillation wave quite differently.

The CHORUS and NOMAD experiments had a specific motivation, which was legitimate at the time but may appear somewhat misguided in retrospect. A quite successful model to explain the structure of the Universe included some neutrino ("hot") dark matter with a preponderance of cold dark matter (see Chap. 12). It was possible to assume, for a hierarchical mass pattern, that the ν_τ would be massive enough to be this ~ 10 eV dark matter. The solar neutrino deficit was assumed to be due to $\nu_e \to \nu_\mu$, and one could invoke some carefully chosen parameters to give such a ν_τ mass utilizing the seesaw mechanism. The latter could be a justification but not really a prediction, as there are three forms of the seesaw mechanism, as described in Chap. 2, and the masses one uses in the seesaw relation have a lot of arbitrariness to them. This motivation also had to ignore the atmospheric ν_μ/ν_e anomaly, assuming either that those observations were not evidence for oscillations or that if they were, the process was $\nu_\mu \to \nu_s$, which conflicted with apparent limits from Big Bang nucleosynthesis.[1] The latter is still a controversial subject, however, as discussed in Chap. 11. In 1993 it was pointed out [7.8] that such one-neutrino dark matter was unlikely. Either all three neutrinos were nearly degenerate in mass ($\nu_e \to \nu_\mu$ for the solar and $\nu_\mu \to \nu_\tau$ for the atmospheric puzzles) and contributed about equally to dark matter, or ν_μ and ν_τ shared that role (with a light $\nu_e \to \nu_s$ for the solar case and a heavier $\nu_\mu \to \nu_\tau$ for the atmospheric

---

[1] An even more unlikely possibility was a solar $\nu_e \to \nu_s$ and an atmospheric $\nu_\mu \to \nu_e$ process, now ruled out.

one).[2] This was prior to the LSND experiment, the results of which would be expected in the latter four-neutrino scheme. In short, if $\nu_\mu \rightarrow \nu_\tau$ explains the atmospheric neutrino observation, requiring $\Delta m^2 \sim 10^{-3}$ eV^2, ν_τ alone cannot provide significant dark matter, and CHORUS and NOMAD were bound to obtain a null result when searching in the large-Δm^2 region.

7.3   Intermediate-Baseline ν_μ Experiments at High-Energy Accelerators

These null results are useful in reinforcing the $\nu_\mu \rightarrow \nu_\tau$ explanation of the atmospheric neutrino results, and the NOMAD limit on $\nu_\mu \rightarrow \nu_e$ provides an upper bound on possible $\Delta m^2_{e\mu}$ values obtained from the LSND experiment. The NOMAD experiment has also found universality in quark fragmentation properties by comparing $\nu_\mu N$ and $\bar{\nu}_\mu N$ deep inelastic scattering events with hadronic jets produced in ep and e^+e^- experiments [7.11]. In addition, the NOMAD group has published a better limit on a new light gauge boson, X, associated with an extra $U(1)$ factor which could be produced in $\pi^0 \rightarrow \gamma + X$ [7.12]. Auxiliary results from CHORUS include the first direct observation of a neutrino-induced diffractive charged-current D_s^{*+} production with subsequent decays $D_s^{*+} \rightarrow D_s^+ \gamma$, $D_s^+ \rightarrow \tau^+ \nu_\tau$, $\tau^+ \rightarrow \mu^+ \nu_\mu \bar{\nu}_\tau$ [7.13].

The NOMAD and CHORUS detectors shared a wide-band neutrino beam from the 450 GeV proton synchrotron (the SPS) at CERN, with one detector behind the other at ~ 800 m from the proton target. The relative abundances and average energies of the neutrinos in the beam are given in Table 7.1, along with the relative effectiveness and average energies of those neutrinos in charged-current interactions. The CHORUS experiment searched for ν_τ charged-current interactions in emulsion via kinks produced in the subsequent τ lepton decay into a negative hadron, electron or muon. The NOMAD experiment also searched for ν_τ charged-current interactions but utilizing kinematic criteria to identify the τ^- decay to $e^- \bar{\nu}_e \nu_\tau$, inclusive decays to one or three charged hadron(s) $+\nu_\tau$ and exclusive decays to $\varrho^- \nu_\tau$.

The hybrid CHORUS (CERN Hybrid Oscillation Research apparatUS) detector [7.14], shown in Fig. 7.1, combined 770 kg of emulsion with several electronic detectors to facilitate the search for τ^- decays in the emulsion. The neutrino target consisted of four stacks of 36 plates each, and each 36 cm $\times$ 72 cm plate had a 90 µm transparent plastic base with 350 µm of emulsion on both sides. The stacks were sandwiched with scintillating fiber trackers and interface sheets having 100 µm thick emulsions, which were changed every few weeks during the run. The fiber trackers were used for vertex and track position prediction, which was then made more precise by the interface emulsion sheets, as shown in Fig. 7.2. This enabled an automatic microscope system to

---

[2] These neutrino schemes were subsequently given theoretical bases in [7.9] and, only for the four-neutrino model, in [7.10].

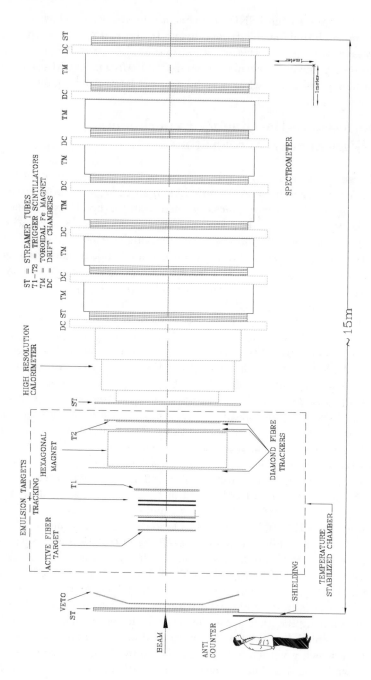

ST = STREAMER TUBES
T1-T2 = TRIGGER SCINTILLATORS
TM = TOROIDAL Fe MAGNET
DC = DRIFT CHAMBERS

DC ST   TM   DC   TM   DC   TM   DC   TM   DC   TM   DC ST

SPECTROMETER

HIGH RESOLUTION
CALORIMETER

ST

EMULSION TARGETS
TRACKING

HEXAGONAL
MAGNET

T2

T1

DIAMOND FIBRE
TRACKERS

ACTIVE FIBER
TARGET

TEMPERATURE
STABILIZED CHAMBER

SHIELDING

VETO
ST

BEAM

ANTI
COUNTER

1 meter

~ 15m

Fig. 7.1. Side view drawing of the CHORUS detector

Table 7.1. Mean energies and relative abundances of the ν fluxes, and the same quantities weighted for the ν-induced charged-current interactions, as given by NOMAD. The integrated ν_μ flux is 1.11×10^{-2} ν_μ per proton on target

	Flux		Charged-current interactions	
ν type	$\langle E_\nu \rangle$ (GeV)	Relative abundance	$\langle E_\nu \rangle$ (GeV)	Relative abundance
ν_μ	23.5	1.00	42.6	1.00
$\bar{\nu}_\mu$	19.2	0.061	41.0	0.0249
ν_e	37.1	0.0094	56.7	0.0148
$\bar{\nu}_e$	31.3	0.0024	53.6	0.0016
ν_τ	~ 35	$\sim 5 \times 10^{-6}$		

find tracks with an efficiency above 98% for track angles up to 0.4 radians in 0.3 s per microscope view (150 μm × 120 μm). The target system was followed by a hadronic spectrometer consisting of an air-core magnet (0.12 T) with one fiber tracker downstream and two upstream to measure track momenta. The energy of the event was then determined by a lead-scintillating-fiber calorimeter having an electromagnetic section and two hadronic sections. This was followed by a muon spectrometer consisting of six magnetized iron disk groups sandwiched with drift chambers and streamer tubes for tracking.

CHORUS took data between May 1994 and the end of 1997 and collected 2.31×10^6 triggers, of which 458 601 had a muon identified in the final state (the so-called 1μ events) and 116 049 were 0μ events at the time information was last released (summer 1999 [7.15]).[3] Both classes of events have a μ^- or a negatively charged hadron which can be followed back to the emulsion target. The reconstruction accuracy of the fiber tracker predicted the position of that track to within a 360 μm × 360 μm area on the interleaved emulsion sheets, where on average there were only five background muon tracks per sheet collected during its short exposure time, and these could be rejected by angular measurements. The search was continued into the emulsion target, involving a 20 μm × 20 μm area, which had a negligible background despite the two-year exposure of the target. With 66% of the 1μ events and 55% of the 0μ events scanned at the last report [7.15], vertices in the emulsion had been located for 126 229 1μ events and 19 426 0μ events.

Once a vertex is located, a search is made for a τ^- decay via $\mu^- \bar{\nu}_\mu \nu_\tau$ (17.4%), $h^-(n\pi^0)\nu_\tau$ (49.8%) and $\pi^- \pi^+ \pi^-(n\pi^0)\nu_\tau$ (14.9%). For the 1μ events the identification is a kink in the negatively charged track, which should have momentum $p < 30$ GeV, no other charged leptons at the primary vertex, and the transverse momentum (p_T) of the selected particle relative to the τ direction greater than 0.25 GeV (to eliminate strong particle decays). If

---

[3] As of April 2000, there was no later information available.

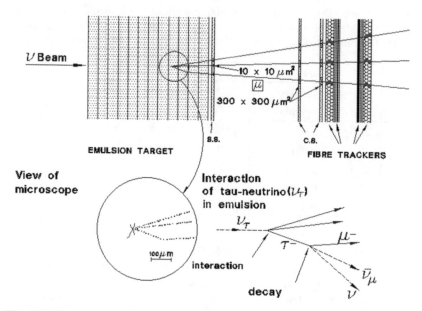

Fig. 7.2. Schematic view of the principle used for detecting a ν_τ charged-current interaction in the CHORUS emulsion. In this case the μ^- from $\tau^- \to \mu^- \nu_\tau \bar{\nu}_\mu$ is visible as a kink after a short τ^- track

observed, the kink must be within 3.95 mm of the interaction point. For the hadronic decays the negative hadron must have $1 < p < 20$ GeV, and the kink has to be within 2.37 mm of the interaction point. Again, $p_T < 0.25$ GeV is required. The emulsion scanning to find the interaction point and to eliminate events incompatible with τ^- decays is fully automatic. The remaining events undergo a computer-assisted eye scan to determine the presence of a τ decay. None has been observed [7.3, 7.15].

The dominant background for 1μ events is the production of charmed particles from $\bar{\nu}_\mu$ interactions:

$$\bar{\nu}_\mu N \to \mu^+ D^- X, \text{ then } D^- \to \mu^- + \text{neutrals}.$$

These events matter only if the μ^+ is not identified, and their contribution to the data sample analyzed so far is 0.24 ± 0.05 of an event. In the same sample, charm production provides a background of 0.075 ± 0.015 of an event for the 0μ channel, but there is a more serious background, the "white kink" events. These result from the interaction of negative hadrons with nuclei in which there is only one outgoing negatively charged particle and no evidence of nuclear breakup, providing a background of 0.73 ± 0.66 of an event.

For an oscillation probability $P_{\mu\tau} = 1$, 4876 1μ and 1137 0μ events would be expected, but since none is observed, the 90% confidence upper limit is

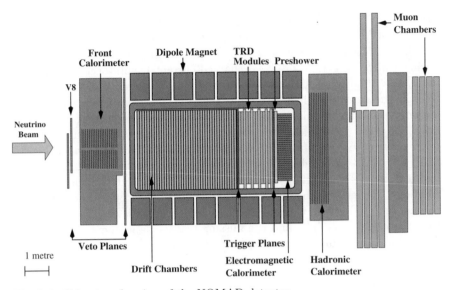

Fig. 7.3. Side view drawing of the NOMAD detector

$P_{\mu\tau} < 4.0\times10^{-4}$.[4] Since the observation is less than the expected background, the now popular Feldman–Cousins method [7.16] would give a more stringent limit of $P_{\mu\tau} < 2.6 \times 10^{-4}$. A direct measurement of the "white kink" process is in progress, and kinematic cuts are planned for its reduction. In addition, a second phase of the analysis has started, with better efficiencies, larger statistics and faster automatic emulsion scanning, with the aim of reaching the design sensitivity of $P_{\mu\tau} \leq 1.0 \times 10^{-4}$.

NOMAD (Neutrino Oscillation MAgnetic Detector [7.17]), shown in Fig. 7.3, was designed to search for $\nu_\mu \to \nu_\tau$ oscillations using kinematic criteria to observe τ^- production. This requires a precise measurement of secondary-particle momenta and good particle identification. To achieve the first aim, drift chambers were located inside a uniform magnetic field of 0.4 T perpendicular to the beam direction. The chambers also served as the neutrino target, with a fiducial mass of $\sim$ 2.7 tons, an average density of 0.1 g/cm^3 and a radiation length of $\sim$ 5 m. Nine independent transition radiation detectors (TRDs) for electron identification were interleaved with additional drift chambers, also inside the magnet. Electron identification was furthered by an electromagnetic calorimeter, which followed a preshower detector behind the TRD but still inside the magnet. Beyond the field was first a hadronic calorimeter, followed by muon chambers. The neutrino interaction trigger consisted of a coincidence between signals from two planes of counters lo-

---

[4] Note added in proof: at Neutrino 2000 the CHORUS Collaboration reported analyzing enough additional events to give a limit ($P_{\mu\tau} < 2.2 \times 10^{-4}$) just equal to NOMAD's final result.

cated after the active target, and no signal from a large-area system of veto counters in front of the magnet and target.

The NOMAD experiment [7.4, 7.5] ran starting in 1995 through 1998, during which time 1.35×10^6 ν_μ charged-current events occurred, providing 1.04×10^6 events with an identified muon, of which 0.95×10^6 events were in the detector fiducial volume. These events were analyzed for both leptonic and hadronic decays of the τ^- to find the presence of ν_τ interactions. Of these decay modes, $\tau^- \to \nu_\tau e^- \bar\nu_e$ was especially useful, since the main background was from ν_e charged-current interactions, which constituted only 1.5% (see Table 7.1) of the neutrino interactions and in addition had a quite different energy spectrum from that expected from ν_μ–ν_τ oscillations. To perform the discrimination against background, the transverse momentum (p_T^e) of the isolated electron, the total transverse momentum of the hadronic system (p_T^H) and the missing transverse momentum (p_T^m) were used. Momentum components perpendicular to the beam had to be employed because the incident neutrino energy was unknown. For ν_e charged-current events, p_T^e was generally opposite to p_T^H, and hence $|p_T^m|$ was small and would be zero if all momenta were measured precisely and the target nucleon were at rest. This was in contrast to the $\tau^- \to \nu_\tau e^- \bar\nu_e$ case, in which there was an appreciable $|p_T^m|$ associated with the two outgoing neutrinos. In addition, in a large fraction of events p_T^m and p_T^H were at opposite azimuthal angles, whereas, for ν_e charged-current interactions, if there were large values of $|p_T^m|$ this resulted mostly from hadrons escaping detection, in which case the azimuthal separation between p_T^m and p_T^H was small. This illustrates the way in which kinematic quantities were utilized for all the observed decays, which constituted a total branching ratio of 82.8%.

For most of the τ^- decay channels the signal was separated from backgrounds using ratios of likelihood functions. These functions were approximated by products of probability density functions of kinematic variables, utilizing large samples of simulated events corrected for differences between real and simulated data [7.4]. Table 7.2 summarizes the NOMAD results and expected backgrounds for various decay channels, separately analyzed for deep inelastic scattering events (DIS) and low-multiplicity (LM) events [7.5]. The separation of LM and DIS events occurs at $|p^H| > 1.5$ GeV. In the table, N_τ^μ is the number of expected τ^- events for an oscillation probability $P_{\mu\tau} = 1$. Note the impressive agreement for all channels between not only the observed number of τ^- events and the background prediction, but also for the control sample of τ^+, which could not come from oscillations. For no observed signal the 90% confidence level upper limit on the oscillation probability (using the method of [7.16]) is $P_{\mu\tau} < 2.2 \times 10^{-4}$, but the sensitivity of the experiment is $P_{\mu\tau} = 4.3 \times 10^{-4}$, since the number of observed events is smaller than the estimated background. The probability to obtain an upper limit of 2.2×10^{-4} or lower is 27%.

Table 7.2. Summary of estimated backgrounds and observed events for each decay type and for both the τ^- candidate events and the τ^+ control sample. Here DIS means deep inelastic scattering events, LM means low-multiplicity events and the N_τs are the numbers of events expected for unit oscillation probability for $\nu_\mu \to \nu_\tau$ (N_τ^μ) and $\nu_e \to \nu_\tau$ (N_τ^e). When the hadron, h, is identified as a ϱ, it is so indicated

Decay channel			Observed	Total background	Observed	Total background	N_τ^μ	N_τ^e
			τ^-		τ^+			
$\tau \to e$		DIS	5	$5.3^{+0.7}_{-0.5}$	9	8.0 ± 2.4	4110	81.5
$\tau \to h(n\pi^0)$	ϱ	DIS	7	9.5 ± 2.5	6	5.6 ± 1.5	3307	78.2
	h	DIS	5	6.8 ± 2.1	19	16.0 ± 4.0	2022	43.7
	h/ϱ	DIS	1	$0.0^{+0.74}_{-0.0}$			210	5.0
$\tau \to 3h(n\pi^0)$		DIS	9	9.6 ± 2.4	6	6.9 ± 3.3	1820	42.6
$\tau \to e$		LM	6	5.4 ± 0.9	3	2.2 ± 0.5	859	8.5
$\tau \to h(n\pi^0)$	ϱ	LM	7	5.2 ± 1.8	21	22.2 ± 6.6	458	8.4
	h	LM	5	6.7 ± 2.3	19	21.9 ± 6.4	357	7.2
$\tau \to 3h(n\pi^0)$		LM	5	3.5 ± 1.2	1	2.2 ± 1.1	288	4.8

The CHORUS and NOMAD results are shown in Fig. 7.4, where the mass-squared difference between ν_μ and ν_τ ($\Delta m_{\mu\tau}^2$) is plotted against the mixing angle, $\sin^2 2\theta$, on the basis of a two-flavor oscillation (but again see footnote 4). The limits can be combined, using the method of [7.16], and the result is also shown in the figure, as are results from earlier experiments [7.18]. The combined oscillation probability limit is $P_{\mu\tau} < 0.8 \times 10^{-4}$, but this is an optimistic result, since both experiments see fewer events than expected from the estimated backgrounds, an all too common occurrence using the method of [7.16].

In the context of a two-flavor oscillation, the NOMAD result can be reinterpreted in terms of a $\nu_e \to \nu_\tau$ oscillation by assuming that any observed ν_τ signal should come from the small ν_e component of the beam [7.5, 7.19]. The observed events, summarized in Table 7.2, are simply compared with N_τ^e, the number of τ^- events for an oscillation probability $P_{e\nu} = 1$, which is obtained from N_τ^μ by reweighting the simulated signal events for the ν_e-to-ν_μ flux ratio, given in Table 7.1. The resulting 90% C.L. limit on the $\nu_e \to \nu_\tau$ oscillation probability is $P_{e\tau} < 1.1 \times 10^{-2}$, for which the sensitivity is 2.0×10^{-2}, and the probability to obtain an upper limit of 1.1×10^{-2} or lower is 31%. This method [7.16] is also used in Fig. 7.5, where the exclusion limit is shown along with those of earlier experiments [7.20].

Before leaving this section, mention needs to be made of other experiments in this category. Some results appear in the plots for E531, CDHS and

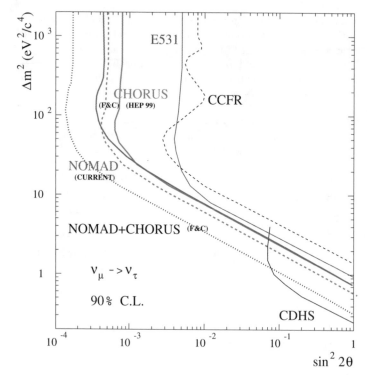

Fig. 7.4. The Δm^2–$\sin^2 2\theta$ plane for ν_μ–ν_τ oscillations. The regions excluded at the 90% C.L. by CHORUS and NOMAD and by their combined result (*dotted line*), by that from NOMAD alone (*solid line*) and by the *dashed-line* version of the CHORUS result were analyzed by the method of [7.16], whereas the *solid-line* version of the CHORUS result (essentially the same as that of NOMAD for $\sin^2 2\theta > 10^{-3}$) is Bayesian

CCFR, for example, but these and others (e.g. CHARM) are not discussed in detail because the emphasis here is on neutrino masses and mixings. Those experiments gave results on deep inelastic scattering, $\sin^2 2\theta_\mathrm{w}$, structure functions, etc., which are not the subject of this book, and for the most part their results on our main topic are not the leading ones at this time.

7.4   Short-Baseline ν_μ and $\bar{\nu}_\mu$ Experiments at Lower-Energy Accelerators

There has been much confusion about the relation between the results of the LSND (Liquid Scintillator Neutrino Detector) experiment [7.1, 7.21] performed at the Los Alamos National Laboratory and those of the KARMEN (KArlsruhe–Rutherford Medium Energy Neutrino) experiment [7.6] operating at the ISIS spallation facility at the Rutherford Appleton Laboratory.

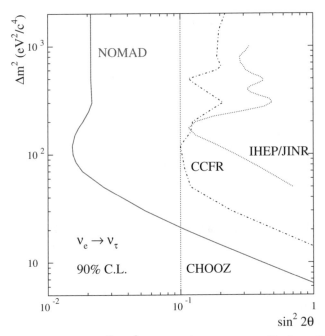

Fig. 7.5. The Δm^2–$\sin^2 2\theta$ plane for $\nu_e \to \nu_\tau$ oscillations. The regions excluded at the 90% C.L. by NOMAD and previous experiments are shown for the large-Δm^2 region

After comparative descriptions of the experiments and their recent analyses, this section will attempt to clarify that confusion, indicating the degree of compatibility of the two results, as well as taking into account limits from other experiments. Since LSND is the only accelerator experiment claiming to detect neutrino oscillations, an understanding of the validity of the results is of obvious importance. As was noted in Sect 7.1, if the LSND conclusion is correct there must be at least one light sterile neutrino, and there is significant hot dark matter.

Both LSND and KARMEN use neutrinos produced in the beamstop of 800 MeV proton accelerators. LSND collected data starting in 1993 through 1995 using a water target followed by a drift space (allowing $\pi^+ \to \mu^+ \nu_\mu$ decay in flight at the 3.4% level for $\nu_\mu \to \nu_e$ [7.2]), but the 1996–1998 data were taken under parasitic conditions using an iron target. KARMEN's target is $Ta + D_2O$ with no drift space, so only 0.1% of its π^+ mesons decayed in flight. The Los Alamos beam intensity was 1 mA, whereas that at Rutherford has been 0.2 mA. Since the two experiments suppressed and measured cosmic-ray backgrounds by comparing data with and without beam, KARMEN's duty factor of 5×10^{-4} is an advantage over the 0.06 value of LSND, but this advantage is diminished by the factor of 5 difference in beam intensities. The KARMEN experiment suffered from bad cosmic-ray backgrounds from 1990 to 1996, however, so additional shielding and a third veto system were added, reducing the neutrons from cosmic muons by a factor of 40. Data from 1997 onwards now totally supplant those of the earlier experiment and are all that are now reported. The LSND experiment was under a 2 kg/cm^2 overburden, eliminating the hadronic cosmic rays, and a liquid-scintillator veto shield enclosing all but the bottom of the detector rejected muons efficiently.

In both experiments the protons produced π^+ mesons, which decayed mainly into $\mu^+\nu_\mu$ (giving, in the LSND case only, a ν_μ beam of up to 200 MeV), and nearly all the μ^+ decayed at rest into $e^+\nu_e\bar{\nu}_\mu$, giving a $\bar{\nu}_\mu$ beam of energies up to 53 MeV. Both experiments searched for $\bar{\nu}_\mu$ oscillations into $\bar{\nu}_e$, which would be detected via the reaction $\bar{\nu}_e p \rightarrow e^+ n$. The e^+ and n were observed in large liquid-scintillator detectors, the n by delayed scintillation light caused by the γ from neutron capture in the detector. Neither experiment measured the sign of the charge of the e^+, so a $\bar{\nu}_e$ background came from π^- production. However, in the LSND case production of π^+ was eight times that of π^-, the decay of which was suppressed by a factor of 20, since π^- generally were absorbed before they were able to decay, and 88% of μ^- were similarly absorbed, giving a relative $\bar{\nu}_e$ yield of only 8×10^{-4}. The corresponding ratio for KARMEN was 6.4×10^{-4}.

Although the detection scheme is the same, there are significant differences between the detectors. The LSND detector was a tank 5.7 m in diameter and 8.3 m long centered 30 m from the target at an angle of 12° to the proton beam, whereas the compact KARMEN detector's center is 17.6 m from the target at an angle of 90° to the proton beam, resulting in rather different sensitivities to Δm^2. The distance gave LSND more shielding from target neutrons. The LSND tank contained 167 tons of mineral oil with a dilute scintillator so that both scintillation and Cerenkov light could be viewed by 1220 8 inch diameter photomultipliers (25% coverage) around the tank walls. A typical 45 MeV electron created in the detector produced ~ 1500 photoelectrons, of which ~ 280 were in the Cerenkov cone. The 56 ton KARMEN detector was segmented into 512 modules, giving better position resolution from both the module size and the timing of light from both of its ends. This helped in the detection of the γ from neutron capture, which was either via Gd(n, γ)Gd (with γs up to 8 MeV) or $p(n, \gamma)d$ (a γ of 2.2 MeV), only the latter being used by LSND. While LSND's dilute scintillator gave poorer energy resolution ($\sim 7\%$) than KARMEN's ($11.5\%/\sqrt{E(\text{MeV})}$), making the distinction between real and accidental γs more difficult, the LSND observation of Cerenkov light provided much better separation of signal e^+ from knock-ons produced by high-energy cosmic-ray neutrons and also allowed measurement of the e^+ direction. The e^+ was required to give a good Cerenkov ring, to have a larger portion of fast light than would come from pure scintillation and to be confined in position.

Besides the common exclusion of charged cosmic rays by shielding and veto counters, LSND also excluded events with evidence of something other than a neutrino entering the detector by using the e^+ direction and position to tell if it entered from outside the tank. To remove the background from $\mu \to e\nu\bar{\nu}$, LSND used information from both before and after the e^+ was detected, whereas KARMEN used only the prior information. Once events with an apparent e^+ were selected, both experiments required a γ to be near in space and time to the e^+ and to have a reasonable energy. KARMEN made cuts on these three quantities, whereas LSND, with its poorer resolution, used a product of position, time and energy distributions for correlated neutron capture events (from cosmic rays) divided by a similar product for accidental γs (from laser-triggered events) to form a likelihood ratio, R. While low R indicates a γ that is accidental, high R corresponds to a γ that appears more like a neutron capture correlated with the production of an e^+. LSND has both cut on high R in order to select clean candidates and also fitted the R distribution to measure and subtract the contribution from accidental γs. The author, not necessarily reflecting the views of the rest of the collaboration, believes the R distribution has been a source of some quantitative errors in some results given at conferences, but not in refereed publications.

In order to exclude e^- from $\nu_e C \to e^- X$, which almost always has a measured e^- energy below 36 MeV, providing the largest background with no correlated γ, LSND, in its publications claiming an oscillation signal, cleaned its sample further by requiring $E_e > 36$ MeV, giving reliable results. Unfortunately, plots were made down to 20 MeV, placing a heavy burden on the correctness of the R distribution, which was especially a problem for the 1996–1998 data taken under parasitic conditions with a lower beam intensity and hence a higher ratio of cosmic-ray (a source of γs) to beam events. With 16 times as much data taken with the beam off as with it on, the beam-off subtraction should have taken care of this, but the R distribution had a statistically poor shape, with an upward fluctuation at high R and a depletion in the mid range. While there were concerns initially about the spatial distribution of events in the tank, these distributions – especially in more complete data sets – are statistically no problem. Rather, the only worrisome distribution has been that of R. The results of a new analysis will be discussed below, with which this difficulty disappears.

The electron energy cut of 20 MeV, which is also used by KARMEN, is necessitated by the endpoint of the $\nu_e {}^{12}C \to {}^{11}Ne^- n$ reaction. The upper energy cut is 50 MeV for KARMEN and 60 MeV for LSND, set by the 52.8 MeV maximum $\bar{\nu}_\mu$ energy from the μ^+ decay at rest. The official KARMEN data are for February 1997 to February 1999 and consist of 8 events with 7.8 ± 0.5 background expected. The preliminary data up to December 1999 are based on 6317 coulombs of protons on target, whereas the LSND final sample results from 28 896 coulombs. The latest KARMEN results [7.22], for data-taking from February 1997 to December 1999, show 10 events with a calculated background of 10.6 ± 0.6 events. The latter number is derived from 2.51 ± 0.2 events from cosmic rays (data between beam pulses), 3.46 ± 0.3 events from $\nu_e {}^{12}C$ charged-current events, 3.18 ± 0.5 events from ν neutral-current events and 1.47 ± 0.2 events from $\bar{\nu}_e$ contamination from the π^- decay chain. The strong claim [7.22] is then that KARMEN sees no oscillations and in particular that $\Delta m^2 > 2$ eV^2 is eliminated, while the parameter space for $\Delta m^2 < 2$ eV^2 is further restricted. The oscillation probability at the 90% C.L. is $P_{\mu e} < 0.85 \times 10^{-3}$ (i.e., for large Δm^2, $\sin^2 2\theta < 1.7 \times 10^{-3}$).

It is interesting, however, to note that KARMEN's γ energy distribution is a rather poor fit, perhaps indicating the presence of more γs than expected, since these would show up around 2 and 4 MeV, where there are upward fluctuations. The electron energy distribution shows a small upward fluctuation of an event or two at 22 MeV, which is near the most probable value if Δm^2 were ~ 6 eV^2. Indeed, only about two such events are expected, on the basis of the LSND rate, since this Δm^2 is unfavorable for KARMEN because it is at half of LSND's distance from the target. Clearly, with such small numbers one cannot say KARMEN is observing a signal, but, by the same token, even if the backgrounds have been determined correctly, fluctuations in those numbers could hide a few signal events. Hence it is quite premature to claim KARMEN excludes $\Delta m^2 > 2$ eV^2, especially as the experiment is so insensitive to part of that range.

The new LSND analysis is an outgrowth of trying to deal with the very difficult decay-in-flight data. The high-energy ν_μ beam would display $\nu_\mu \to \nu_e$ oscillations via $\nu_e {}^{12}C \to e^- X$, providing only one observable instead of the two for $\bar{\nu}_\mu \to \bar{\nu}_e$. A further difficulty is that there is no test system for the 60–200 MeV e^-, which must be above 60 MeV to avoid background from μ decays at rest. In the $\bar{\nu}_\mu \to \bar{\nu}_e$ case the very large sample of decay $e^\pm$ from stopped cosmic-ray muons covers exactly the right energy range to provide an energy calibration, a measurement of energy resolution and a means of tuning cuts in an unbiased manner. Therefore, particle identification can be checked by comparing these decay electrons with the muons from which they were created and with cosmic-ray neutrons identified by their 2.2 MeV capture γ rays and initial small signal. The $\nu_\mu \to \nu_e$ data thus not only have far worse backgrounds, but also require very difficult extrapolations in energy. Nevertheless, a result has been obtained and published [7.2] and provides some support for the $\bar{\nu}_\mu \to \bar{\nu}_e$ result, although the errors are large.

To obtain any result for $\nu_\mu \to \nu_e$ it was necessary to improve the analysis techniques. The biggest improvement was introducing a maximum-likelihood fit of all of the photomultiplier hit times and pulse heights to the hypothesis of an electron with unknown position, energy, direction, track length and amount of Cerenkov radiation. Formerly, this information had been used in pieces, e.g. pulse height for energy and timing for position. The new analysis not only reduced errors in all the fit quantities but also now provides

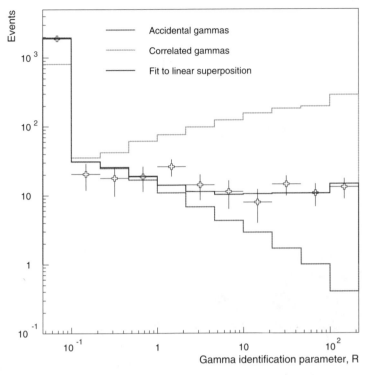

Fig. 7.6. Fit of the beam-on minus beam-off data (*middle curve* at large R) to a linear superposition of accidental γs (from laser-triggered events), shown as the *lower curve*, and correlated γs (from cosmic neutrons), shown as the *upper* (*gray*) *curve*. The larger the value of R, the more likely the positron in the event is to have a correlated γ

an excellent R distribution, as has already been mentioned, which is shown in Fig. 7.6. As an example of the improvement, at an R cut which previously gave a 0.6% accidental rate, the new analysis now gave 0.3% while increasing the efficiency for e–γ correlated events from 0.23 to 0.39. With this evidence that the e–γ correlation has now been done well, the electron energy distribution once more looks like what one would predict for oscillations plus known backgrounds, as may be seen in Fig. 7.7. While the published data (1993–1995 in [7.1]) also show a good energy distribution, those given at conferences utilizing the parasitic 1996–1998 data did not. Now, using all the data and the new analysis, the energy distribution provides good evidence for oscillations, as does the distribution in angle between the incoming neutrino and the produced electron. The results can be expressed in different ways, and Table 7.3 gives the size of the effect for different cuts on R. Recall that the beam-off subtraction has good accuracy because there are 16 times as much beam-off data as beam-on.

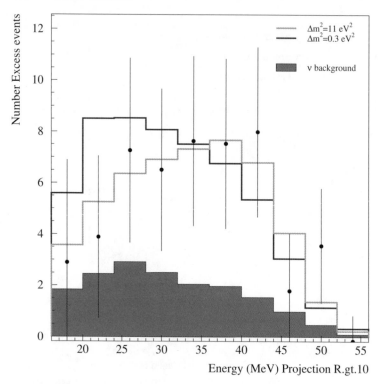

Fig. 7.7. The energy distribution for positron events with a tight requirement ($R >$ 10) to have a correlated γ. Beam-on minus beam-off data are shown with the estimated neutrino background (*shaded area*) and the expected neutrino oscillations at $\Delta m^2 = 11$ eV^2 (*light line*) and $\Delta m^2 = 0.3$ eV^2 (*solid line*), with the estimated neutrino background added to the oscillation curves

Table 7.3. LSND $\bar{\nu}_\mu \to \bar{\nu}_e$ results as a function of e–γ accidentals for the electron energy interval 20–60 MeV

Accidentals cut	Beam on	Beam off	ν background	Excess events
$R > 100$	27	8.3 ± 0.7	5.4	13.3 ± 5.2
$R > 10$	86	36.9 ± 1.5	16.9	32.2 ± 9.4
$R > 1$	205	106.8 ± 2.5	39.2	59.0 ± 14.5

Another way is to make a fit to the R distribution, so as to get a single, R-independent, result. Using the 20 to 60 MeV electron energy interval, the beam-on minus beam-off excess is 117.9 ± 22.4 events, of which 30.0 ± 6.0 are expected to be neutrino-induced background. The fitted oscillation excess of $87.9 \pm 22.4 \pm 12.2$ events gives an oscillation probability $P_{e\mu} = (2.64 \pm 0.67 \pm 0.41) \times 10^{-3}$.

Table 7.4. Cross sections in units of cm^2 for conventional neutrino processes as measured by LSND [7.23] and KARMEN [7.24] and predicted theoretically [7.25–7.27]

Reaction	LSND σ_{meas}	KARMEN σ_{meas}	σ_{theo}
$^{12}C(\nu_e, e^-)^{12}N$	$(9.1 \pm 0.4 \pm 0.9) \times 10^{-42}$	$(9.4 \pm 0.4 \pm 0.8) \times 10^{-42}$	9.1×10^{-42}
$^{12}C(\nu_e, e^-)^{12}N^*$	$(5.7 \pm 0.6 \pm 0.6) \times 10^{-42}$	$(5.1 \pm 0.6 \pm 0.5) \times 10^{-42}$	6.3×10^{-42}
$^{12}C(\nu_\mu, \mu^-)^{12}N$	$(6.6 \pm 1.0 \pm 1.0) \times 10^{-41}$		6.4×10^{-41}
$^{12}C(\nu_\mu, \mu^-)^{12}N^*$	$(12.4 \pm 0.3 \pm 1.8) \times 10^{-40}$		See text

While the existence of an effect is statistically solid – the probability (for $R > 10$) of an expected 53.8 (36.9 + 16.9) events to fluctuate up to 86 is $< 1 \times 10^{-4}$ (taking into account all errors), for example – there is the important issue of systematic effects. An excellent check is provided by the simultaneous measurements of conventional neutrino processes [7.23], comparisons for which are provided by KARMEN determinations [7.24] of some of the same cross sections and by theoretical predictions [7.25]. These values are shown in Table 7.4. One comparison is not given, that for $\nu_\mu + {}^{12}C \rightarrow \mu^- + {}^{12}N^*$, where $^{12}N^*$ indicates in this case that the final state includes all energy levels up to some tens of MeV. Whereas the decay-at-rest reaction cross section is dominated by the Gamow–Teller transition to the ^{12}N ground state, for which transition probabilities can be obtained from other measured processes, the decay-in-flight high energy populates a multitude of levels for which there is little information, making calculations difficult. Originally calculations were a factor of two higher than the measured value. A more recent charge-exchange random-phase-approximation calculation [7.26] gave 17.5×10^{-40} cm^2, whereas very recent work [7.27] using a shell model gave 15.2×10^{-40} cm^2, and 19.2×10^{-40} cm^2 using a random-phase approximation. Clearly this is too difficult a calculation to provide more than an approximate check, but where calculations are reliable, the agreements (as with KARMEN) are excellent, giving evidence that a lot of potential systematic problems are under control.

If LSND is seeing an effect and KARMEN and other experiments are not, is there a conflict? Part of the confusion on this issue arises from LSND's use of a likelihood plot and KARMEN's use of frequentist confidence levels from [7.16]. The likelihood plot was intended to show the favored regions of Δm^2, and all information about each event was used. The likelihood analysis applied did not have a Gaussian likelihood distribution, since its integral is infinite, but the likelihood contour labeled "90%" was obtained by going down a factor of 10 from the maximum, as in the Gaussian case. The contours in the LSND plot have been widely misinterpreted as confidence levels – which they certainly are not – because they were plotted along with confidence-level limits from other experiments.

Recently the difficult, computer-intensive analysis in terms of real confidence levels has been done [7.28]. The likelihood for a grid in $(\sin^2 2\theta, \Delta m^2)$ space, including backgrounds, has been computed and compared with numerous Monte Carlo experiments to obtain a 90% confidence region. While the equivalency varies from point to point in the Δm^2–$\sin^2 2\theta$ plane, a typical value for the 90% confidence level is down a factor of 20 from the likelihood maximum. Thus the LSND allowed regions are considerably broader in $\sin^2 2\theta$ than in the plots published so far, and other experiments constrain allowed Δm^2 regions less. The problem has been exacerbated by using the 20–36 MeV region for the LSND data, especially when the 1996–1998 runs were included, for reasons already discussed.

Nevertheless, when a joint analysis [7.28] is made of the LSND and KARMEN experiments, even using the 20–36 MeV region, including the 1996–1998 data, and employing the old LSND analysis (with the "official" eight-event KARMEN data) there are considerable regions of compatibility between the experiments at the 90% and 95% C.L., as shown in Fig. 7.8. Note that in some places in the figure the 90% C.L. exclusion regions of KARMEN do not look compatible with the 90% C.L. acceptance regions of LSND, but in fact they are when a joint analysis is done properly. Thus one is easily misled when looking at this type of plot, and this is also the reason results from some earlier $\bar{\nu}_\mu \to \bar{\nu}_e$ experiments, such as E776 [7.29], are not plotted here, as their exclusion curves were obtained in a very different way. Limits, which in view of the above considerations should be considered as approximate upper and lower bounds on Δm^2, are also shown from NOMAD [7.30] (lack of extra ν_e appearing beyond those expected from the weak ν_e beam) and Bugey [7.31] (lack of ν_e disappearance).

Figure 7.8 represents the current state of evidence for neutrino oscillations at accelerators, although it ought to be updated using the new LSND analysis and current KARMEN (ten-event) data, which is not an easy task. Such an updated Fig. 7.8, however, would look much the same but shifted to slightly smaller mixing angles, since LSND's new analysis reduced the oscillation probability from 0.31% to 0.26%, and KARMEN's added events provide a similar leftward shift. While the LSND experiment is finished, the new analysis was shown first at Neutrino 2000 and will soon be published. A word of caution should be given about those results, however, since it has been decided to present a global analysis, using all the data from 20 to 200 MeV. If only the region where the $\bar{\nu}_\mu \to \bar{\nu}_e$ signal can exist (20–60 MeV) is used, then higher values of Δm^2 are favored, as can be deduced from Fig. 7.7. On the other hand, there is lack of a significant $\nu_\mu \to \nu_e$ signal in the 60–200 MeV region observed in this global analysis ($8.1 \pm 12.2 \pm 1.8$ events), unlike the more thorough analysis [7.2] specifically looking for $\nu_\mu \to \nu_e$, which detected $18.1 \pm 6.6 \pm 4.0$ oscillation events. The global analysis optimized parameters for the 20–60 MeV region and included 1996–1998 data with its worse backgrounds, whereas the published [7.2] $\nu_\mu \to \nu_e$ result optimized for the

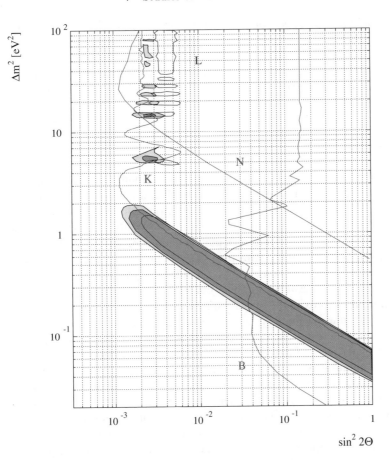

Fig. 7.8. The *shaded areas* are the 90% and 95% confidence regions based on the product of the KARMEN and LSND likelihood ratios, obtained with the method of [7.16]. The 90% C.L. regions for LSND alone (L), obtained using this method and the old analysis, and exclusion regions for KARMEN (K) for the eight-event sample and for NOMAD (N) and Bugey (B) are also shown

60–200 MeV region and used the 1993–1995 data only with their better signal/background ratio. The reduced $\nu_\mu \to \nu_e$ result is more compatible with smaller Δm^2 values at smaller mixing angles than with larger Δm^2, and hence including it makes the ~ 6 eV^2 region in particular seem less probable. In the author's opinion the 60–200 MeV tail of small, systematics-prone signal is wagging the dog of the 20–60 MeV region where the much larger and better-determined signal is found. Thus this global result is not shown here.

The KARMEN experiment will continue into 2001, and two or three years later results should be available from the MiniBooNE experiment at Fermilab. The latter experiment plans to probe this parameter space more completely and is the only real prospect for testing the LSND result fully in the foreseeable future, since CERN's management turned down the excellent I216 (or P231) proposal, although the $\bar{\nu}_\mu \to \bar{\nu}_e$ experiment is without doubt the most important neutrino experiment which can be done at this time. There are prospects for higher-neutrino-flux experiments, especially at possible muon storage rings, and this is addressed in Chapter 9.

7.5   Conclusions

With the final results from NOMAD and only a small remaining improvement expected from further analysis by CHORUS, it seems rather clear that $\nu_\mu \to \nu_\tau$ oscillations do not correspond to the large-Δm^2 region explored by those experiments, but rather that Δm^2 ($\sim 10^{-3}$ eV^2) lies in the domain of atmospheric neutrinos. Thus hot dark matter can no longer be attributed to a single active neutrino, making two- (or possibly three-) neutrino dark matter likely. Limits from NOMAD on $\nu_e \to \nu_\tau$ provide nearly an order of magnitude improvement at large Δm^2, and their limit on $\nu_\mu \to \nu_e$ is quite important in providing an approximate upper bound on the Δm^2 range observed by LSND. Further improvement in the $\nu_\mu \to \nu_e$ limit can be expected.

The only evidence for neutrino oscillations from accelerators comes from the LSND experiment, for which the $\nu_\mu \to \nu_e$ Δm^2 is in the range from about 0.2 to about 10 eV^2, with some significant gaps. KARMEN's null result somewhat limits the LSND parameter space, but when the two experiments are properly analyzed together, much commonly acceptable space remains throughout the Δm^2 range. LSND's new analysis of all its data now shows excellent distributions in all parameters, strongly reinforcing the observation of oscillations. This makes more likely the need for at least one light sterile neutrino and for appreciable hot dark matter.

Acknowledgments

Several people helped in providing information, but I want to thank especially Luigi DiLella, Pierre Loverre and Eric Church. Debbie Ceder provided excellent help in putting this in electronic form. Partial support was provided by the U.S. Department of Energy.

References

7.1 C. Athanassopoulos et al., Phys. Rev. Lett. **75**, 2650 (1995); Phys. Rev. C **54**, 2685 (1996); Phys. Rev. Lett. **77**, 3082 (1996).

7.2 C. Athanassopoulos et al., Phys. Rev. Lett. **81**, 1774 (1998); Phys. Rev. C **58**, 2489 (1998).

7.3 E. Eskut et al., Phys. Lett. B **424**, 202 (1998); **B**434, 205 (1998).

7.4 J. Altegoer et al., Phys. Lett. B **431**, 219 (1998); P. Astier et al., Phys. Lett. B **453**, 169 (1999).

7.5 P. Astier et al., Phys. Lett. B **483**, 387 (2000).

7.6 G. Drexlin et al., Prog. Part. Nucl. Phys. **32**, 375 (1994); B. Zeitnitz et al., Prog. Part. Nucl. Phys. **40**, 169 (1998).

7.7 S. J. Freedman et al., Phys. Rev. D **47**, 811 (1993).

7.8 D. O. Caldwell, in *Perspectives in Neutrinos, Atomic Physics and Gravitation* (Editions Frontières, Gif-sur-Yvette, 1993), p. 187.

7.9 D. O. Caldwell and R. N. Mohapatra, Phys. Rev. D **48**, 3259 (1993).

7.10 J. T. Peltoniemi and J. W. F. Valle, Nucl. Phys. B **406**, 409 (1993).

7.11 J. Altegoer et al., Phys. Lett. B **445**, 439 (1999).

7.12 J. Altegoer et al., Phys. Lett. B **428**, 197 (1998).

7.13 P. Annis et al., Phys. Lett. B **435**, 458 (1998).

7.14 E. Eskut et al., Nucl. Instrum. Meth. A **401**, 7 (1997).

7.15 CHORUS collaboration, contributed to *19th Int. Symp. on Lepton and Photon Interactions at High Energies*, August 1999, Stanford University; L. Dihella, Rev. Proc. 19th Int. Symp. on Lepton and Photon Interactions at High Energies (World Scientific, Singapore, 2000), p. 257.

7.16 G. J. Feldman and R. D. Cousins, Phys. Rev. D **57**, 3873 (1998).

7.17 J. Altegoer et al., Nucl. Instrum. Meth. A **404**, 96 (1998).

7.18 N. Ushida et al. (E531 collaboration), Phys. Rev. Lett. **57**, 2897 (1986); K. S. McFarland et al. (CCFR collaboration), Phys. Rev. Lett. **75**, 3993 (1995); F. Dydak et al. (CDHS collaboration) Phys. Lett. B **134**, 281 (1984).

7.19 P. Astier et al., Phys. Lett. B **471**, 406 (2000).

7.20 M. Apollonio et al. (Chooz collaboration), Phys. Lett. B **466**, 415 (1999); D. Naples et al. (CCFR Collaboration), Phys. Rev. D **59**, 031101 (1999); A. A. Borisov et al. (IHEP/JINR collaboration), Phys. Lett. B **369**, 39 (1996).

7.21 C. Athanassopoulos et al., Nucl. Instrum. Meth. A **388**, 149 (1997).

7.22 M. Steidl, talk at *Les Recontres de Physique de la Vallée d'Aoste*, Feb. 2000; http://www.pi.infn.it/lathuile/2000/programme.html.

7.23 C. Athanassopoulos et al., Phys. Rev. C **55**, 2078 (1997) Phys. Rev. C **56**, 2806 (1997); R.L. Imlay et al., Nucl. Phys. A **629**, C531 (1998).

7.24 B. Armbruster et al., Phys. Rev. Lett. **81**, 520 (1998); Phys. Rev. C **57**, 3414 (1998); G. Drexlin et al., Prog. Part. Nucl. Phys. **40**, 183 (1998).

7.25 E. Kolbe et al., Phys. Rev. C **52**, 3437 (1995); N. Auerbach, N. Van Giai, and O. K. Vorov, Phys. Rev. C **56**, R2368 (1997); A. C. Hayes and I. S. Towner, Phys. Rev. C **61**, 044603 (2000).

7.26 E. Kolbe, K. Langanke and P. Vogel, Nucl. Phys. A **652**, 91 (1999).

7.27 C. Volpe et al., Phys. Rev. C **62**, 015501 (2000).

7.28 K. Eitel, New J. Phys. **2**, 1.1 (2000); hep-ex/9909036.

7.29 L. Borodovsky et al., Phys. Rev. Lett. **68**, 274 (1997).

7.30 M. Mezzetto, Nucl. Phys. B, Proc. Suppl. **70**, 214 (1999); Proc. 5th Int. Workshop on Topics in Astroparticle and Underground Physics, Laboratori Nazionali del Gran Sasso, Italy (Elsevier/North-Holland).

7.31 B. Achkar et al., Nucl. Phys. B **434**, 503 (1995).

8 Double Beta Decay:
Theory, Experiment and Implications

Petr Vogel

8.1 Introduction

Double beta decay is a rare spontaneous nuclear transition in which the nuclear charge changes by two units while the mass number remains the same. It has been long recognized as a powerful tool to study lepton number conservation in general, and neutrino properties in particular. Since the lifetimes of double beta decays are so long, the experiments on double beta decay are very challenging and have led to the development of many generally valuable techniques to achieve extremely low backgrounds.

For $\beta\beta$ decay to proceed, the initial nucleus must be less bound than the final one, but more bound than the intermediate nucleus. These conditions are realized in nature for a number of even–even nuclei (and never for nuclei with an odd number of protons or neutrons). Since the lifetime of $\beta\beta$ decay is always much longer than the age of the Universe, both the initial and the final nuclei exist in nature (some of the actinides being the only exceptions). In many of the "candidates" the transition of two neutrons into two protons is energetically possible, with the largest Q value just above 4 MeV. In a few cases the opposite transition, which decreases the nuclear charge, is also possible, but the Q values are typically smaller.

A nuclear $\beta\beta$ transition can proceed in two ways. One of them, the 2ν decay

$$(Z, A) \rightarrow (Z + 2, A) + e_1^- + e_2^- + \bar{\nu}_{e_1} + \bar{\nu}_{e_2} , \qquad (8.1)$$

conserves the lepton number, while the other one, the 0ν decay

$$(Z, A) \rightarrow (Z + 2, A) + e_1^- + e_2^- , \qquad (8.2)$$

violates lepton number conservation and is therefore forbidden in the standard electroweak theory. The prospect of discovering this neutrinoless double-beta-decay mode is the driving force of most of the interest in this field. It is a possible window into physics "beyond the Standard Model".

Double beta decay has been and continues to be a popular topic since it was first discussed by Maria Goeppert-Mayer in the 1930s. There have been numerous earlier reviews, beginning with the "classics" by Primakoff and Rosen [8.1], Haxton and Stephenson [8.2] and Doi, Kotani and Takasugi [8.3], and continuing to the more recent ones, often devoted to particular aspects of $\beta\beta$ decay [8.4–8.10]. Many details are also described in the monograph [8.11]. The "Review of particle physics" [8.12] regularly summarizes the most recent experimental data.

Double beta decay, in all its modes, is a second-order weak semileptonic process, hence its lifetime, proportional to $(G_F \cos\theta_C)^{-4}$, is so very long. (Here $G_F = 1.166 \times 10^{-5}$ GeV^{-2} is the Fermi coupling constant, and θ_C is the Cabibbo angle.) The neutrinoless decay can be mediated by a variety of virtual particles, in particular by the exchange of light or heavy Majorana neutrinos. The decay amplitude then depends on the masses and coupling constants of these virtual particles. Independently of the actual mechanism of the 0ν $\beta\beta$ decay, its observation would imply that neutrinos necessarily have a nonvanishing Majorana mass [8.14]. In fact, if the 0ν decay is actually observed and its rate measured, one can obtain, at least in principle, a lower limit on that mass [8.15].

However, so far no 0ν decay has been observed. This means (barring artificial complete cancellation of the amplitudes, which we dismiss as unnatural) that the upper limit of the decay rate can be interpreted as an independent limit for each of the possible amplitudes of the decay. In particular, we can obtain a limit on the properties of light and heavy virtual Majorana neutrinos. Below we concentrate on the decays mediated by these particles. (Other possibilities, e.g. the decays mediated by the new particles predicted by supersymmetry, are discussed in [8.9, 8.10].)

Double beta decay with majoron emission, the $0\nu\chi$ mode,

$$(Z, A) \to (Z + 2, A) + e_1^- + e_2^- + \chi \,, \tag{8.3}$$

belongs to the category of the lepton-number-violating decays, even though the lepton number is formally conserved when χ is assigned the lepton number -2. The hypothetical scalar particle χ, which must in this case be light enough to be emitted in the $\beta\beta$ decay, is usually associated with spontaneous breaking of the $B - L$ symmetry [8.16, 8.17].

Empirically, it is easy to distinguish between the three decay modes listed above, provided the electron energies are measured. The electron sum-energy spectra are determined by the phase space of the outgoing leptons and clearly characterize the decay mode, as schematically illustrated in Fig. 8.1. (Geochemical or milking experiments, however, cannot distinguish between the different $\beta\beta$ modes as they determine only the total decay rate.)

There are two distinct groups of theoretical issues associated with the interpretation of $\beta\beta$ decay experiments. The particle physics issues deal with the expression of the decay rate in terms of the fundamental parameters, such as the neutrino masses and mixing angles, the coupling constants in the weak-interaction Hamiltonian, etc. This group of problems involves also the relation of $\beta\beta$ decay to other processes, such as neutrino oscillations, direct mass measurements and searches for other lepton-number-violating processes.

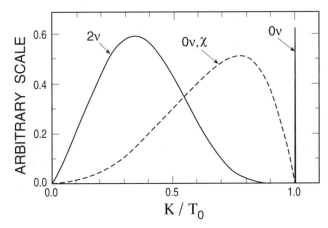

Fig. 8.1. Schematic electron sum-energy spectra of the three $\beta\beta$ decay modes. Each is normalized arbitrarily and independently of the others. The abscissa is the ratio K/T_0 of the sum of the electron kinetic energies divided by its maximum value

The other, essentially decoupled, set of problems involves the nuclear-structure issues associated with $\beta\beta$ decay. The decay rate is expressed in terms of nuclear matrix elements (NMEs) which have to be evaluated. One would like to know, first of all, their values and the uncertainties of those values. This area of research has attracted a lot of attention, and there are many, often conflicting, evaluations available in the literature. Unfortunately, there is no simple way of judging the correctness and accuracy of the evaluations of the nuclear matrix elements for neutrinoless decay. Comparison with the experimentally known rate of the 2ν $\beta\beta$ decay rate is often invoked in that context as a test of the ability of the nuclear model to describe the relevant phenomena. It is not clear, however, if this is indeed a valid test. For example, if one assumes that the 0ν decay is mediated by the exchange of a heavy particle (whether this exchanged particle is a heavy neutrino or not), the corresponding internucleon potential is of short range, and additional issues involving nucleon structure, irrelevant to the 2ν decay, play an important role. Another, less fundamental but in practice perhaps more important example deals with the dependence of the NME on the number of single-nucleon subshells included in the calculation. For the 2ν decay, where only the Gamow–Teller operator $\sigma\tau$ plays a role, it is clearly sufficient to include just the states within the valence oscillator shell. It is less clear that the same truncation is sufficient for the correct description of the 0ν decay.

The experimental study of $\beta\beta$ decay presents a formidable challenge since the goal is to detect a process with a half-life in excess of 10^{25} years (the present best limit for the 0ν decay). The $\beta\beta$ decay must be detected in the presence of an inevitable background of similar energy caused by trace radioisotopes with half-lives 15 or more orders of magnitude shorter. Thus, the optimum separation of the signal from the background, combined with the requirement of kilogram quantities of the source isotopes, characterizes the present-day experiments.

The past and current experiments are still relatively modest in size, and therefore also in complexity and cost. Given the importance of the search for neutrinoless decay, ambitious plans, involving much larger amounts of the source nuclei, are being considered. Naturally, the larger source mass will be beneficial only if it is accompanied by a corresponding reduction of the background. The future projects will therefore be inevitably much more complex, and will involve larger groups of researchers. With them, the field of $\beta\beta$ decay, which already competes with the other experiments described in this book in importance, will also compete in size and cost.

8.2   Lepton Number Violation

With the usual assignment of the lepton number, $L(l^-) = L(\nu) = -L(l^+) = -L(\bar{\nu}) = +1$, 0ν $\beta\beta$ decay represents a change in the global lepton number by two units, $\Delta L = 2$. In that respect its observation would be related to the attempts to detect $\bar{\nu}_e$ from the sun or to detect ν_e from nuclear reactors. Both of these latter processes represent a kind of "$\nu \leftrightarrow \bar{\nu}$ oscillations", and also are possible only for massive Majorana neutrinos.

For light Majorana neutrinos lepton number conservation is irrelevant and the 0ν decay is hindered only by the helicity mismatch. However, the "antineutrino" born in association with one of the e^- in the 0ν decay is not fully right-handed, but has a left-handed component of amplitude $\sim m_\nu/E_\nu$. This left-handed piece can be absorbed by another neutron, which is converted into a proton, and the second e^- is emitted. Similar considerations would govern the above-mentioned $\nu \leftrightarrow \bar{\nu}$ oscillations. The word "oscillations" in this context is a misnomer, however, since the process (if it exists) would proceed without an oscillatory behavior [8.18].

The expected branching ratio for the "wrong" neutrinos at low energies, those relevant to the case of the sun or nuclear reactors, is [8.19]

$$R \sim \frac{m_\nu^2}{2E_\nu^2} \frac{\sigma^{\bar{\nu}N}}{\sigma^{\nu N}} \sim 10^{-14} \,, \tag{8.4}$$

where the numerical factor has been derived for $m_\nu \sim 1$ eV, $E_\nu \sim 5$ MeV, and with the ratio of cross sections taken as unity. Since the 0ν $\beta\beta$ decay is at present sensitive to such neutrino masses, one cannot expect a signal for this kind of $\nu \leftrightarrow \bar{\nu}$ oscillations until similar sensitivity is achieved, i.e. not any time soon, if ever.

However, it is also possible that $\bar{\nu}_e$ from the sun are produced in a more complicated, but possibly more efficient way. Let us assume that a transition magnetic moment $\mu_{e,l}$ connects the left-handed $\nu_{e,\mathrm{L}}$ with a right-handed $\bar{\nu}_{l,\mathrm{R}}$

of a different flavor, which can subsequently oscillate (by vacuum or matter-enhanced oscillations) into the right-handed and thus observable $\bar{\nu}_{e,\mathrm{R}}$, i.e., when neutrinos propagate in a transverse solar magnetic field $B_{\perp}$, one or both of the sequences $\nu_{e,\mathrm{L}} \to \bar{\nu}_{l,\mathrm{R}} \to \bar{\nu}_{e,\mathrm{R}}$ or $\nu_{e,\mathrm{L}} \to \nu_{l,\mathrm{L}} \to \bar{\nu}_{e,\mathrm{R}}$ occurs. Such a process requires that magnetic conversion, which is possible only for massive Majorana neutrinos and which depends on the product $\mu_{e,l}B_{\perp}$, and flavor oscillation, which depends on Δm^2 and $\sin^2 2\theta$, are both present. There is no obvious relation between this process and neutrinoless $\beta\beta$ decay, except that both require the existence of a neutrino Majorana mass term. (This brief discussion of magnetic conversion is highly simplified. In reality, the transition magnetic moments ought to be written in terms of mass eigenstates [8.20].)

Finally, tight experimental limits exist on total-lepton-number violating processes which involve both electrons and muons (see [8.12]), such as the muon conversion

$$\mu^- + (Z, A) \to (Z - 2, A) + e^+ \tag{8.5}$$

and the muonium–antimuonium conversion

$$\mu^+ e^- \to \mu^- e^+ \ . \tag{8.6}$$

The relation of these processes to $\beta\beta$ decay is, however, not well established. (See, however, [8.13] for a discussion of this problem.).

8.3   Particle Physics Aspects

In this section we shall consider how the rate of neutrinoless $\beta\beta$ decay is related to the unknown parameters of the neutrino mass matrix and to the phenomenological parameters describing generalized semileptonic charged-current weak interactions H_{W}:

$$H_{\mathrm{W}} = \frac{G_{\mathrm{F}}}{\sqrt{2}} \left[J_{\mathrm{L}}^{\alpha}(M_{\mathrm{L}\alpha}^+ + \kappa M_{\mathrm{R}\alpha}^+) + J_{\mathrm{R}}^{\alpha}(\eta M_{\mathrm{L}\alpha}^+ + \lambda M_{\mathrm{R}\alpha}^+) \right] + \mathrm{H.c.} \ , \tag{8.7}$$

where J_{L} and M_{L} are the lepton and quark left-handed current four-vectors, respectively. and J_{R} and M_{R} are the corresponding right-handed four-vectors. The dimensionless parameters η, λ and κ characterize deviations from the Standard Model. (Since κ gives a negligible contribution to double beta decay, we shall not consider it from now on.) The coupling parameters η and λ, modified by neutrino mixing, and denoted then usually as $\langle \eta \rangle$ and $\langle \lambda \rangle$, are unknown (and presumably small).

The lepton sector of the theory contains in general n generations of charged leptons as well as n left- and n right-handed neutrinos. The neutrino mass matrix is the $2n \times 2n$ matrix

$$M = \begin{pmatrix} M_{\mathrm{L}} & M_{\mathrm{D}}^{\mathrm{T}} \\ M_{\mathrm{D}} & M_{\mathrm{R}} \end{pmatrix} \ , \tag{8.8}$$

where M_D is the $n \times n$ lepton-number-conserving Dirac mass term, and the symmetric $n \times n$ matrices M_L and M_R are the lepton-number-violating Majorana mass terms. The matrix M has $2n$ real, but not necessarily positive, eigenvalues. Writing the eigenvalues as $m_j \epsilon_j$, we can impose the physically reasonable condition that $m_j \geq 0$. The sign of the eigenvalues of the mass matrix is contained in the phases $\epsilon_j = \pm 1$, which are the intrinsic CP parities of the neutrinos j.

Neutrino oscillation phenomena arise because the "mass eigenstates" of M or, more precisely, their chiral projections N_j^L and N_j^R, are not necessarily the familiar weak-interaction neutrinos that couple to the known intermediate vector boson W_L and to the hypothetical right-handed boson W_R. The physical "weak eigenstate" or current neutrinos, the n left-handed neutrinos ν_L and the n right-handed ones ν_R' (the prime has been added in order to stress that they are *different* particles), are related to the neutrinos of definite mass by the $n \times 2n$ mixing matrices U and V:

$$\nu_L = U N^L \ , \quad \nu_R' = V N^R \ . \tag{8.9}$$

The mixing matrices U and V obey the normalization and orthogonality conditions

$$\sum_{j=1}^{2n} U_{lj}^* U_{l'j} = \delta_{ll'} \ , \quad \sum_{j=1}^{2n} V_{lj}^* V_{l'j} = \delta_{ll'} \ , \quad \sum_{j=1}^{2n} U_{lj}^* V_{l'j} = 0 \ . \tag{8.10}$$

In neutrinoless $\beta\beta$ decay the rate depends on the following effective parameters, which are expressed in terms of the mixing matrices U and V:

$$\langle m_\nu \rangle = \sum_j{}' \epsilon_j m_j U_{e,j}^2 \ ,$$

$$\langle \lambda \rangle = \lambda \sum_j{}' \epsilon_j U_{e,j} V_{e,j} \ ,$$

$$\langle \eta \rangle = \eta \sum_j{}' \epsilon_j U_{e,j} V_{e,j} \ ,$$

$$\langle g_{\nu,\chi} \rangle = \frac{1}{2} \sum_{i,j}{}' (g_{i,j} \epsilon_i + g_{j,i} \epsilon_j) U_{e,i} U_{e,j} \ . \tag{8.11}$$

Here the prime indicates that the summation is over only relatively light neutrinos. Also, λ and η are the dimensionless coupling constants for the right-handed-current weak interaction, (8.7), and $g_{i,j}$ are the coupling constants of interaction between the majoron χ and the Majorana neutrinos N_i and N_j. For the heavy neutrino one obtains

$$\langle m_\nu^{-1} \rangle_H = \sum_j{}'' \epsilon_j m_j^{-1} U_{e,j}^2 \ , \tag{8.12}$$

where the double prime indicates that the summation, involving the inverse neutrino masses m_j^{-1}, is over only the heavy neutrino mass eigenstates ($m_j \geq 1$ GeV).

It is now clear that, within the mechanism considered so far, there is no neutrinoless double beta decay if all neutrinos are massless. Not only does $\langle m_\nu \rangle$ vanish in such a case, but also $\langle \lambda \rangle$ and $\langle \eta \rangle$ vanish owing to the orthogonality condition (8.10). Moreover, $\langle \lambda \rangle$ and $\langle \eta \rangle$ vanish for the same reason even if some or all neutrinos are massive but light, and therefore the summation in (8.11) contains all neutrino mass eigenstates. In that case, however, there is a smaller next-order contribution from the mass dependence of the neutrino propagator, which for this purpose can be written as

$$\frac{\gamma_\mu q^\mu}{q^2 + m_j^2} \approx \frac{\gamma_\mu q^\mu}{q^2} \left(1 - \frac{m_j^2}{q^2} \right) . \tag{8.13}$$

The expression for $\langle \lambda \rangle$, for example, now contains $\sum_j{}' \epsilon_j U_{e,j} V_{e,j} m_j^2$, which clearly shows that a nonvanishing neutrino mass is required.

The presence of the phases ϵ_j in the expression for $\langle m_\nu \rangle$ means that cancellations are possible. In particular, for every Dirac neutrino there is an exact cancellation, since the Dirac neutrino is equivalent to a pair of Majorana neutrinos with opposite signs of the phases ϵ_j and degenerate masses.

In the general case the neutrinoless double-beta-decay rate is a quadratic polynomial in unknown parameters:

$$[T_{1/2}^{0\nu}(0^+ \to 0^+)]^{-1} = C_1 \frac{\langle m_\nu \rangle^2}{m_e^2} + C_2 \langle \lambda \rangle \frac{\langle m_\nu \rangle}{m_e} \cos \psi_1 + C_3 \langle \eta \rangle \frac{\langle m_\nu \rangle}{m_e} \cos \psi_2$$
$$+ C_4 \langle \lambda \rangle^2 + C_5 \langle \eta \rangle^2 + C_6 \langle \lambda \rangle \langle \eta \rangle \cos(\psi_1 - \psi_2) . \tag{8.14}$$

Here ψ_1 and ψ_2 are the phase angles between the generally complex numbers m_ν, λ and η. (However, when CP invariance is assumed, $\psi_{1,2}$ are either 0 or π.) The phase space integrals *and* the nuclear matrix elements are combined in the factors C_i. Assuming that we can calculate the C_i, (8.14) represents an ellipsoid which restricts the allowed range of the unknown parameters $\langle m_\nu \rangle$, $\langle \lambda \rangle$ and $\langle \eta \rangle$ for a given value (or limit) of the 0ν double-beta-decay lifetime.

In order to evaluate the nuclear matrix elements, we must consider the neutrino propagator. Assuming that $\langle m_\nu \rangle^2$ is the only relevant quantity, one can perform an integration over the four-momentum of the exchanged particle and obtain the "neutrino potential", which for $m_\nu < 10$ MeV has the form

$$H(r, \Delta E) = \frac{2R}{\pi r} \int_0^\infty dq \frac{\sin(qr)}{q + \Delta E} , \tag{8.15}$$

where $\Delta E = \langle E_N \rangle - (1/2)(M_i + M_f)$ is the average excitation energy of the intermediate odd–odd nucleus and the factor R (the nuclear radius) has been added to make the neutrino potential dimensionless.

When the 0ν decay is mediated by the right-handed weak-current interaction the evaluation of the decay rate becomes more complicated, since many more terms must be included [8.2, 8.3, 8.21]. If the four-momentum of the virtual neutrino is $q_\mu \equiv \omega, q$, the neutrino propagator contains

$$\omega\gamma_0 - q \cdot \gamma + m_j .$$

The part of the propagator proportional to m_j is responsible for the neutrino potential (8.15). The part containing q leads to a new potential related to the derivative of $H(r, \Delta E)$, and the part containing ω leads to yet another potential, which is a combination of $H(r, \Delta E)$ and its derivative.

Similarly, there are now also more nuclear matrix elements, which contain in addition the nucleon momenta (i.e. the gradient operators) and depend on the nucleon spins and radii in a more complicated way (e.g. they contain tensor operators). The outgoing electrons are no longer just in $s_{1/2}$ states, because for some of the operators one of the electrons will be in a $p_{1/2}$ state. The recoil matrix element, which originates from the recoil term in the nuclear vector current, is numerically relatively large [8.21], resulting in more sensitivity to the parameter $\langle\eta\rangle$. The current best limits on $\langle\eta\rangle$ and $\langle\lambda\rangle$ are listed in [8.12].

8.4   Experimental Techniques and Results

It is beyond the scope of this review to describe in detail the experimental techniques developed to meet the challenge of background suppression and signal recognition needed to determine the rate of (or an interesting limit on) $\beta\beta$ decay. Thus, only the briefest outline is given, and the most important experimental results are summarized in tables.

Historically, the existence of $\beta\beta$ decay was first established using the *geochemical* method. Here one takes advantage of geologic integration times by searching for daughter products accumulated in ancient minerals that are rich in the parent isotope. (The related *radiochemical* method is applicable if the daughter isotope is radioactive.) Since the energy information is long lost, the mode of $\beta\beta$ decay responsible is not directly determined. Instead, the total decay rate is determined, and thus an upper limit on each mode as well.

Only by measuring the energies of the electrons released in the decay in direct counting experiments can one distinguish directly the mode of decay. The 2ν and $0\nu\chi$ decay modes both result in a rather generic-looking electron spectrum (Fig. 8.1), and the observation of these decays requires either an extremely efficient background suppression or additional information, such as a tracking capability.

The measured half-lives of the 2ν mode are collected in Table 8.1. Many of them have been measured by several groups; only the results with the smallest claimed errors are shown. (The case of ^{130}Te, where the two competing results have the same error but exclude each other, is the only exception.) Also, the numerous half-life limits have been omitted. The 2ν mode is now well

Table 8.1. Recent $\beta\beta_{2\nu}$ results. (Only positive results are listed. The most accurate published values are given, except for ^{130}Te, where two conflicting results with the same claimed errors are quoted)

Isotope	$T_{1/2}^{2\nu}$ (y)	Reference
^{48}Ca	$(4.3^{+2.4}_{-1.1} \pm 1.4) \times 10^{19}$	[8.22]
^{76}Ge	$(1.77 \pm 0.01 \ ^{+0.13}_{-0.11}) \times 10^{21}$	[8.23]
^{82}Se	$(8.3 \pm 1.0 \pm 0.7) \times 10^{19}$	[8.24]
^{96}Zr	$(3.9 \pm 0.9) \times 10^{19}$ [a]	[8.25]
^{100}Mo	$(6.82^{+0.38}_{-0.53} \pm 0.68) \times 10^{18}$	[8.26]
^{116}Cd	$(3.75 \pm 0.35 \pm 0.21) \times 10^{19}$	[8.27]
^{128}Te	$(7.2 \pm 0.4) \times 10^{24}$ [a]	[8.28]
^{130}Te	$(2.7 \pm 0.1) \times 10^{21}$ [a]	[8.28]
	$(7.9 \pm 1.0) \times 10^{20}$	[8.29]
^{150}Nd	$(6.75^{+0.37}_{-0.42} \pm 0.68) \times 10^{18}$	[8.26]
^{238}U	$(2.0 \pm 0.6) \times 10^{21}$ [b]	[8.30]

[a] geochemical determination; total decay rate.
[b] radiochemical determination; total decay rate.

established; no doubt many more and more accurate results will become available soon.

In fact, $\beta\beta$ decay is becoming a valuable tool of nuclear spectroscopy. The decay of ^{100}Mo into the excited 0^+ state of ^{100}Ru at 1130 keV has been observed [8.31, 8.32]. The technique used, the observation of the subsequent γ decay cascade, can be readily applied to other nuclei as well. This development not only expands the scope of the experimental study of $\beta\beta$ decay, but also allows more detailed comparison between theory and experiment (for an early attempt, see [8.33]).

The 0ν mode can be approached quite differently from the 2ν and $0\nu, \chi$ modes because of the distinctive character of the 0ν electron sum spectrum – a monoenergetic line at the full Q value (Fig. 8.1). Obviously, a sharp energy resolution of the detector is a big advantage which helps to isolate the line from the background. As in the case of the 2ν decay, other features, such as tracking, naturally help as well.

The best reported limits for the neutrinoless $\beta\beta$ decay modes are collected in Tables 8.2 and 8.3. Again, only the most restrictive limits for the transitions considered are shown. The longest half-life limit, reported for ^{76}Ge by the Heidelberg–Moscow collaboration [8.34], is based on 24.16 kg y of exposure and uses pulse shape discrimination to suppress the background (in the relevant energy region the background is a mere (0.06 ± 0.02) events/(kg y keV)).

Table 8.2. The best reported limits on $T_{1/2}^{0\nu}$ and $\langle m_\nu \rangle$. The experimental result is listed first, with its reference. This is followed by the limit on $\langle m_\nu \rangle$, followed by the reference for the nuclear matrix element (NME) employed. Whenever possible, the choice of the authors of the experimental paper regarding the NME has been respected. See the text for discussion of uncertainties associated with the evaluation of NMEs

Isotope	$T_{1/2}^{0\nu}$ (10^{22} y) (C.L.(%))	Reference for experiment	$\langle m_\nu \rangle$ (eV)	Reference for NME
^{48}Ca	> 0.95 (76)	[8.35]	< 18.3	[8.2]
^{76}Ge	$> 1600\,(5700)^{\mathrm{a}}$ (90)	[8.34]	$< 0.4\,(0.2)^{\mathrm{a}}$	[8.36]
^{82}Se	> 2.7 (68)	[8.37]	< 5	[8.2]
^{100}Mo	> 5.2 (68)	[8.38]	< 6.6	[8.21]
^{116}Cd	> 2.9 (90)	[8.39]	< 4.6	[8.36]
$T_{1/2}(^{130}$Te$)/T_{1/2}(^{128}$Te$)^{\mathrm{b}}$	$(3.52 \pm 0.11) \times 10^{-4}$	[8.28]	$< 1.1\text{--}1.5$	[8.36, 8.40]
^{136}Xe	> 44 (90)	[8.41]	$< 2.3\text{--}2.8$	[8.40]
^{150}Nd	> 0.12 (90)	[8.26]	< 4.0	[8.36]

[a] The first entry is based on the average background, and the second entry (in parentheses) is based on the apparent lack of background counts in the corresponding energy interval.

[b] Geochemical determination of the lifetime ratio.

Table 8.3. The most restrictive (C.L. (%)) majoron limits

Isotope	$T_{1/2}^{0\nu,\chi}$ (y) (C.L. (%))	$\langle g_{\nu,\chi} \rangle$	Reference
^{48}Ca	$> 7.2 \times 10^{20}$ (90)	$< 5.3 \times 10^{-4}$	[8.42]
^{76}Ge	$> 1.66 \times 10^{22}$ (90)	$< 1.8 \times 10^{-4}$,	[8.43]
^{82}Se	$> 2.4 \times 10^{21}$ (68)	$< 2.3 \times 10^{-4}$	[8.24]
^{100}Mo	$> 5.4 \times 10^{21}$ (68)	$< 7.3 \times 10^{-5}$	[8.38]
^{116}Cd	$> 1.2 \times 10^{21}$ (90)	$< 2.1 \times 10^{-4}$	[8.44]
^{128}Te	$> 7.7 \times 10^{24}$ ^{a} (90)	$< 3 \times 10^{-5}$	[8.28]
^{136}Xe	$> 7.2 \times 10^{21}$ (90)	$< 1.6 \times 10^{-4}$	[8.41]
^{150}Nd	$> 2.8 \times 10^{20}$ (90)	$< 1 \times 10^{-4}$	[8.26]

[a] Geochemical determination; from total decay rate.

In that experiment seven events were observed in the 3σ region around the 0ν decay Q value, while from the background extrapolation one expects 13 events. Using this lack of background events, an even more stringent limit (the entry in parentheses in Table 8.2) is obtained.

The limit based on the Te lifetime ratio in Table 8.2 is based on the different Q value dependences of the 0ν and 2ν modes. That this offers a valuable tool was recognized already in the prophetic early paper by Pontecorvo [8.45]. Even though the corresponding NMEs are not exactly equal, they are close enough to allow one to use the geochemical lifetime determination here and in Table 8.3.

8.5 Nuclear-Structure Aspects

The rate of the 2ν $\beta\beta$ decay is simply

$$1/T^{2\nu}_{1/2} = G_{2\nu}(E_0, Z)|M_{2\nu}|^2 ,$$

(8.16)

while for the neutrinoless decay (assuming that it is mediated by a light Majorana neutrino and that there are no right-handed weak interactions) and for the decay with majoron emission, it is given by

$$1/T^{0\nu}_{1/2} = G_{0\nu}(E_0, Z)|M_{0\nu}|^2 \langle m_\nu \rangle^2 ,$$
$$1/T^{0\nu,\chi}_{1/2} = G_{0\nu,\chi}(E_0, Z)|M_{0\nu,\chi}|^2 \langle g_{\nu,\chi} \rangle^2 .$$

(8.17)

Here the phase space functions $G(E_0, Z)$ are accurately calculable, and the nuclear matrix elements M are the topic of this section. Obviously, the accuracy with which the fundamental particle physics parameters $\langle m_\nu \rangle$ and $\langle g_{\nu,\chi} \rangle$ can be determined is limited by our ability to evaluate these nuclear matrix elements.

In that context there are three distinct set of problems:

- 2ν decay: the physics of the Gamow–Teller amplitudes
- 0ν decay with the exchange of light massive Majorana neutrinos: no selection rules on multipoles, the role of nucleon correlations and sensitivity to nuclear models
- 0ν decay with the exchange of heavy neutrinos: the physics of the nucleon–nucleon states at short distances.

8.5.1 Two-Neutrino Decay

Since the energies involved are modest, the allowed approximation should be applicable, and the rate is governed by the double Gamow–Teller (GT) matrix element

$$M^{2\nu}_{\text{GT}} = \sum_m \frac{\langle f||\sigma\tau_+||m\rangle \times \langle m||\sigma\tau_+||i\rangle}{E_m - (M_i + M_f)/2} ,$$

(8.18)

where i, f are the ground states in the initial and final nuclei, and m are the intermediate 1^+ (virtual) states in the odd–odd nucleus. The first factor in the numerator represents the β^+ (or (n,p)) amplitude for the final nucleus, while the second one represents the β^- (or (p,n)) amplitude for the initial nucleus. Thus, in order to correctly evaluate the 2ν decay rate, we have to know, at least in principle, *all* GT amplitudes for both β^- and β^+ processes, including their signs. The difficulty is that the 2ν matrix element exhausts a very small fraction $(10^{-5}\text{–}10^{-7})$ of the double GT sum rule [8.46], and hence it is sensitive to details of nuclear structure.

Various approaches used in the evaluation of the 2ν decay rate have been reviewed recently in [8.6]. The quasiparticle random-phase approximation (QRPA) has been the most popular theoretical tool in the recent past. Its main ingredients, the repulsive particle–hole spin–isospin interaction, and the attractive particle–particle interaction, clearly play a decisive role in the concentration of the β^- strength in the giant GT resonance, and in the relative suppression of the β^+ strength and its concentration at low excitation energies. Together, these two ingredients are able to explain the suppression of the 2ν matrix element when expressed in terms of the corresponding sum rule.

Yet, the QRPA is often criticized. Two "undesirable", and to some extent unrelated, features are usually quoted. One is the extreme sensitivity of the decay rate to the strength of the particle–particle force (often denoted by g_{pp}). This decreases the predictive power of the method. The other one is the fact that, for a realistic value of g_{pp}, the QRPA solutions are close to their critical value (so-called collapse). This indicates a phase transition, i.e. a rearrangement of the nuclear ground state. The QRPA is meant to describe small deviations from the unperturbed ground state, and thus is not fully applicable near the point of collapse. Numerous attempts have been made to extend the range of validity of the QRPA (see e.g. [8.6]). Altogether, the QRPA and its various extensions, with their ability to adjust at least one free parameter, are typically able to explain the observed 2ν decay rates.

At the same time, detailed calculations show that the sum over the excited states in (8.18) converges quite rapidly [8.47]. In fact, a few low-lying states usually exhaust the whole matrix element. Thus, it is not really necessary to describe all GT amplitudes; it is enough to describe correctly the β^+ and β^- amplitudes of the low-lying states, and include everything else in the overall renormalization (quenching) of the GT strength.

Nuclear shell model methods are now capable of handling much larger configuration spaces than even a few years ago. Thus, for many nuclei the evaluation of the 2ν rates within the $0\hbar\omega$ shell model space is feasible. (Heavy nuclei with permanent deformation, such as ^{150}Nd and ^{238}U, remain, however, beyond the reach of the shell model techniques.) Using the shell model avoids, naturally, the above difficulties of the QRPA. At the same time, the shell model can describe, using the same method and the same residual interaction, a wealth of spectroscopic data, allowing much better tests of its predictive power.

8.5.2 Neutrinoless Decay: Light Majorana Neutrino

If one assumes that 0ν decay is caused by the exchange of a virtual light Majorana neutrino between the two nucleons, then several new features arise: (a) the exchanged neutrino has a momentum $q \sim 1/r_{nn} \simeq 50\text{--}100$ MeV (where r_{nn} is the distance between the decaying nucleons). Hence, the dependence on the energy in the intermediate state is weak, the closure approximation is applicable and one does not have to sum explicitly over the nuclear intermediate states. (b) Since $qR > 1$ (where R is the nuclear radius), the expansion in multipoles is not convergent, unlike the case in 2ν decay. In fact, all possible multipoles contribute by a comparable amount. (c) The neutrino propagator results in a neutrino potential of relatively long range (see (8.15)).

Thus, in order to evaluate the rate of the 0ν decay, we need to evaluate only the matrix element connecting the ground states 0^+ of the initial and final nuclei. Again, we can use the QRPA or the shell model. Both calculations show that the features enumerated above are indeed present. In addition, the QRPA typically shows a less extreme dependence on the particle–particle coupling constant g_{pp} than it does for the 2ν decay, since the contribution of the 1^+ multipole is relatively small. The calculations also suggest that for quantitatively correct results one has to treat the short-range nucleon–nucleon repulsion carefully, despite the long range of the neutrino potential.

Does that mean that the calculated matrix elements are insensitive to nuclear structure? An answer to that question obviously has great importance, since unlike the situation for 2ν decay, we cannot directly test whether the calculation is correct or not.

For simplicity, let us assume that 0ν $\beta\beta$ decay is mediated only by the exchange of a light Majorana neutrino. The relevant nuclear matrix element is then the combination $M_{\mathrm{GT}}^{0\nu} - M_{\mathrm{F}}^{0\nu}$, where the GT and F operators change two neutrons into two protons, and contain the corresponding operator plus the neutrino potential. One can express these matrix elements either in terms of the proton particle–neutron hole multipoles (i.e. the usual beta decay operators) or in terms of the multipole coupling of the exchanged pair, nn and pp.

When using the decomposition into the proton particle–neutron hole multipoles, one finds that all possible multipoles (given the one-nucleon states near the Fermi level) contribute, and the contributions have typically equal signs. Hence, there does not seem to be much cancellation.

However, perhaps more physical is the decomposition into the exchanged-pair multipoles. There one finds, first of all, that only natural-parity multipoles ($\pi = (-1)^I$) contribute noticeably. And there is a rather severe cancellation. The biggest contribution comes from the 0^+ multipole, i.e. the pairing part. All other multipoles, related to higher-seniority states, contribute with an opposite sign. The final matrix element is then a difference of the pairing

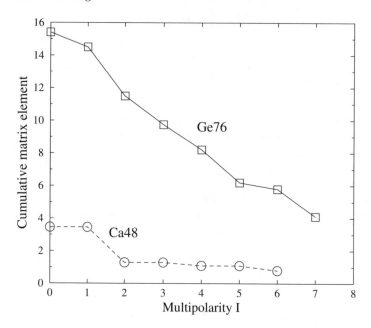

Fig. 8.2. The cumulative contribution, i.e. the summed contribution of all natural-parity multipoles up to I of the exchanged nn and pp pair, to the 0ν nuclear-matrix-element combination $M_{GT}^{0\nu} - M_F^{0\nu}$. The *full line* is for ^{76}Ge and the *dashed line* for ^{48}Ca

and higher-multipole (or broken-pair $\equiv$ higher-seniority) parts, and is considerably smaller than either of them. This is illustrated in Fig. 8.2, where the cumulative effect is shown, i.e. the quantity $M(I) = \sum_J^I \left[M_{GT}^{0\nu}(J) - M_F^{0\nu}(J) \right]$ is displayed for ^{76}Ge (from [8.48]) and ^{48}Ca (from [8.49]). Thus, the final result depends sensitively on both the correct description of the pairing and on the admixtures of higher-seniority configurations in the corresponding initial and final nuclei. It appears, moreover, that the final result might depend on the size of the single-particle space included. That important question requires further study.

Since there is no objective way to judge which calculation is correct, one often uses the spread between the calculated values as a measure of the theoretical uncertainty. This is illustrated in Fig. 8.3. There, I have chosen two representative QRPA sets of results, the highly truncated "classical" shell model result of Haxton and Stephenson [8.2] and the result of a more recent shell model calculation which is convergent for the set of single-particle states chosen (essentially the $0\hbar\omega$ space).

For the most important case of ^{76}Ge the calculated rates differ by a factor of 6–7. Since the effective neutrino mass $\langle m_\nu \rangle$ is inversely proportional to the square root of the lifetime, the experimental limit of 1.6×10^{25} y translates into limits of about 1 eV using the NME of [8.40, 8.50], and about 0.4 eV with

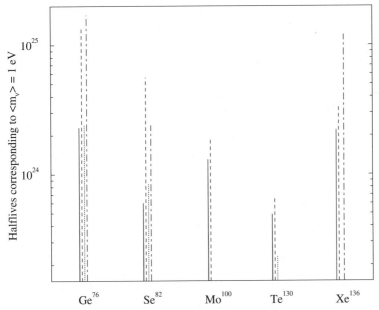

Fig. 8.3. Half-lives (in years) calculated for $\langle m_\nu \rangle = 1$ eV by various representative methods and different authors for the most popular double-beta-decay candidate nuclei. *Solid lines*, QRPA from [8.36]; *dashed lines*, QRPA from [8.40] (recalculated for $g_A = 1.25$ and $\alpha' = -390$ MeV fm^3); *dotted lines*, shell model [8.2]; *dot–dashed lines*, shell model [8.50]

the NME of [8.2, 8.36]. On the other hand, if one accepts the more stringent limit of 5.7×10^{25} y [8.34], even the more pessimistic matrix elements restrict $\langle m_\nu \rangle$ to less than 0.5 eV. Needless to say, a more objective measure of the theoretical uncertainty would be highly desirable.

In Tables 8.2 and 8.3 we list the deduced limits on the fundamental parameters, namely the effective neutrino Majorana mass $\langle m_\nu \rangle$ and the majoron coupling constant $\langle g_{\nu,\chi} \rangle$. The references to the sources of the corresponding nuclear matrix elements, used to translate the experimental half-life limit into the listed limits on $\langle m_\nu \rangle$ and $\langle g_{\nu,\chi} \rangle$, are also given. When using the tables one has to keep in mind the uncertainties illustrated in Fig. 8.3.

8.5.3 Neutrinoless Decay: Very Heavy Majorana Neutrino

Neutrinoless $\beta\beta$ decay can be also mediated by the exchange of a heavy neutrino. The decay rate is then inversely proportional to the square of the effective neutrino mass [8.51]. In this context it is particularly interesting to consider the left–right symmetric model proposed by Mohapatra [8.52]. In it, one can find a relation between the mass of the heavy neutrino M_N and the mass of the right-handed vector boson W_R. Thus, the limit on the $\beta\beta$ rate provides, within that specific model, a stringent lower limit on the mass of the W_R.

The process then involves the emission of a heavy W_R^- by the first neutron and its virtual decay into an electron and a heavy Majorana neutrino, $W_R^- \to e^- + \nu_N$. This is followed by the transition $\nu_N \to e^- + W_R^+$ and the absorption of the W_R^+ in the second neutron, changing it into the second proton. Since all particles exchanged between the two neutrons are very heavy, the corresponding "neutrino potential" is of essentially zero range. Hence, when calculating the nuclear matrix element, one has to take into account carefully the short-range nucleon–nucleon repulsion.

As long as we treat the nucleus as an ensemble of nucleons only, the only way to have nonvanishing nuclear matrix elements for the above process is to treat the nucleons as finite-size particles. In fact, that is the standard way to approach the problem [8.51]; the nucleon size is described by a dipole form factor with a cutoff parameter $\Lambda \simeq 0.85$ GeV. Using such a treatment of the nucleon size, and the half-life limit for the ^{76}Ge 0ν decay listed in Table 8.2, one obtains a very interesting limit on the mass of the vector boson W_R [8.53]

$$m_{W_R} \geq 1.6 \text{ TeV} . \tag{8.19}$$

However, another way of treating the problem is possible, and has already been mentioned in [8.51]. Let us recall how the analogous situation is treated in the description of the parity-violating nucleon–nucleon force [8.54]. There, instead of the weak (i.e. very short range) interaction of two nucleons, one assumes that a meson (π, ω, ρ) is emitted by one nucleon and absorbed by another one. One of the vertices is the parity-violating one, and the other one is the usual parity-conserving strong one. The corresponding range is then just the meson exchange range, easily treated. The situation is schematically depicted in the left-hand panel of Fig. 8.4. The analogy for $\beta\beta$ decay is shown in the right-hand graph. It involves two pions, and the "elementary" lepton-number-violating $\beta\beta$ decay then involves a transformation of two pions into two electrons. Again, the range is just the pion exchange range. It would be interesting to see if a detailed treatment of this graph would lead to a more or less stringent limit on the mass of the W_R than the treatment with form factors. The relation to the claim in [8.55] that an analogous graph contributing to the lepton-number-violating muon capture vanishes identically should be further investigated; in fact that claim is probably not valid.

8.6   Prospects

As stated earlier, the present best limits on the rate of 0ν $\beta\beta$ decay, or equivalently on the neutrino effective Majorana mass $\langle m_\nu \rangle$, have been obtained with an exposure of about 20 kg yr. Several experiments (Heidelberg–Moscow, ^{76}Ge [8.34]; IGEX, ^{76}Ge [8.56]; Caltech–Neuchatel TPC, ^{136}Xe [8.41]) are currently at or near that level. The other limits in Tables 8.2 and 8.3 were

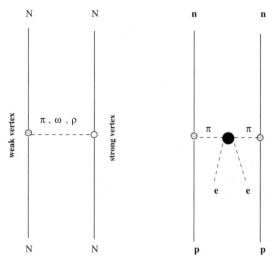

N   N   n   n

weak vertex   π, ω, ρ   strong vertex   π   π   e   e

N   N   p   p

Fig. 8.4. The Feynman graph description of the parity-violating nucleon–nucleon force (*left graph*) and of $\beta\beta$ decay with the exchange of a heavy neutrino mediated by pion exchange. The short-range lepton-number-violating amplitude is symbolically described by the *filled blob* in the *right graph*

obtained with smaller exposures of ~ 1 kg yr. The detector NEMO-3, with a planned source mass of 10 kg, is being built and should be operational soon [8.57]. In a few years of operation it should reach a half-life limit of $\sim 10^{25}$ y for the 0ν decay of ^{100}Mo, and perhaps other nuclei as well. However, further improvements with the existing detectors become increasingly difficult, since the sensitivity to $\langle m_\nu \rangle$ is proportional to only the 1/4 power of the source mass and exposure time. Thus, much larger sources are clearly needed.

What are the prospects for a radical improvement in the search for 0ν $\beta\beta$ decay? To achieve that, one would have to build a detector capable of using hundreds of kilograms or even several tons of the source material. At the same time, the background per unit mass has to be correspondingly improved so that one can benefit from the larger mass. Obviously, such a program is very challenging.

The difficulty begins with the problem of acquiring such a large mass of the isotopically separated and radioactively clean material. Here, the principal obstacle is the cost of the isotope separation. (This can be avoided only if the source isotope has a large abundance; in practice that is true only for ^{130}Te, with 34% abundance.)

The second unavoidable difficulty is the background caused by the 2ν decay. One can observe the 0ν decay only if its rate exceeds the fluctuations of the 2ν events at the same energy, i.e. near the decay Q value. The number of 2ν decays in an energy interval ΔE near Q depends on these quantities as $\sim (\Delta E/Q)^6$, provided $\Delta E \ll Q$. (If the energy resolution is folded in, this dependence is somewhat modified.) Thus, good energy resolution, which determines how wide an interval ΔE one must consider, is again crucial in order to reduce the effect of this "ultimate" background.

One of the proposals for such a large $\beta\beta$ experiment has been extensively discussed in the literature (e.g. [8.9]). The project, with the acronym GE-NIUS, would use a large amount of "naked" enriched ^{76}Ge, in the form of an array of about 300 detectors, suspended in liquid nitrogen, which simultaneously provides cooling and shielding. It is envisioned that the detector would consist of one ton of enriched ^{76}Ge. The anticipated background is 0.04 counts/(keV y t), i.e. about 1000 times lower than the best existing backgrounds. Such a detector could reach a half-life limit of about 6×10^{27} y within one year of operation, thus improving the neutrino mass limit by an order of magnitude.

Another large project, CUORE, [8.58] is a cryogenic setup consisting of 17 towers, each containing 60 cubic crystals of TeO$_2$. It would be housed in a single, specially constructed dilution refrigerator and would contain about 800 kg of the sensitive material. A prototype system, CUORICINO, consisting of one of the towers, is being developed now.

An experiment with a large amount (100 tons) of natural or enriched molybdenum (the abundance of the $\beta\beta$ candidate ^{100}Mo is 9.6%), with good energy and position resolution, MOON, is proposed in [8.59].

In order to radically suppress the background, Ba ions, the final products of ^{136}Xe double beta decay, could be identified by laser tagging. That approach, EXO, described in [8.60], would allow the use of a large time projection chamber with perhaps ton quantities of ^{136}Xe, reaching sensitivities to half-lives $\sim 10^{28}$ years.

This still incomplete list of proposed very large $\beta\beta$ decay experiments shows that the field is entering a critical phase. If the new techniques mentioned above can be developed in conjunction with the large source mass, the background caused by radioactivity can be essentially eliminated. However, as stated above, the ultimate background due to the tail of the 2ν decay can be compensated only by a superior energy resolution.

Given the importance of neutrinoless decay, it is likely that several of these large and costly projects, involving ton years of exposure and a correspondingly reduced background, will be realized in the foreseeable future. Thus, sensitivity to a neutrino Majorana mass $\langle m_\nu \rangle$ approaching 0.01 eV may be in sight. Whether neutrinoless decay will be discovered is unknown, but the reasons to look for it are so compelling that the search will undoubtedly continue.

8.7   Implications

The study of 0ν $\beta\beta$ decay provides at present an upper limit well below 1 eV for the effective electron neutrino Majorana mass $\langle m_\nu \rangle$ even if the most pessimistic nuclear matrix elements are used. What are the consequences of that limit when combined with the manifestations of neutrino oscillations?

Recall that the atmospheric neutrino anomaly (with its zenith angle dependence) implies nearly maximum mixing of μ and τ neutrinos (or μ and sterile neutrinos) with $\Delta m^2 \sim 10^{-3}$ eV^2 (see Chap. 5). There is, so far, no unique neutrino oscillation solution to the solar neutrino deficit (see Chap. 4). However, all of the acceptable solutions have $\Delta m^2 < 10^{-4}$ eV^2 and involve electron neutrinos. Both large- and small-mixing-angle solutions are currently compatible with the data. Finally, the third piece of "positive" evidence comes from the LSND experiment (see Chap. 7), and implies relatively small mixing between the electron and muon neutrinos and $\Delta m^2 \geq 0.1$ eV^2. A full analysis must contain, in addition, all experimental results which exclude various parts of the possible regions of the quantities Δm^2 and the mixing angles. Taking all these findings together would necessarily imply the existence of a fourth neutrino, which must be "sterile" given the constraint on the invisible width of the Z. At the same time, it is well known that oscillation experiments are not able to furnish the overall scale of the neutrino masses.

This absolute neutrino mass scale is essential not only as a matter of principle, but also, in particular, if one wants to ascribe part of the dark matter, namely its "hot" component, to massive neutrinos. Doing that would mean that the sum of the neutrino masses $\sum m_\nu$ is one or several eV. Tritium beta decay gives an upper limit of a similar magnitude for any mass eigenstate with a large electron flavor component. Clearly, if light neutrinos are responsible for a nonnegligible part of the dark matter, the oscillation data mean that at least two and possibly all neutrino masses are nearly degenerate. (Such scenario was discussed for the first time in [8.61].) The relation of $0\nu\,\beta\beta$ decay to the oscillation scenarios, in particular the scenarios involving degenerate neutrinos, has been a topic of several recent papers [8.62–8.66].

The consequences are particularly dramatic if one assumes that only three massive Majorana neutrinos exist with nearly degenerate masses $m_i \simeq \bar{m} \sim O(\text{eV})$ (hence discarding for this purpose the LSND experimental result, even though there is no evidence against it). The $\beta\beta$ decay constraint can be expressed as [8.62]

$$\langle m_\nu \rangle = |m_1 c_2^2 c_3^2 e^{i\phi} + m_2 c_2^2 s_3^2 e^{i\phi'} + m_3 s_2^2 e^{i2\delta}| \,, \tag{8.20}$$

where c_i, s_i denote $\cos\theta_i, \sin\theta_i$ in the 3×3 mixing matrix, δ is the CP-violating phase in that matrix, and ϕ, ϕ' are the CP-violating phases in the diagonal mass matrix. Clearly, the differences in m_i can be neglected in this case. Moreover, the reactor long-baseline experiments have established that ν_e do not mix very much with anything else near $\Delta m^2 \sim 10^{-3}$ eV^2, which means that the angle θ_2 is small. At the same time, the angle θ_1, which controls the atmospheric neutrino oscillations, is near its maximum value, with $\sin^2 2\theta_1 \simeq 1$. Thus

$$|\cos^2 \theta_3 + \sin^2 \theta_3 e^{i(\phi'-\phi)}| < \langle m_\nu \rangle / \bar{m} \ll 1 \,. \tag{8.21}$$

Hence θ_3 must also be near maximum mixing, $\sin 2\theta_3 \simeq 1$, and the CP phases in the above equation (8.21) are such that the two terms cancel each other.

That would be a very unexpected result. We would have three massive, highly degenerate neutrinos with bimaximal mixing. Moreover, the electron neutrino would be "quasi-Dirac", with its two components essentially canceling each other in their contribution to 0ν $\beta\beta$ decay. While such a scenario is rather problematic (see [8.63]), and it does not accommodate the LSND result at all, it illustrates the power of neutrinoless $\beta\beta$ decay in constraining the choice of neutrino oscillation scenarios.

8.8   Conclusions

The quest for neutrino mass is at a critical stage at present. The evidence for neutrino mixing is getting stronger and stronger, and the basic parameters describing the neutrino oscillation phenomena are being constrained more and more. At the same time, the searches for oscillation cannot give us the scale of the neutrino masses, but only their differences. Among the experiments that are sensitive to the masses themselves, albeit to different aspects of them (the endpoint of ordinary beta decay, observation of supernova neutrinos and neutrinoless double beta decay), only the 0ν decay is able to reach the sub-eV region, and in the foreseeable future to extend this region by a substantial margin.

In this review the present status of $\beta\beta$ decay has been described. The unpleasant uncertainty related to the nuclear-structure aspect of the problem is estimated to be at the level of a factor of 2–3 for the effective neutrino mass. However, the experimental progress has been such that even using the most conservative nuclear matrix elements allows us to push the limit well below that obtained from the competing techniques. The nuclear-structure uncertainty can be reduced by further development of the corresponding nuclear models. At the same time, by reaching comparable experimental limits in several nuclei, the chances of a severe error in the NMEs will be substantially reduced.

Several projects are under way that will improve the lifetime limit substantially or find 0ν decay. Already, the search for $\beta\beta$ decay gives important constraints on the fundamental properties of neutrinos and their interactions. The role of $\beta\beta$ decay in the whole enterprise described in the various chapters of this book will be substantially strengthened once these ambitious projects are under way.

Acknowledgment

This work was supported by the US Department of Energy under grant no. DE-FG03-88ER-40397.

References

8.1 H. Primakoff and S. P. Rosen, Rep. Prog. Phys. **22**, 121 (1959).

8.2 W. C. Haxton and G. J. Stephenson Jr., Prog. Part. Nucl. Phys. **12**, 409 (1984).

8.3 M. Doi, T. Kotani and E. Takasugi, Prog. Theor. Phys. Suppl. No. 83, 1 (1985).

8.4 M. Moe and P. Vogel, Annu. Rev. Nucl. Part. Sci. **44**, 247 (1994).

8.5 V. I. Tretyak and Yu. Zdesenko, Atom. Data Nucl. Data Tables **61**, 43 (1995).

8.6 J. Suhonen and O. Civitarese, Phys. Rep. **300**, 123 (1998).

8.7 F. Šimkovic, G. Pantis and A. Faessler, Phys. Atom. Nucl. **61**, 218 (1998).

8.8 A. Morales, Nucl. Phys. B, Proc. Suppl. **77**, 335 (1999).

8.9 H. V. Klapdor-Kleingrothaus, hep-ex/9907040.

8.10 J. D. Vergados, Phys. Atom. Nucl. **63**, 1137 (2000); hep-ph/9907316.

8.11 F. Boehm and P. Vogel, *Physics of Massive Neutrinos*, 2nd edn. (Cambridge University Press, Cambridge, 1992).

8.12 C. Caso et al., Eur. Phys. J. **3**, 1 (1998).

8.13 S. Bergmann, H. V. Klapdor-Kleingrothaus and H. Päs, hep-ph/0004048.

8.14 J. Schechter and J. W. F. Valle, Phys. Rev. D **25**, 2951 (1982).

8.15 B. Kayser, Nucl. Phys. B, Proc. Suppl. **19**, 177 (1991).

8.16 Y. Chikashige, R. N. Mohapatra and R. D. Peccei, Phys. Rev. Lett. **45**, 1926 (1980).

8.17 G. Gelmini and M. Rondacelli, Phys. Lett. B **99**, 411 (1981).

8.18 P. Fisher, B. Kayser and K. D. McFarland, Ann. Res. Nucl. Part. Sci. **49**, 481 (1999); hep-ph/9906244.

8.19 P. Langacker and J. Wang, Phys. Rev. D **58**, 093004 (1998).

8.20 J. Beacom and P. Vogel, Phys. Rev. Lett. **83**, 5222 (1999).

8.21 T. Tomoda, Rep. Prog. Part. Phys. **54**, 53 (1991).

8.22 A. Balysh et al., Phys. Rev. Lett. **77**, 5186 (1996).

8.23 M. Günther et al., Phys. Rev. D **55**, 54 (1997).

8.24 R. Arnold et al., Nucl. Phys. A **636**, 209 (1998).

8.25 A. Kawashima, K. Takahashi and A. Masuda, Phys. Rev. C **47**, 2452 (1993).

8.26 A. De Silva et al., Phys. Rev. C **56**, 2451 (1997).

8.27 R. Arnold et al., Z. Phys. C **72**, 239 (1996).

8.28 T. Bernatowicz et al., Phys. Rev. Lett. **69**, 2341 (1992).

8.29 N. Takaoka, Y. Motomura and K. Nagao, Phys. Rev. C **53**, 1557 (1996).

8.30 A. L. Turkevich, T. E. Economou and G. A. Cowan, Phys. Rev. Lett. **67**, 3211 (1991).

8.31 A. S. Barabash et al., Phys. Lett. B **345**, 408 (1995).

8.32 L. De Braeckeleer et al., preprint, September 1999.

8.33 A. Griffiths and P. Vogel, Phys. Rev. C **46**, 181 (1992).

8.34 L. Baudis et al., Phys. Rev. Lett. **83**, 41 (1999).

8.35 K. E. You et al., Phys. Lett. B **265** 53 (1991).

8.36 A. Staudt, K. Muto and H. V. Klapdor-Kleingrothaus, Europhys. Lett. **13**, 31 (1990).

8.37 S. R. Elliott et al., Phys. Rev. C **46**, 1535 (1992).

8.38 H. Ejiri et al., Nucl. Phys. A **611**, 85 (1996).

8.39 A. Sh. Georgadze et al., Phys. Atom. Nucl. **58**, 1093 (1995).

8.40 J. Engel, P. Vogel and M. R. Zirnbauer, Phys. Rev. C **37**, 731 (1988).

8.41 R. Luescher et al., Phys. Lett. B **434**, 407 (1998).

8.42 A. S. Barabash, Phys. Lett. B **216**, 257 (1989).

8.43 M. Beck et al., Phys. Rev. Lett. **70**, 2853 (1993).

8.44 F. A. Danevich et al., Nucl. Phys. A **643**, 317 (1998).

8.45 B. Pontecorvo, Phys. Lett. B **26**, 630 (1968).

8.46 P. Vogel, M. Ericson and J. D. Vergados, Phys. Lett. B **212**, 259 (1988); K. Muto, Phys. Lett. B **277**, 13 (1992).

8.47 M. Ericson, T. Ericson and P. Vogel, Phys. Lett. B **328**, 259 (1994).

8.48 K. Muto, E. Bender and H. V. Klapdor, Z. Phys. A **334**, 187 (1989).

8.49 E. Caurier, A. Poves, and A. P. Zuker, Phys. Lett. B **252**, 13 (1990).

8.50 E. Caurier, F. Nowacki, A. Poves and J. Retamosa, Phys. Rev. Lett. **77**, 1954 (1996); E. Caurier, private communication.

8.51 J. D. Vergados, Phys. Rev. C **24**, 640 (1981).

8.52 R. N. Mohapatra, Phys. Rev. D **34**, 909 (1986).

8.53 M. Hirsch, H. V. Klapdor-Kleingrothaus and O. Panella, Phys. Lett. B **374**, 7 (1996).

8.54 E. G. Adelberger and W. C. Haxton, Ann. Rev. Nucl. Part. Sci. **35**, 501 (1985).

8.55 M. D. Shuster and M. Rho, Phys. Lett. B **42**, 54 (1972).

8.56 C. E. Aalseth et al., Phys. Rev. C **59**, 2108 (1999).

8.57 X. Sarazin et al., Nucl. Phys. B, Proc. Suppl. **70**, 239 (1999).

8.58 E. Fiorini, Phys. Rep. **307**, 309 (1998).

8.59 H. Ejiri et al., nucl-ex/9911008.

8.60 M. Danilov et al., Phys. Lett. B **480**, 12 (2000).

8.61 D. O. Caldwell and R. N. Mohapatra, Phys. Rev. D **48**, 3259 (1993).

8.62 H. Georgi and S. L. Glashow, hep-ph/9808293.

8.63 J. Ellis and S. Lola, Phys. Lett. B **458**, 310 (1999).

8.64 V. Barger and K. Whisnant, Phys. Lett. B **456**, 194 (1999).

8.65 S. M. Bilenky et al., hep-ph/9907234.

8.66 H. V. Klapdor-Kleingrothaus, H. Päs and A. Yu. Smirnov, hep-ph/0003219.

9 Neutrino Mixing Schemes

V. Barger and K. Whisnant

9.1 Introduction

A revolution in our understanding of the neutrino sector is under way, driven by observations that are interpreted in terms of changes in the flavors of neutrinos as they propagate. Since neutrino oscillations occur only if neutrinos are massive, these phenomena indicate physics beyond the Standard Model. With the present evidence for oscillations from atmospheric, solar and accelerator data, we are already able to begin to make strong inferences about the mass spectrum and the mixings of neutrinos. Theoretical efforts to achieve a synthesis have produced a variety of models with differing testable consequences. A combination of particle physics, nuclear physics and astrophysics is needed for a full determination of the fundamental properties of neutrinos. This chapter reviews what has been achieved thus far and the future prospects for understanding the nature of neutrino masses and mixing.

9.2 Two-Neutrino Analyses

In a model with two neutrinos, the probability of a given neutrino flavor ν_α oscillating into ν_β in a vacuum is

$$P(\nu_\alpha \to \nu_\beta) = \sin^2 2\theta \sin^2 \left(1.27 \frac{\delta m^2 L}{E}\right) , \qquad (9.1)$$

where θ is the two-neutrino mixing angle, δm^2 is the mass-squared difference of the two mass eigenstates in eV^2, L is the distance from the neutrino source to the detector in kilometers and E is the neutrino energy in GeV.

9.2.1 Atmospheric Neutrinos

The atmospheric neutrino experiments determine the ratios

$$\frac{N_\mu}{N_\mu^0} = \alpha \left[\langle P(\nu_\mu \to \nu_\mu)\rangle + r \langle P(\nu_e \to \nu_\mu)\rangle\right] , \qquad (9.2)$$

$$\frac{N_e}{N_e^0} = \alpha \left[\langle P(\nu_e \to \nu_e)\rangle + r^{-1} \langle P(\nu_\mu \to \nu_e)\rangle\right] , \qquad (9.3)$$

where N_e^0 and N_μ^0 are the expected numbers of atmospheric e and μ events, respectively, in the absence of oscillations; $r \equiv N_e^0/N_\mu^0$; $\langle\,\rangle$ indicates an average over the neutrino spectrum; and α is an overall neutrino flux normalization. Atmospheric neutrino data have generally indicated that the expected number of muons detected is suppressed relative to the expected number of electrons [9.1]; this suppression can be explained via neutrino oscillations [9.2].

The atmospheric data also indicate that N_e/N_e^0 is relatively flat with respect to zenith angle, while N_μ/N_μ^0 decreases with increasing zenith angle (i.e. with longer oscillation distance). Assuming $\nu_\mu \to \nu_\tau$ oscillations, the favored two-neutrino parameters are [9.3]

$$\delta m^2 = 3.5 \times 10^{-3}\,\mathrm{eV}^2 \quad (1.5\text{–}7 \times 10^{-3}\,\mathrm{eV}^2)\,, \tag{9.4}$$

$$\sin^2 2\theta = 1.00 \quad (0.80\text{–}1.00)\,; \tag{9.5}$$

the 90% C.L. allowed ranges are given in parentheses. The absolute normalization of the electron data indicates $\alpha \simeq 1.18$, which is within the theoretical uncertainties [9.4]. The flatness versus L of the electron data implies that simple $\nu_\mu \to \nu_e$ oscillations are strongly disfavored. Large-amplitude ($\sin^2 2\theta > 0.2$) $\nu_\mu \to \nu_e$ oscillations are also excluded by the Chooz reactor data [9.5] for $\delta m_{\mathrm{atm}}^2 \gtrsim 10^{-3}\,\mathrm{eV}^2$.

It is also possible that atmospheric ν_μ are oscillating into sterile neutrinos. However, measurements of the upgoing zenith angle distribution and π^0 production disfavor this possibility ([9.6] and Chap. 5 of this book).

9.2.2   Solar Neutrinos

For the ^{37}Cl [9.7] and ^{71}Ga [9.8] experiments, the expected number of neutrino events is

$$N = \int \sigma P(\nu_e \to \nu_e)(\beta\phi_{\mathrm{B}} + \phi_{\mathrm{non\text{-}B}})\,\mathrm{d}E\,, \tag{9.6}$$

where we allow an arbitrary normalization factor β for the ^{8}B neutrino flux since its normalization is not certain. For the Kamiokande [9.9] and Super-Kamiokande [9.10] experiments the interaction is $\nu e \to \nu e$ and the outgoing electron energy is measured. The number of events per unit of electron energy is then

$$\frac{\mathrm{d}N}{\mathrm{d}E_e} = \beta \int \left(\frac{\mathrm{d}\sigma_{\mathrm{CC}}}{\mathrm{d}E_e'} P(\nu_e \to \nu_e) + \frac{\mathrm{d}\sigma_{\mathrm{NC}}}{\mathrm{d}E_e'}\left[1 - P(\nu_e \to \nu_e)\right]\right)$$
$$\times G\left(E_e', E_e\right) \phi_{\mathrm{B}}\,\mathrm{d}E_\nu\,\mathrm{d}E_e'\,, \tag{9.7}$$

where $\mathrm{d}\sigma_{\mathrm{CC}}/\mathrm{d}E_e'$ and $\mathrm{d}\sigma_{\mathrm{NC}}/\mathrm{d}E_e'$ are the charged-current and neutral-current differential cross sections, respectively, for an incident neutrino energy E_ν, and $G(E_e', E_e)$ is the probability that an electron of energy E_e' is measured as having energy E_e. The neutrino fluxes are taken from the standard solar model (SSM) [9.11]. If ν_e oscillates into a sterile neutrino, $\sigma_{\mathrm{NC}} = 0$.

For two-neutrino vacuum oscillations (VOs) of ν_e into either ν_μ or ν_τ [9.12], the solar data indicate [9.13, 9.14]

$$\delta m^2 = 7.5 \times 10^{-11}\,\mathrm{eV}^2\,, \tag{9.8}$$

$$\sin^2 2\theta = 0.91\,, \tag{9.9}$$

although there are also regions near $\delta m^2 = 2.5 \times 10^{-10}$, 4.4×10^{-10} and 6.4×10^{-10} eV^2 that give acceptable fits.

For two-neutrino MSW oscillations [9.15] of ν_e into either ν_μ or ν_τ, there are three possible solution regions [9.14]. The small-angle MSW (SAM) solution is

$$\delta m^2 = 7.5 \times 10^{-6}\,\mathrm{eV}^2\,, \tag{9.10}$$

$$\sin^2 2\theta = 0.01\,, \tag{9.11}$$

and the large-angle MSW (LAM) solution is

$$\delta m^2 \sim 10^{-5}\,\mathrm{eV}^2\,, \tag{9.12}$$

$$\sin^2 2\theta = 0.6\text{--}0.9\,. \tag{9.13}$$

There is also a low-δm^2 MSW (LOW) solution,

$$\delta m^2 \sim 10^{-7}\,\mathrm{eV}^2\,, \tag{9.14}$$

$$\sin^2 2\theta = 0.6\text{--}0.9\,, \tag{9.15}$$

although it is less favored. Note that all solutions except for the small-angle MSW one have large mixing in the solar sector.

Two-neutrino oscillations of ν_e into a sterile neutrino are somewhat disfavored because sterile neutrinos do not have an NC interaction, which tends to make the oscillation predictions for the ^{37}Cl and SuperK experiments similar, contrary to experimental evidence. Future measurements of NC interactions in the SNO detector [9.16] will provide a more thorough test for oscillations of solar ν_e into sterile neutrinos.

SuperK and SNO will also provide an improved measurement of the ^{8}B neutrino spectrum in the near future, which should help distinguish the various solar scenarios. The Borexino [9.17] and ICARUS [9.18] experiments will provide a measurement of the ^{7}Be neutrinos, and could detect the strong seasonal dependence that exists for ^{7}Be neutrinos in many VO models [9.19].

9.2.3   LSND

There are also indications of neutrino oscillations from the LSND accelerator experiment [9.20]. The data from this experiment suggest $\nu_\mu \to \nu_e$ oscillations with the following two-neutrino parameters:

$$0.3\,\mathrm{eV}^2 \leq \delta m^2 = \frac{0.03\,\mathrm{eV}^2}{(\sin^2 2\theta)^{0.7}} \leq 2.0\,\mathrm{eV}^2\,. \tag{9.16}$$

Values of δm^2 above 10 eV^2 are also allowed for $\sin^2 2\theta \simeq 0.0025$ [9.21]. The future MiniBooNE experiment [9.22] is expected to either confirm or refute the LSND result.

9.3   Global Analyses

A complete description of all neutrino data requires a model that can explain all of the oscillation phenomena simultaneously. Since each of the three types of oscillation evidence (atmospheric, solar and LSND) requires a distinct δm^2 scale, and since N neutrinos have only $N-1$ independent mass-squared differences, four neutrinos are needed to completely explain all of the data. If one of the types of neutrino data is ignored, then it is in principle possible to explain the remaining data with a three-neutrino model. Because the LSND results have yet to be confirmed by another experiment, three-neutrino models are generally used in the context of describing the atmospheric and solar data.

9.3.1   Three-Neutrino Models

If the atmospheric and solar data are to be described by a three-neutrino model, there are two distinct mass-squared difference scales (compare (9.4) with (9.8), (9.10), (9.12) and (9.14)). The two possible patterns of three neutrino masses are illustrated in Fig. 9.1. We assume without loss of generality that $|m_1| < |m_2| < |m_3|$, and that $\delta m_{21}^2 \equiv \delta m_{\text{sun}}^2$ and $\delta m_{31}^2 \simeq \delta m_{32}^2 \equiv \delta m_{\text{atm}}^2 \gg \delta m_{\text{sun}}^2$. Then the off-diagonal vacuum oscillation probabilities in a three-neutrino model are [9.23]

$$P(\nu_e \to \nu_\mu) = 4\left|U_{e3}U_{\mu3}^*\right|^2 \sin^2 \Delta_{\text{atm}} - 4\,\text{Re}\left(U_{e1}U_{e2}^*U_{\mu1}^*U_{\mu2}\right)\sin^2 \Delta_{\text{sun}}$$
$$-2J\sin 2\Delta_{\text{sun}}\,, \tag{9.17}$$

$$P(\nu_e \to \nu_\tau) = 4\left|U_{e3}U_{\tau2}^*\right|^2 \sin^2 \Delta_{\text{atm}} - 4\,\text{Re}\left(U_{e1}U_{e2}^*U_{\tau1}^*U_{\tau2}\right)\sin^2 \Delta_{\text{sun}}$$
$$+2J\sin 2\Delta_{\text{sun}}\,, \tag{9.18}$$

$$P(\nu_\mu \to \nu_\tau) = 4\left|U_{\mu3}U_{\tau3}^*\right|^2 \sin^2 \Delta_{\text{atm}} - 4\,\text{Re}\left(U_{\mu1}U_{\mu2}^*U_{\tau1}^*U_{\tau2}\right)\sin^2 \Delta_{\text{sun}}$$
$$-2J\sin 2\Delta_{\text{sun}}\,, \tag{9.19}$$

where U is the neutrino mixing matrix (in the basis where the charged-lepton mass matrix is diagonal), $\Delta_j \equiv 1.27(\delta m_j^2/\text{eV}^2)(L/\text{km})(E/\text{GeV})$ and

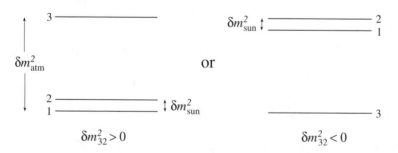

Fig. 9.1. Two possible patterns of neutrino masses that can explain the atmospheric and solar anomalies

$J = \mathrm{Im}\left(U_{e2}U_{e3}^{*}U_{\mu2}^{*}U_{\mu3}\right)$ is the CP-violating invariant [9.24]. For a discussion of CP-violating effects in neutrino oscillations see [9.25–9.27]

The matrix elements $U_{\alpha j}$ are the mixings between the flavor ($\alpha = e, \mu, \tau$) and mass ($j = 1, 2, 3$) eigenstates. The CP-odd term changes sign under reversal of the oscillating flavors, or if neutrinos are replaced by antineutrinos. For either Dirac or Majorana neutrinos, we choose the following parametrization for U to describe neutrino oscillations:

$$U = \begin{pmatrix} c_{13}c_{12} & c_{13}s_{12} & s_{13}e^{-i\delta} \\ -c_{23}s_{12} - s_{13}s_{23}c_{12}e^{i\delta} & c_{23}c_{12} - s_{13}s_{23}s_{12}e^{i\delta} & c_{13}s_{23} \\ s_{23}s_{12} - s_{13}c_{23}c_{12}e^{i\delta} & -s_{23}c_{12} - s_{13}c_{23}s_{12}e^{i\delta} & c_{13}c_{23} \end{pmatrix}, \quad (9.20)$$

where $c_{jk} \equiv \cos\theta_{jk}$ and $s_{jk} \equiv \sin\theta_{jk}$. For Majorana neutrinos, U contains two further phase factors, but these do not enter into oscillation phenomena.

For the oscillation of neutrinos in atmospheric and long-baseline experiments with $L/E \gtrsim 10^2\,\mathrm{km/GeV}$, the Δ_{sun} terms are negligible and the relevant vacuum oscillation probabilities are

$$P(\nu_e \to \nu_\mu) = s_{23}^2 \sin^2 2\theta_{13} \sin^2 \Delta_{\mathrm{atm}}, \tag{9.21}$$

$$P(\nu_e \to \nu_\tau) = c_{23}^2 \sin^2 2\theta_{13} \sin^2 \Delta_{\mathrm{atm}}, \tag{9.22}$$

$$P(\nu_\mu \to \nu_\tau) = c_{13}^4 \sin^2 2\theta_{23} \sin^2 \Delta_{\mathrm{atm}}. \tag{9.23}$$

The diagonal oscillation probabilities for a given neutrino species can be found by subtracting all the off-diagonal probabilities involving that species from unity.

For neutrinos from the sun, $L/E \sim 10^{10}\,\mathrm{km/GeV}$, and the $\sin^2\Delta_{\mathrm{atm}}$ terms oscillate very rapidly, averaging to $1/2$. Then the observable oscillation probability in a vacuum is

$$P(\nu_e \to \nu_e) = 1 - \frac{1}{2}\sin^2 2\theta_{13} - c_{13}^4 \sin^2 2\theta_{12} \sin^2 \Delta_{\mathrm{sun}}. \tag{9.24}$$

When $\theta_{13} = 0$ (i.e. $U_{13} = 0$), the oscillations of atmospheric and long-baseline neutrinos decouple from those of solar neutrinos: at the $\delta m_{\mathrm{atm}}^2$ scale, there are pure $\nu_\mu \to \nu_\tau$ oscillations with amplitude $\sin^2 2\theta_{23}$, with no admixture of ν_e, and at the $\delta m_{\mathrm{sun}}^2$ scale the ν_e oscillations are described by a simple two-neutrino formula with amplitude $\sin^2 2\theta_{12}$. Then the two-neutrino parameters for atmospheric and solar oscillations can be adopted directly in the three-neutrino case.

If $\theta_{13} \neq 0$, then ν_e will participate in atmospheric and long-baseline oscillations. As discussed earlier, pure $\nu_\mu \to \nu_e$ oscillations at the $\delta m_{\mathrm{atm}}^2$ scale are strongly disfavored by the atmospheric data, but some small amount of $\nu_\mu \to \nu_e$ is still allowed. The Chooz reactor experiment measures $\bar{\nu}_e$ survival, and is sensitive to oscillations between $\bar{\nu}_e$ and $\bar{\nu}_\mu$ for $\delta m_{\mathrm{atm}}^2 > 10^{-3}\,\mathrm{eV}^2$. The combined data from atmospheric experiments and Chooz favor $\sin\theta_{13} = 0$ (i.e. pure $\nu_\mu \to \nu_\tau$ oscillations in the atmosphere) and suggest that $\sin\theta_{13} < 0.3$ [9.23].

The solar data also allow solar neutrinos to mix maximally, or nearly maximally. If we require both atmospheric and solar oscillations to be maximal, there is a unique three-neutrino solution to the neutrino mixing matrix [9.28]. This "bimaximal" mixing corresponds to $\theta_{13} = 0$ and $|\theta_{12}| = |\theta_{23}| = \pi/4$, and is a special case of the decoupled solution for atmospheric and solar neutrinos. One interesting aspect of the bimaximal solution is that the solar ν_e oscillations are 50% into ν_μ and 50% into ν_τ, although the flavor content of the ν_e oscillation is not observable in solar experiments. Further properties of the bimaximal and nearly bimaximal solutions and models that give rise to such solutions are discussed in [9.28].

9.3.2 Four-Neutrino Models

As discussed earlier, four neutrinos are necessary to completely describe the atmospheric, solar and LSND results. A fourth light neutrino must be sterile since it is not observed in Z decays [9.29]. There must be three separate mass-squared scales, which must satisfy the hierarchy $\delta m_{\mathrm{sun}}^2 \ll \delta m_{\mathrm{atm}}^2 \ll \delta m_{\mathrm{LSND}}^2$. In the three-neutrino case the relation $\delta m_{\mathrm{sun}}^2 \ll \delta m_{\mathrm{atm}}^2$ leads to only one fundamental type of mass structure, in which one heavier mass is separated from two lighter, nearly degenerate states (or vice versa). In the four-neutrino case, however, there are two distinct types of mass structure: one heavier mass separated from three lighter, nearly degenerate states, or vice versa (which we shall refer to as the $1 + 3$ spectrum), and two nearly degenerate mass pairs separated from each other (the $2 + 2$ spectrum). In each case, the largest separation scale is determined by the LSND scale. It can be shown [9.30, 9.31] that only the $2 + 2$ spectrum is completely consistent with the positive oscillation signals in the solar, atmospheric and LSND experiments, and with the negative results from the Bugey reactor [9.32] and CDHS accelerator [9.33] experiments. Therefore our discussions below assume the $2 + 2$ case, which is illustrated in Fig. 9.2. Here we shall assume that the mass splitting of the heavier pair drives the atmospheric oscillations, the mass splitting of the lighter pair drives the solar oscillations, and the separation of the two mass

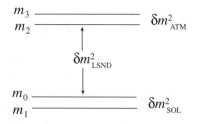

Fig. 9.2. The mass pattern of two separated pairs required to account for the LSND, atmospheric and solar data; the locations of $\delta m_{\mathrm{ATM}}^2$ and $\delta m_{\mathrm{SUN}}^2$ may be interchanged

pairs drives the LSND oscillations. Sterile neutrinos are also of interest in explaining r-process nucleosynthesis via supernova explosions (see e.g. [9.21]).

The vacuum neutrino flavor oscillation probabilities, for oscillation of an initially produced ν_α to a finally detected ν_β, can be written [9.25]

$$P(\nu_\alpha \to \nu_\beta) = \delta_{\alpha\beta} - \sum_{j<k}\left[4\,\mathrm{Re}(W_{\alpha\beta}^{jk})\sin^2\Delta_{kj} - 2\,\mathrm{Im}(W_{\alpha\beta}^{jk})\sin 2\Delta_{kj}\right],$$

$$(9.25)$$

where

$$W_{\alpha\beta}^{jk} = U_{\alpha j}U_{\alpha k}^* U_{\beta j}^* U_{\beta k}, \tag{9.26}$$

and $\Delta_{kj} \equiv 1.27\,\delta m_{kj}^2 L/E$. We assume that there are four mass eigenstates m_0, m_1, m_2 and m_3, which are most closely associated with the flavor states ν_s, ν_e, ν_μ and ν_τ, respectively. The solar oscillations are driven by δm_{01}^2, the atmospheric oscillations by δm_{32}^2 and the LSND oscillations by $\delta m_{02}^2 \simeq \delta m_{03}^2 \simeq \delta m_{12}^2 \simeq \delta m_{13}^2$. Hence, the solar oscillations are primarily $\nu_e \to \nu_s$ and the atmospheric oscillations are primarily $\nu_\mu \to \nu_\tau$. For oscillations of solar ν_e to sterile neutrinos, the solar data disfavor large mixing; hence the most likely solar solution is MSW small mixing. It is also possible to reverse the roles of ν_s and ν_τ; however, current data disfavor oscillations of atmospheric ν_μ to sterile neutrinos [9.6].

The 4×4 mixing matrix U may be parameterized by six mixing angles (θ_{01}, θ_{02}, θ_{03}, θ_{12}, θ_{13}, θ_{23}) and six phases (δ_{01}, δ_{02}, δ_{03}, δ_{12}, δ_{13}, δ_{23}); only three of the phases are observable in neutrino oscillations. A complete analysis of the four-neutrino mixing matrix is quite complicated. However, the smallness of the mixing indicated by the LSND results suggests that ν_e does not mix strongly with the two heavier states, i.e. that θ_{12} and θ_{13} are small. Similarly, one can assume that the other light state does not mix strongly with the heavier states, i.e. θ_{02} and θ_{03} are also small. This situation occurs naturally in the explicit four-neutrino models in the literature. Then, after dropping terms of second order in small mixing angles, U takes the form [9.25]

$$U = \begin{pmatrix} c_{01} & s_{01}^* & s_{02}^* & s_{03}^* \\[4pt] -s_{01} & c_{01} & s_{12}^* & s_{13}^* \\[4pt] \begin{matrix}-c_{01}(s_{23}^*s_{03}+c_{23}s_{02}) \\ +s_{01}(s_{23}^*s_{13}+c_{23}s_{12})\end{matrix} & \begin{matrix}-s_{01}^*(s_{23}^*s_{03}+c_{23}s_{02}) \\ -c_{01}(s_{23}^*s_{13}+c_{23}s_{12})\end{matrix} & c_{23} & s_{23}^* \\[4pt] \begin{matrix}c_{01}(s_{23}s_{02}-c_{23}s_{03}) \\ -s_{01}(s_{23}s_{12}-c_{23}s_{13})\end{matrix} & \begin{matrix}s_{01}^*(s_{23}s_{02}c_{23}s_{03}) \\ +c_{01}(s_{23}s_{12}-c_{23}s_{13})\end{matrix} & -s_{23} & c_{23} \end{pmatrix}, \tag{9.27}$$

where $c_{jk} \equiv \cos\theta_{jk}$ and $s_{jk} \equiv \sin\theta_{jk}e^{i\delta_{jk}}$. We see that under these conditions, U has approximately block diagonal form. The mixing of solar neutrinos is due to θ_{01} and the mixing of atmospheric neutrinos is due to θ_{23}; the values for these mixing angles are essentially given by the two-neutrino fits in Sect. 9.2. Both vacuum and MSW solar oscillation solutions are allowed; for MSW oscillations to occur in the sun, $m_0 > m_1$ is required. The mixing that drives the LSND oscillations is due to θ_{12} and θ_{13}; the effective amplitude of the $\nu_\mu \to \nu_e$ oscillations in the LSND case is

$$\sin^2 2\theta_{\mathrm{LSND}} \simeq 4|s_{12}c_{23} + s_{13}s_{23}^*|^2 \ . \tag{9.28}$$

The assumption that θ_{02} and θ_{03} are small is not required by current data. However, most explicit models have the approximate form given by (9.27); see Sect. 9.4.2. If in fact θ_{02} and θ_{03} are not small, ν_e oscillations to the flavor associated with the other light state (ν_s or ν_τ) are possible at the LSND L/E scale with an amplitude of the same order as the LSND oscillation amplitude.

9.4   Consequences for Masses and Mixings

9.4.1   Three-Neutrino Models

The atmospheric and solar data put restrictions on the neutrino mixing angles and mass-squared differences, but do not at all constrain the absolute neutrino mass scale, which must be determined by other methods. The freedom to choose the mass scale allows a wide variety of possible mass matrix structures, even for the same mass-squared differences and mixing.

The matrix U that relates neutrino flavor eigenstates to neutrino mass eigenstates depends in general on mixing in both the neutrino and the charged-lepton sectors. If U_ℓ diagonalizes the charged-lepton mass matrix and U_ν the neutrino mass matrix, then $U = U_\ell^\dagger U_\nu$. Except where stated otherwise, in our discussions here we shall work in the basis where the charged-lepton mass matrix is diagonal, so that $U = U_\nu$. In general, the neutrino mass matrix in the flavor basis can then be written $M_{\alpha\beta} = \sum_j U_{\alpha j} m_j U_{\beta j}$ for Majorana neutrinos or $M_{\alpha\beta} = \sum_j U_{\alpha j} m_j U_{\beta j}^*$ for Dirac neutrinos (these are the same if CP is conserved, i.e. when U is real).

As an example of the different mass matrix structures that are possible, we consider the case where at least one of the masses is much smaller than the largest mass. Then there is one type of mass matrix of the form $M = M_0 + O(\delta m_{jk}^2)$ (up to trivial sign changes) that can lead to maximal mixing of atmospheric neutrinos:

$$M_0 = \frac{m}{2} \begin{pmatrix} 0 & 0 & 0 \\ 0 & a & b \\ 0 & b & a \end{pmatrix} , \tag{9.29}$$

where $a, b \sim 1$. If $a = b$, then there is only one large mass ($m_1, m_2 \ll m_3 \equiv m$) and the form of (9.29) automatically fixes $\theta_{23} \simeq \pi/4$ and $\theta_{13} \simeq 0$; the $O(\delta m_{jk}^2)$

terms determine θ_{12}. Bimaximal mixing ($\theta_{12} \simeq \pi/4$) is obtained if the leading $O(\delta m_{jk}^2)$ terms have the form

$$\Delta M = \epsilon \begin{pmatrix} 0 & -1 & 1 \\ -1 & 0 & 0 \\ 1 & 0 & 0 \end{pmatrix}, \tag{9.30}$$

where $\epsilon \ll m$; subleading $O(\delta m_{jk}^2)$ terms are then needed to split m_1 and m_2. If $a \neq b$ in (9.29), then there are two large masses, with $\theta_{23} \simeq \pi/4$ and $\theta_{13} \simeq 0$. Small θ_{12}, appropriate for the small-angle MSW solar solution, is achieved if the leading $O(\delta m_{jk}^2)$ terms have the form

$$\Delta M = \epsilon \begin{pmatrix} 0 & 1 & 1 \\ 1 & 0 & 0 \\ 1 & 0 & 0 \end{pmatrix}; \tag{9.31}$$

see [9.34] for an example of a GUT model that has this form.

In the situation where all masses are approximately degenerate, $m \equiv |m_1| \simeq |m_2| \simeq |m_3| \gg \delta m_{jk}^2$, there are three different types of mass matrices of the form $M = M_0 + O(\delta m_{jk}^2)$ (up to trivial changes in sign) that can lead to bimaximal mixing, depending on the relative signs of the m_j:

$$M_0 = m \begin{pmatrix} 0 & -1/\sqrt{2} & 1/\sqrt{2} \\ -1/\sqrt{2} & 1/2 & 1/2 \\ 1/\sqrt{2} & 1/2 & 1/2 \end{pmatrix} \text{ or } m \begin{pmatrix} 1 & 0 & 0 \\ 0 & 0 & 1 \\ 0 & 1 & 0 \end{pmatrix} \text{ or } m \begin{pmatrix} 1 & 0 & 0 \\ 0 & 1 & 0 \\ 0 & 0 & 1 \end{pmatrix}. \tag{9.32}$$

The requirement from neutrinoless double beta decay that the $\nu_e \nu_e$ element of the neutrino mass matrix be small (described below) implies that only the first case is allowed for Majorana neutrinos. The form of M_0 gives three degenerate neutrinos of mass m and fixes two combinations of mixing angles ($c_{13}s_{12} \simeq -1/\sqrt{2}$ and $c_{23}c_{12} - s_{13}s_{23}s_{12} \simeq 1/2$), while the remaining degrees of freedom among the mixing angles and the neutrino mass splittings are determined by the $O(\delta m_{jk}^2)$ terms. If the leading $O(\delta m_{jk}^2)$ terms are proportional to the mass matrix in (9.29) with $a = b$, m_3 is split from m_1 and m_2, $\theta_{13} \simeq 0$ and bimaximal mixing is obtained. Subleading $O(\delta m_{jk}^2)$ terms then provide the splitting between m_1 and m_2.

A different mixing scheme occurs if the neutrino mass matrix is approximately proportional to unity and the charged-lepton mass matrix is close to the so-called democratic form [9.35],

$$M_\ell = \frac{m_\ell}{3} \begin{pmatrix} 1 & 1 & 1 \\ 1 & 1 & 1 \\ 1 & 1 & 1 \end{pmatrix}; \tag{9.33}$$

then there is one large eigenvalue $m_\ell \simeq m_\tau$ and two constraints on the flavor mixing angles, $c_{13}c_{23} \simeq 1/\sqrt{3}$ and $s_{23}s_{12} - s_{13}c_{23}c_{12} \simeq 1/\sqrt{3}$. If a small perturbation is added to the lower right element of M_ℓ, the muon acquires mass and θ_{13} is constrained to be approximately zero; then there is maximal mixing in the solar sector ($\sin^2 2\theta_{12} = 1$) and nearly maximal mixing in the atmospheric sector ($\sin^2 2\theta_{23} = 8/9$) [9.35]. Additional perturbations to the diagonal elements of the charged-lepton and neutrino mass matrices are then needed to give the electron–muon and neutrino mass splittings, respectively.

An $SO(10)$ SUSY GUT model with a minimal Higgs sector can provide large ν_μ–ν_τ mixing for atmospheric neutrinos, and can accommodate either small or large mixing of solar neutrinos [9.36]. Other possible neutrino mass matrix textures are discussed in [9.37], [9.38] and references therein, [9.39] and Chap. 10.

Although neutrino oscillations are not sensitive to the overall neutrino mass scale, there are other processes that do depend on absolute masses. For example, studies of the tritium beta decay spectrum put an upper limit on the effective mass of the electron neutrino,

$$m_{\nu_e} \equiv \sum_j |U_{ej}|^2 m_j \,, \tag{9.34}$$

of about 2.5 eV [9.40]; in a three-neutrino model this implies an upper limit of 2.5 eV on the heaviest neutrino [9.41].

The current limit on the magnitude of the ν_e–ν_e element of the neutrino mass matrix for Majorana neutrinos obtained from neutrinoless double beta ($0\nu\beta\beta$) decay [9.42] is of order 0.5 eV [9.43], taking into account the imprecise knowledge of the nuclear matrix element and the sensitivity of a background fluctuation analysis [9.44]. For the three-neutrino model this implies

$$|M_{\nu_e\nu_e}| = |\sum_j U_{ej} m_j U_{ej}|$$

$$= |(c_{13}c_{12})^2 m_1 + (c_{13}s_{12})^2 m_2 e^{i\phi_2} + s_{13}^2 m_3 e^{i\phi_3}|$$

$$\leq M_{\max} = 0.2 \text{ eV} \,, \tag{9.35}$$

where ϕ_2 and ϕ_3 are extra phases present for Majorana neutrinos that are not observable in neutrino oscillations. For models with one large mass, $m_1, m_2 \ll m_3 \simeq \sqrt{\delta m_{\mathrm{atm}}^2} \simeq 0.05$ eV, and (9.35) does not provide any additional constraint. However, if all three masses are nearly degenerate ($m_1 \simeq m_2 \simeq m_3 \equiv m$), the $0\nu\beta\beta$ decay limit becomes $s_{12}^2 \geq [1 - 2s_{13}^2 - (M_{\max}/m)]/(2c_{13}^2)$, which in turn implies that the solar $\nu_e \to \nu_e$ oscillation amplitude (see (9.24)) has the constraint [9.45]

$$4c_{13}^4 s_{12}^2 c_{12}^2 \geq 1 - \left(\frac{M_{\max}}{m}\right)^2 - 2s_{13}^2 \left(1 + \frac{M_{\max}}{m}\right) \,. \tag{9.36}$$

For any value of $m > M_{\max}/(1 - 2s_{13}^2)$ there will be a lower limit on the size of the solar $\nu_e \to \nu_e$ oscillation amplitude; for example, given the current limit on s_{13}, the small-angle MSW solar solution is ruled out for nearly degenerate

Majorana neutrinos if $m > 0.25$ eV [9.45]. Large-angle MSW and vacuum solar solutions, which have large mixing, are still allowed; any solar solution with maximal mixing can never be excluded by this bound.

Neutrino mass also provides an ideal hot-dark-matter component [9.46]; the contribution of neutrinos to the mass density of the Universe is given by $\Omega_\nu = \sum m_\nu/(h^2 \times 93 \text{ eV})$, where h is the Hubble expansion parameter in units of 100 km/(s Mpc) [9.47]. For $h = 0.65$, the model with three nearly degenerate neutrinos has $\Omega_\nu \simeq 0.075(m/\text{eV})$. In three-neutrino models with hierarchical masses, in which the largest mass is of order $\sqrt{\delta m_{\text{atm}}^2}$, Ω_ν is much smaller, on the order of 0.001. Future measurements of the hot-dark-matter component may be sensitive to $\sum m_\nu$ down to 0.4 eV [9.48].

9.4.2 Four-Neutrino Models

As described in Sect. 9.3.2, four-neutrino models must have the $2 + 2$ mass structure, i.e. two nearly degenerate pairs separated from each other by the LSND scale. One possible class of mass matrices that can give this situation is

$$
M = m \begin{pmatrix} \epsilon_1 & \epsilon_2 & 0 & 0 \\ \epsilon_2 & 0 & 0 & \epsilon_3 \\ 0 & 0 & \epsilon_4 & 1 \\ 0 & \epsilon_3 & 1 & \epsilon_5 \end{pmatrix} , \tag{9.37}
$$

presented in the $(\nu_{\text{s}}, \nu_e, \nu_\mu, \nu_\tau)$ basis (i.e. the basis in which the charged-lepton mass matrix is diagonal).

When $\epsilon_5 = \epsilon_4$, the mass matrix in (9.37) can accommodate any of the three solar solutions, depending on the hierarchy of the mass matrix elements [9.31]

$$\text{SAM}: \quad \epsilon_2 \ll \epsilon_1, \epsilon_4 \ll \epsilon_3 \ll 1 , \tag{9.38}$$

$$\text{LAM}: \quad \epsilon_1, \epsilon_2, \epsilon_4 \ll \epsilon_3 \ll 1 , \tag{9.39}$$

$$\text{VO}: \epsilon_1 \ll \epsilon_2 \ll \epsilon_4 \ll \epsilon_3 \ll 1 . \tag{9.40}$$

In each case, the mass eigenvalues have the hierarchy $m_1 < m_0 \ll m_2, m_3$, as required for the MSW solution. The mass matrix contains five parameters, just enough to incorporate the required three mass-squared differences and the oscillation amplitudes for the solar and LSND neutrinos. The large amplitude for atmospheric oscillations does not require an additional parameter, since the mass matrix naturally gives nearly maximal mixing of ν_μ with ν_τ. For the VO case, ϵ_1 does not contribute to the phenomenology and can be set to zero, so that there are only four parameters; the large amplitude for solar oscillations also occurs naturally in this case [9.31, 9.49].

Another variant with five parameters is $\epsilon_5 = -\epsilon_4$ and $\epsilon_2 \ll \epsilon_1 \ll \epsilon_3 \ll \epsilon_4 \sim 1$ [9.50]. In this case, ϵ_3 both determines the amplitude of the LSND oscillations *and* causes the splitting between m_2 and m_3 that drives the atmospheric oscillations. Two testable consequences of this model are that $\delta m^2_{\text{atm}} \lesssim 1.3 \times 10^{-3}$ eV^2 and that there should be observable $\nu_e \to \nu_\tau$ oscillations in short-baseline experiments.

For both of the previous cases ($\epsilon_5 = \epsilon_4 \ll 1$ and $-\epsilon_5 = \epsilon_4 \sim 1$), the heavier-mass pair is much heavier than the lighter-mass pair, i.e. $m_1 < m_0 \ll m_2, m_3$. Yet another possibility is to have $\epsilon_1 = \epsilon_5 = 0$ and $\epsilon_3, \epsilon_4 \ll \epsilon_2 < 1$, where ϵ_2 is not small compared with unity [9.49]. In this case, the two degenerate pairs of masses have mass eigenvalues that are the same order of magnitude; there are only four parameters as both the solar and the atmospheric mixings are naturally close to maximal. However, the m_0–m_1 splitting in this model can give only MSW solar oscillations, which for large mixing are disfavored when $\nu_e \to \nu_s$.

Other four-neutrino mass matrix ansatzes have been presented in the literature [9.51], but they generally have characteristics similar to those discussed here. In all of the explicit four-neutrino models discussed above, since the ν_e–ν_e element of the neutrino mass matrix vanishes, there is no contribution to neutrinoless double beta decay. However, because m_2 and m_3 are always of order 1 eV or more (to provide the necessary mass-squared splitting for the LSND oscillations), these models always contribute to the hot dark matter in the Universe [9.52].

9.5   Long-Baseline Experiments

Long-baseline experiments (with $L/E_\nu \sim 10^2$–10^3 km/GeV) are expected to confirm the $\nu_\mu \to \nu_\mu$ disappearance oscillations at the δm^2_{atm} scale. The K2K experiment [9.53] from KEK to SuperK, with a baseline of $L \simeq 250$ km and a mean neutrino energy of $\langle E_\nu \rangle \sim 1.4$ GeV, is under way. The MINOS experiment from Fermilab to Soudan [9.54], with a longer baseline $L \simeq 730$ km and higher mean energies $\langle E_\nu \rangle = 3$, 6 or 12 GeV, is under construction, and the ICANOE [9.55] and OPERA [9.56] experiments, with similar baselines from CERN to Gran Sasso, have been approved. These experiments with dominant ν_μ and $\bar{\nu}_\mu$ beams will securely establish the oscillation phenomena and may measure δm^2_{atm} to a precision of order 10% [9.57]. They will also measure the dominant oscillation amplitude $\sin^2 2\theta_{23}$.

Further tests of the neutrino parameters are likely to require higher-intensity neutrino beams, and ν_e, $\bar{\nu}_e$ as well as ν_μ, $\bar{\nu}_\mu$ beams, such as those generated in a neutrino factory [9.26, 9.58–9.60]. The ν_e, $\bar{\nu}_e$ components of the beam allow one to search for $\nu_e \to \nu_\mu$ and $\bar{\nu}_e \to \bar{\nu}_\mu$ appearance, which will occur in the leading δm^2_{atm} oscillation if $\sin^2 2\theta_{13} \neq 0$. Depending on the machine parameters, δm^2_{atm} and $\sin^2 2\theta_{23}$ can be measured to an accuracy of 1–2%, and $\sin^2 2\theta_{13}$ can be tested down to 0.01 or below [9.60]. If the baseline is long enough ($L \gtrsim 1000$ km), matter effects in the earth will also play an important role in an appearance experiment: for $\delta m^2_{\text{atm}} > 0$ and $\delta m^2_{\text{atm}} < 0$, $\nu_e \to \nu_\mu$ oscillations are enhanced and suppressed, respectively, and $\bar{\nu}_e \to \bar{\nu}_\mu$ oscillations are suppressed and enhanced, respectively. Therefore by comparing $\nu_e \to \nu_\mu$ with $\bar{\nu}_e \to \bar{\nu}_\mu$ oscillations it may be possible to determine the sign of δm^2_{atm} [9.60]. The enhancement due to matter of either $\nu_e \to \nu_\mu$ or $\bar{\nu}_e \to \bar{\nu}_\mu$ may also improve the $\sin^2 2\theta_{13}$ sensitivity for appropriate choices of L and E_ν [9.60]. In a four-neutrino model, both short-baseline experiments with $L/E_\nu \sim 1$ km/GeV (which probe δm^2_{LSND}) and long-baseline experiments will be required to obtain maximal information on the neutrino mixing parameters [9.25].

CP-violating effects due to the phase δ only become appreciable at the subleading δm^2 scale, and only then if the mixing angles are large enough.[1] In a three-neutrino model with δm^2_{sun} and δm^2_{atm}, CP violation will be observable only for the large-angle MSW solar solution; a long-baseline experiment with a high-intensity neutrino beam from a neutrino factory may be able to give information on δ in this case [9.62]. In a four-neutrino model, potentially large CP-violating effects are possible at the δm^2_{atm} scale [9.25, 9.63].

9.6   Summary and Outlook

In a three-neutrino world, it may be possible to completely determine the neutrino mixing matrix and two independent mass-squared differences using existing and planned neutrino oscillation experiments. Future measurements of solar neutrinos should pin down the neutrino mass and mixing parameters δm^2_{21} and $\sin^2 2\theta_{12}$ that are predominantly responsible for the solar neutrino deficit. Long-baseline experiments can provide a more precise determination of δm^2_{32} and $\sin^2 2\theta_{23}$, which drive the atmospheric neutrino anomaly, and also measure the size of $\sin^2 2\theta_{13}$ and determine the sign of δm^2_{32}. If the solar solution is the large-angle MSW solution, long-baseline experiments may also be sensitive to the CP-violating phase δ. Future measurements of beta decay spectra, double beta decay and hot dark matter may then help determine the last remaining neutrino mass parameter, the absolute neutrino mass scale.

In a four-neutrino world, there are three additional mixing angles and two additional CP phases. Since the extra neutrino is sterile, it may be difficult to determine some of the additional mixing angles, especially if they are small, as is the case in many existing models. Short-baseline experiments that probe the δm^2_{LSND} oscillation will be helpful in making sense of the larger parameter space. Furthermore, in four-neutrino models CP violation may become observable at the δm^2_{atm} scale (rather than the δm^2_{sun} scale, as is the case in three-neutrino models), and there will be a contribution to hot dark matter on the order of $\sum m_\nu \sim 1$ eV.

---

[1] This was first discussed in the context of three neutrinos in [9.61].

References

9.1 K. S. Hirata et al. (Kamiokande collaboration), Phys. Lett. B **280**, 146 (1992); Y. Fukuda et al., Phys. Lett. B **335**, 237 (1994); R. Becker-Szendy et al. (IMB collaboration), Nucl. Phys. B, Proc. Suppl. **38**, 331 (1995); W. W. M. Allison et al. (Soudan-2 collaboration), Phys. Lett. b **391**, 491 (1997); M. Ambrosio et al. (MACRO collaboration), Phys. Lett. B **434**, 451 (1998).

9.2 J. G. Learned, S. Pakvasa and T. J. Weiler, Phys. Lett. B **207**, 79 (1988); V. Barger and K. Whisnant, Phys. Lett. B **209**, 365 (1988); K. Hidaka, M. Honda and S. Midorikawa, Phys. Rev. Lett. **61**, 1537 (1988).

9.3 Y. Fukuda et al. (SuperKamiokande collaboration), Phys. Rev. Lett. **81**, 1562 (1998); Phys. Rev. Lett. **82**, 2644 (1999).

9.4 G. Barr, T. K. Gaisser and T. Stanev, Phys. Rev. D **39**, 3532 (1989); M. Honda, T. Kajita, K. Kasahara and S. Midorikawa, Phys. Rev. D **52**, 4985 (1995); V. Agrawal, T. K. Gaisser, P. Lipari and T. Stanev, Phys. Rev. D **53**, 1314 (1996); T. K. Gaisser et al., Phys. Rev. D **54**, 5578 (1996); T. K. Gaisser and T. Stanev, Phys. Rev. D **57**, 1977 (1998).

9.5 M. Apollonio et al. (Chooz collaboration), Phys. Lett. B **420**, 397 (1998).

9.6 T. Kajita, talk presented at *7th International Symposium on Particles, Strings and Cosmology (PASCOS 99)*, Granlibakken, CA, Dec. 1999, http://pc90.ucdavis.edu/talks/plenary/Kajita/.

9.7 B. T. Cleveland et al., Nucl. Phys. B, Proc. Suppl. **38**, 47 (1995).

9.8 W. Hampel et al. (GALLEX collaboration, Phys. Lett. B **388**, 384 (1996); J. N. Abdurashitov et al. (SAGE collaboration), Phys. Rev. Lett. **77**, 4708 (1996).

9.9 Y. Fukuda et al. (Kamiokande collaboration), Phys. Rev. Lett. **77**, 1683 (1996).

9.10 Y. Fukuda et al. (SuperKamiokande collaboration), Phys. Rev. Lett. **82**, 1810 (1999); Phys. Rev. Lett. **82**, 2430 (1999); M. B. Smy, hep-ex/9903034, in *Proceedings of the APS Meeting of the Division of Particles and Fields (DPF-99)*, Los Angeles, January 1999, in press; G. Sullivan, Aspen Winter Conference on Particle Physics, "Advances in Particle Physics: Recent Results and Open Questions", January 1999, Aspen, CO; Transparencies available at http://pheno.physics.wisc.edu/AWC99/program.html

9.11 J. N. Bahcall and M. H. Pinsonneault, Rev. Mod. Phys. **67**, 781 (1995); J. N. Bahcall, S. Basu and M.H. Pinsonneault, Phys. Lett. B **433**, 1 (1998).

9.12 V. Barger, R.J.N. Phillips and K. Whisnant, Phys. Rev. D **24**, 538 (1981); S. L. Glashow and L. M. Krauss, Phys. Lett. B **190**, 199 (1987); V. Barger, R. J. N. Phillips and K. Whisnant, Phys. Rev. Lett. **65**, 3084 (1990); Phys. Rev. Lett. **69**, 3135 (1992); A. Acker, S. Pakvasa and J. Pantaleone, Phys. Rev. D **43**, 1754 (1991); P. I. Krastev and S. T. Petcov, Phys. Lett. B **285**, 85 (1992); Phys. Rev. D **53**, 1665 (1996); N. Hata and P. Langacker, Phys. Rev. D **56**, 6107 (1997); S. L. Glashow, P. J. Kernan and L. M. Krauss, Phys. Lett. B **445**, 412 (1999).

9.13 V. Barger and K. Whisnant, Phys. Rev. D **59**, 093007 (1999).

9.14 J. N. Bahcall, P. I. Krastev and A. Yu. Smirnov, Phys. Rev. D **58**, 096016 (1998).

9.15 L. Wolfenstein, Phys. Rev. D **17**, 2369 (1978); V. Barger, R. J. N. Phillips and K. Whisnant, Phys. Rev. D **22**, 2718 (1980); P. Langacker, J. P. Leveille and J. Sheiman, Phys. Rev. D **27**, 1228 (1983); S. P. Mikheyev and A. Smirnov, Yad. Fiz. **42**, 1441 (1985); Nuovo Cim. C **9**, 17 (1986); S. J. Parke, Phys. Rev. Lett. **57**, 1275 (1986); S. J. Parke and T. P. Walker, Phys. Rev. Lett. **57**, 2322 (1986); W. C. Haxton, Phys. Rev. Lett. **57**, 1271 (1986).

9.16 A. McDonald (Solar Neutrino Observatory (SNO) collaboration), talk at *Neutrino-98*, Takayama, Japan, June 1998.

9.17 L. Oberauer (Borexino collaboration), talk at *Neutrino-98*, Takayama, Japan, June 1998.

9.18 F. Arneodo et al. (ICARUS collaboration), Nucl. Phys. Proc. Suppl. **70**, 453 (1999).

9.19 V. Barger, R. J. N. Phillips and K. Whisnant, Phys. Rev. Lett. **65**, 3084 (1990); Phys. Rev. Lett. **69**, 3135 (1992); S. Pakvasa and J. Pantaleone, Phys. Rev. Lett. **65**, 2479 (1990); A. Acker, S. Pakvasa and J. Pantaleone, Phys. Rev. D **43**, 1754 (1991); P. I. Krastev and S.T. Petcov, Phys. Lett. B **285**, 85 (1992); Nucl. Phys. B **449**, 605 (1995); S. P. Mikheyev and A.Yu. Smirnov, Phys. Lett. B **429**, 343 (1998); B. Faid, G. L. Fogli, E. Lisi, and D. Montanino, Astropart. Phys. **10**, 93 (1999); S. L. Glashow, P. J. Kernan and L.M. Krauss, Phys. Lett. B **445**, 412 (1999); J. M. Gelb and S. P. Rosen, Phys. Rev. D **60**, 011301 (1999); V. Berezinsky, G. Fiorentini and M. Lissia, hep-ph/9811352; V. Barger and K. Whisnant, Phys. Lett. B **456**, 54 (1999).

9.20 C. Athanassopoulos et al. (Liquid Scintillator Neutrino Detector (LSND) collaboration), Phys. Rev. Lett. **75**, 2650 (1995); Phys. Rev. Lett. **77**, 3082 (1996); Phys. Rev. C **58**, 2489 (1998).

9.21 D. O. Caldwell, G. M. Fuller and Y.-Z. Qian, astro-ph/9910175, Phys. Rev. D, published in Phys. Rev. D **61**, 123005 (2000).

9.22 E. Church et al. (BooNE collaboration), "A letter of intent for an experiment to measure $\nu_\mu \rightarrow \nu_e$ oscillations and ν_μ at the Fermilab Booster", May 16, 1997, unpublished.

9.23 V. Barger, T. J. Weiler and K. Whisnant, Phys. Lett. B **440**, 1 (1998).

9.24 L.-L. Chau and W.-Y. Keung, Phys. Rev. Lett. **53**, 1802 (1984); C. Jarlskog, Z. Phys. C **29**, 491 (1985); Phys. Rev. D **35**, 1685 (1987).

9.25 V. Barger, Y.-B. Dai, K. Whisnant and B.-.L. Young, Phys. Rev. D **59**, 113010 (1999).

9.26 A. De Rujula, M. B. Gavela and P. Hernandez, Nucl. Phys. B **547**, 21 (1999).

9.27 V. Barger, K. Whisnant, and R. J. N. Phillips, Phys. Rev. Lett. **45**, 2084 (1980); S. Pakvasa, in *Proceedings of the 20th International Conference on High Energy Physics*, ed. by L. Durand and L. G. Pondrom, AIP Conf. Proc. No. 68 (American Institute of Physics, New York, 1981), Vol. 2, p. 1164; D. J. Wagner and T. J. Weiler, Phys. Rev. D **59**, 113007 (1999); A. M. Gago, V. Pleitez and R.Z. Funchal, Phys. Rev. D **61**, 016004 (2000); K. R. Schubert, hep-ph/9902215; K. Dick, M. Freund, M. Lindner and A. Romanino, Nucl. Phys. B **562**, 29 (1999); J. Bernabeu, hep-ph/9904474, in Proc. 17th Int. Workshop on Weak Interactions and Neutrinos (WIN99), Cape Town, Jan. 1999, ed. by C. Dominguey and R. Viollier (World Scientific, Singapore, 2000), p. 227; S. M. Bilenky, C. Giunti and W. Grimus, Phys. Rev. D **58**, 033001 (1998); M. Tanimoto, Prog. Theor. Phys. **97**, 901 (1997); J. Arafune and J. Sato, Phys. Rev. D **55**, 1653 (1997); T. Fukuyama, K. Matasuda and H. Nishiura, Phys. Rev. D **57**, 5844 (1998); M. Koike and J. Sato, hep-ph/9707203, in *Proceedings of the KEK Meetings on CP Violation and its Origin*; H. Minakata and H. Nunokawa, Phys. Lett. B **413**, 369 (1997); H. Minakata and H. Nunokawa, Phys. Rev. D **57**, 4403 (1998); J. Arafune, M. Koike and J. Sato, Phys. Rev. D **56**, 3093 (1997); M. Tanimoto, Phys. Lett. B **435**, 373 (1998); H. Fritzsch and Z.Z. Xing, Phys. Rev. D **61**, 073016 (2000); Z.Z. Xing, hep-ph/0002246, published in Phys. Lett. B **487**, 327 (2000).

9.28 V. Barger, S. Pakvasa, T.J. Weiler and K. Whisnant, Phys. Lett. B **437**, 107 (1998); A.J. Baltz, A. Goldhaber and M. Goldhaber, Phys. Rev. Lett. **81**, 5730 (1998); F. Vissani, hep-ph/9708483; M. Jezabek and Y. Sumino, Phys. Lett. B **440**, 327 (1998); Y. Nomura and T. Yanagida, Phys. Rev. D **59**, 017303 (1999); G. Altarelli and F. Feruglio, Phys. Lett. B **439**, 112 (1998); JHEP (J. High-Energy Physics) **9811**, 021 (1998); H. Georgi and S. Glashow, Phys. Rev. D **61**, 097301 (2000); S. Davidson and S. F. King, Phys. Lett. B **445**, 191 (1998); R. N. Mohapatra and S. Nussinov, Phys. Lett. B **441**, 299 (1998); Phys. Rev. D **60**, 013002 (1999). S. K. Kang and C. S. Kim, Phys. Rev. D **59**, 091302 (1999); C. Jarlskog, M. Matsuda, S. Skadhauge and M. Tanimoto, Phys. Lett. B **449**, 240 (1999); E. Ma, Phys. Lett. B **456**, 48 (1999). D. V. Ahluwalia, Mod. Phys. Lett. A **13**, 2249 (1998); I. Stancu and D .V. Ahluwalia, Phys. Lett. B **460**, 431 (1999).

9.29 D. Abbaneo et al. (LEP Electroweak Working Group and SLD Heavy Flavor Group), CERN-PPE-96-183, December 1996.

9.30 S. M. Bilenky, C. Giunti and W. Grimus, Eur. Phys. J. C **1**, 247 (1998).

9.31 V. Barger, S. Pakvasa, T.J. Weiler and K. Whisnant, Phys. Rev. D **58**, 093016 (1998).

9.32 Y. Declais et al., Nucl. Phys. B **434**, 503 (1995).

9.33 F. Dydak et al., Phys. Lett. B **134**, 281 (1984).

9.34 J. K. Elwood, N. Irges and P. Ramond, Phys. Rev. Lett. **81**, 5064 (1998).

9.35 H. Fritzsch and Z. Xing, Phys. Lett. B **372**, 265 (1996); **440**, 313 (1998); E. Torrente-Lujan, Phys. Lett. B **389**, 557 (1996); M. Fukugita, M. Tanimoto and T. Yanagida, Phys. Rev. D **57**, 4429 (1998); M. Tanimoto, Phys. Rev. D **59**, 017304 (1999).

9.36 C. H. Albright and S. M. Barr, Phys. Rev. D **58**, 013002 (1998); Phys. Lett. B **461**, 218 (1999).

9.37 R. Barbieri, L. J. Hall, and A. Strumia, Phys. Lett. B **445**, 407 (1999); L. J. Hall and D. Smith, Phys. Rev. D **59**, 113013 (1999).

9.38 G. Altarelli and F. Feruglio, Phys. Rep. **320**, 295 (1999); P. Ramond, hep-ph/0001009, in *Proceedings of the 6th International Workshop on Topics in Astroparticle and Underground Physics (TAUP 99)*, Paris, Sept. 1999, in press; S.M. Barr and I. Dorsner, hep-ph/0003058, published in Nucl. Phys. B **585**, 79 (2000).

9.39 S.F. King, Phys. Lett. B **439**, 350 (1998); Nucl. Phys. **562**, 57 (1999); C. H. Albright and S. M. Barr, Phys. Rev. D **58**, 013002 (1998).

9.40 A. I. Belesev et al., Phys. Lett. B **350**, 263 (1995); V. M. Lobashev et al., Phys. Lett. B **460**, 227 (1999).

9.41 V. Barger, T. J. Weiler and K. Whisnant, Phys. Lett. B **442**, 255 (1998).

9.42 M. Doi et al., Phys. Lett. B **102**, 323 (1981); H. Minakata and O. Yasuda, Phys. Rev. D **56**, 1692 (1997); H. Georgi and S. L. Glashow, Phys. Rev.

D **61**, 097301 (2000); J. Ellis and S. Lola, Phys. Lett. B **458**, 310 (1999); F. Vissani, JHEP (J. High-Energy Physics) **9906**, 022 (1999); S. M. Bilenki, C. Giunti, W. Grimus, B. Kayser and S.T. Petcov, Phys. Lett. B **465**, 193 (1999).

9.43 L. Baudis et al., Phys. Rev. Lett. **83**, 41 (1999).

9.44 D. Caldwell, private communication.

9.45 V. Barger and K. Whisnant, Phys. Lett. B **456**, 194 (1999).

9.46 J. R. Primack, Science **280**, 1398 (1998); E. Gawiser and J. Silk, Science **280**, 1405 (1998); J. R. Primack and M. A. K. Gross, astro-ph/9810204, in *Proceedings of the 10th Rencontres de Blois: The Birth of Galaxies*, Blois, France, June 1998, in press.

9.47 E. W. Kolb and M. S. Turner, *The Early Universe* (Addison-Wesley, Reading, 1990).

9.48 D. J. Eisenstein, W. Hu and M. Tegmark, Phys. Rev. Lett. **80**, 5255 (1998); astro-ph/9807130.

9.49 S. Mohanty, D. P. Roy and U. Sarkar, Phys. Lett. B **445**, 185 (1998).

9.50 S. C. Gibbons, R. N. Mohapatra, S. Nandi and A. Raychaudhuri, Phys. Lett. B **430**, 296 (1998).

9.51 D. O. Caldwell and R. N. Mohapatra, Phys. Rev. D **48**, 3259 (1993); J. T. Peltoniemi and J. W. F. Valle, Nucl. Phys. B **406**, 409 (1993); R. Foot and R. R. Volkas, Phys. Rev. D **52**, 6595 (1995); E. Ma and P. Roy, Phys. Rev. D **52**, R4780 (1995); J. J. Gomez-Cadenas and M. C. Gonzalez-Garcia, Z. Phys. C **71**, 443 (1996); N. Okada and O. Yasuda, Int. J. Mod. Phys. A **12**, 3669 (1997); R. N. Mohapatra, hep-ph/9711444, in *Proceedings of the Workshop on Physics at the First Muon Collider and at the Front End of a Muon Collider*, Fermilab, Nov. 1997, ed. by S. Geer and R. Raja (American Institute of Physics, Woodbury, 1998), AIP Conf. Proc. Vol. 435, p. 358; N. Gaur, A. Ghosal, Ernest Ma and P. Roy, Phys. Rev. D **58**, 071301 (1998).

9.52 D. O. Caldwell and J. R. Primack, Phys. Rev. Lett. **74**, 2160 (1995).

9.53 K. Nishikawa et al. (KEK-PS E362 collaboration), *Proposal for a Long Baseline Neutrino Oscillation Experiment, using KEK-PS and Super-Kamiokande*, 1995, unpublished; Tokyo Univ. Report INS-924, April 1992, (unpublished); Y. Oyama, in *Proceedings of the YITP Workshop on Flavor Physics*, Kyoto, 1998, hep-ex/9803014.

9.54 MINOS collaboration, *Neutrino Oscillation Physics at Fermilab: The NuMI-MINOS Project*, NuMI-L-375 (Fermilab, Chicago, 1998).

9.55 ICANOE web page, http://pcnometh4.cern.ch/

9.56 OPERA web page, http://www.cern.ch/opera/

9.57 D. A. Petyt, "A study of parameter measurement in a long-baseline neutrino oscillation experiment", PhD Thesis, University of Oxford, 1998.

9.58 S. Geer, "Physics Potential of Neutrino Beams from Muon Storage Rings", FERMILAB-CONF-97-417; S. Geer, Phys. Rev. D **57**, 6989 (1998); C. Ankenbrandt et al. (Muon Collider collaboration), Phys. Rev. (Special Topics) Accel. Beams **2**, 081001 (1999); R.B. Palmer et al. (Muon Collider collababoration), http://pubweb/bnl.gov/people/palmer/nu/params.ps.

9.59 M. Campanelli, A. Bueno and A. Rubbia, hep-ph/9905240; I. Mocioiu and R. Shrock, hep-ph/9910554, hep-ph/0002149; A. Cervera, A. Donini, M. B. Gavela, J. J. Gomez Cadenas, P. Hernandez, O. Mena and S. Rigolin, hep-ph/0002108.

9.60 V. Barger, S. Geer and K. Whisnant, Phys. Rev. D **61**, 053004 (2000); V. Barger, S. Geer, R. Raja and K. Whisnant, hep-ph/9911524, Phys. Rev. D (in press).

9.61 V. Barger, K. Whisnant and R. J. N. Phillips, Phys. Rev. Lett. **45**, 2084 (1980).

9.62 M. Freund, M. Lindner, S. T. Petcov and A. Romanino, hep-ph/9912457; V. Barger, S. Geer, R. Raja and K. Whisnant, hep-ph/0003184.

9.63 S. M. Bilenky, C. Giunti and W. Grimus, Phys. Rev. D **58**, 033001 (1998); Prog. Part. Nucl. Phys. **43**, 1 (1999); T. Hattori, T. Hasuike and S. Wakaizumi, hep-ph/0002096, published in Phys. Rev. D **62**, 033006 (2000).

10 Theories of Neutrino Masses and Mixings

Rabindra N. Mohapatra

10.1 Introduction

The history of weak-interaction physics has to a large extent been a history of our understanding of the properties of the elusive spin-half particles called neutrinos. Evidence for only left-handed neutrinos being emitted in beta decay was the cornerstone of the successful V–A theory of weak interactions suggested by Sudarshan, Marshak, Feynman and Gell-Mann; evidence for the neutral-current interactions in the early 1970s provided brilliant confirmation of the successful gauge unification of the weak and electromagnetic interactions proposed by Glashow, Salam and Weinberg.

Today, as we enter a new millennium, we again have evidence for a very important new property of neutrinos, i.e. they have mass and, as a result, like the quarks, they mix with each other and lead to the phenomenon of neutrino oscillation. This is contrary to the expectations based on the Standard Model as well as on the old V–A theory (in fact one may recall that one way to make the V–A theory plausible was to use invariance of the weak Lagrangian under the so-called γ_5 invariance of all fermions, a principle which was motivated by the assumption that neutrinos have zero mass). The simple fact that neutrino masses vanish in the Standard Model is proof that the nonzero neutrino mass is an indication of new physics at some higher scale (or shorter distances). Study of details of neutrino masses and mixings is therefore going to open up new vistas in our journey towards a deeper understanding of the properties of the weak interactions at very short distances. This, no doubt, will have profound implications for the nature of the final theory of particles, forces and the Universe.

We are, of course, far from a complete picture of the masses and mixings of the various neutrinos and cannot therefore have a full outline of the theory that explains them. However, there exist enough information and indirect indications that constrain the masses and mixings among the neutrinos that we can see a narrowing of the possibilities for the theories. Many clever experiments now under way will soon clarify or rule out many of the allowed models. It is one of the goals of this chapter to give a panoramic view of the most likely scenarios for new physics that explain what is now known about neutrino masses.[1] We hope to emphasize the several interesting ideas for understanding the small neutrino masses and discuss in general terms how they can lead to the scenarios for neutrinos currently being discussed in order to understand the observations. These ideas have a very good chance of being part of the final theory of neutrino masses. We then touch briefly on some specific models that are based on the above general framework but attempt to provide an understanding of the detailed mass and mixing patterns. These works are instructive for several reasons: first, they provide proof of the detailed workability of the general ideas described above (they provide a sort of existence proofs that things will work); second, they often illustrate the kind of assumptions needed and through doing so a unique insight into which direction the next step should be in; and finally of course, nature may be generous in picking one of those models as the final message bearer.

As discussed in an earlier section, the neutrino mass can be of either Dirac or Majorana type. In this article we shall discuss our understanding of neutrino masses assuming that they are of Majorana type.

10.2   Experimental Indications of Neutrino Masses

As has been extensively discussed elsewhere in this book, while the direct-search experiments for neutrino masses using tritium beta decay and neutrinoless double beta decay have only yielded upper limits, the searches for neutrino oscillation, which can occur only if neutrinos have masses and mixings, have yielded positive evidence. There is now clear evidence from one experiment and strong indications from other experiments of neutrino oscillations and hence neutrino masses. The evidence comes from the atmospheric neutrino data in the SuperKamiokande experiment [10.4], which confirms the indications of oscillations in earlier data from the Kamiokande [10.5] and IMB [10.6] experiments. More recent data from the Soudan II [10.7] and MACRO [10.8] experiments provide further confirmation of this evidence.

From the existing data, several important conclusions can be drawn: (i) the data cannot be fitted assuming oscillation between ν_μ and ν_e; (ii) two oscillation scenarios that fit the data are ν_μ–ν_τ and ν_μ–ν_s oscillations (where ν_s is a sterile neutrino that does not couple to the W or Z bosons in the basic Lagrangian), although, at the 2σ level, the first scenario is a better fit than the latter. The more precise values of the oscillation parameters at a 90% confidence level are

$$\Delta m^2_{\nu_\mu \nu_\tau} \simeq (2\text{--}8) \times 10^{-3} \text{ eV}^2 \, , \tag{10.1}$$

$$\sin^2 2\theta_{\mu\tau} \simeq 0.8\text{--}1 \, .$$

The second evidence for neutrino oscillation comes from the five experiments that have observed a deficit in the flux of neutrinos from the sun as

---

[1] See [10.1] for a summary of theoretical and phenomenological ideas relating to massive neutrinos; for a summary of recent developments, see [10.2, 10.3].

compared with the predictions of the standard solar model, championed by Bahcall and his collaborators[2] and, more recently, studied by many groups. The experiments responsible for this discovery are the Chlorine, Kamiokande, Gallex, SAGE and SuperKamiokande experiments [10.10], conducted at the Homestake mine in the USA, Kamioka in Japan, Gran Sasso in Italy and Baksan in Russia. The different experiments see different parts of the solar neutrino spectrum. The details of these considerations are discussed in other chapters. The oscillation interpretation of the solar neutrino deficit has more facets to it than does the atmospheric case: first, the final-state particle that the ν_e oscillates into, and second, what kind of Δm^2 and mixings fit the data. At the moment there is a multitude of possibilities. Let us summarize them now.

As far as the final state goes, either it can be one of the two remaining active neutrinos, ν_μ and ν_τ, or it can be the sterile neutrino ν_s as in the case of atmospheric neutrinos. Both possibilities are open at the moment. The SNO experiment, which is expected to measure the solar neutrino flux via neutral-current interactions, will settle the issue of whether the final state of solar neutrino oscillation is the active neutrino or the sterile neutrino; in the former case, the ratio r of the charged-current flux (Φ_{CC}) to the neutral-current flux (Φ_{NC}) is nearly one-half, whereas in the latter case, it is one. As far as the Δm^2 and $\sin^2 2\theta$ parameters go, there are three possibilities: if the oscillation proceeds without any help from the matter in the dense core of the sun, it is a vacuum oscillation (VO); if the oscillation is enhanced by the solar core, it is attributed to the MSW mechanism, in which case there are two ranges of parameters that can explain the deficit – the small-angle range (SMA-MSW) and the large-angle range (LMA-MSW). The parameter ranges, taken from [10.9], are

$$\text{VO}: \Delta m^2 \simeq 6.5 \times 10^{-11}\,\text{eV}^2, \;\; \sin^2 2\theta \simeq 0.75\text{--}1\,,$$
$$\text{SMA-MSW}: \Delta m^2 \simeq 5 \times 10^{-6}\,\text{eV}^2, \;\; \sin^2 2\theta \simeq 5 \times 10^{-3}\,,$$
$$\text{LMA-MSW}: \Delta m^2 \simeq 1.2 \times 10^{-5}\text{--}3.1 \times 10^{-4}\,\text{eV}^2,$$
$$\sin^2 2\theta \simeq 0.58\text{--}1.00\,. \tag{10.2}$$

A relevant point to note is that there is no large-angle MSW fit for the case of sterile neutrinos, owing to absence of a matter effect for them. The situation lately, however, has been quite fluid in the sense that there are measurements from the SuperKamiokande experiment of the electron energy distribution which appear to contradict the MSW-SMA solution; similarly, there are day–night effects which seem to prefer the MSW-LMA solution, although VO solutions also lead to day–night effects owing to matter effects for certain ranges of Δm^2 [10.11]. There are also indications of seasonal variation of the solar neutrino flux above and beyond that expected from the position of the earth in its orbit.

---

[2] For a recent summary and update see [10.9].

Finally, we come to the last indication of neutrino oscillation from the Los Alamos Liquid Scintillation Detector (LSND) experiment [10.12], where neutrino oscillations of both the ν_μ from a stopped-muon decay (DAR) and the ν_μ accompanying the muon in pion decay (DIF) have been observed. The evidence from the DAR is statistically more significant and is an oscillation from $\bar\nu_\mu$ to $\bar\nu_e$. The mass and mixing parameter range that fits the data is

$$\text{LSND} : \Delta m^2 \simeq 0.2\text{-}2 \text{ eV}^2, \quad \sin^2 2\theta \simeq 0.003\text{-}0.03 \ . \tag{10.3}$$

There are also points at higher masses, specifically at 6 eV^2, which are allowed by the present LSND data for small mixings [10.13]. At present, the KARMEN experiment at the Rutherford Laboratory is also searching for the same oscillation. While about eight events have been found at the time of writing, this is consistent with the expected background [10.14]. The proposed MiniBooNE experiment at Fermilab [10.13] will provide more definitive information on this very important process in the next five years.

Our goal now is to study the theoretical implications of these discoveries. We shall proceed towards this goal in the following manner: we shall isolate the mass patterns that fit the above data and then look for plausible models that can, first, lead to the general feature that neutrinos have tiny masses; then we shall try to understand in a simple manner some of the features indicated by the data in the hope that these general ideas will be part of our final understanding of the neutrino masses. As mentioned earlier, to understand the neutrino masses one has to go beyond the Standard Model. First, we shall be more specific about what we mean by this statement. Then we shall present some ideas which may form the basic framework for constructing detailed models. We shall refrain from discussing any specific models except to give examples in the course of illustrating the theoretical ideas.

10.3   Patterns and Textures for Neutrinos

As already mentioned, we shall assume two-component neutrinos and therefore their masses will in general be of Majorana type. Let us also specify our notation to facilitate further discussion: the neutrinos emitted in weak processes such as beta decay or muon decay are weak eigenstates and are not mass eigenstates. The latter determine how a neutrino state evolves in time. Similarly, in the detection process, it is the weak eigenstate that is picked out. This is of course the key idea behind neutrino oscillations [10.15]. It is therefore important to express the weak eigenstates in terms of the mass eigenstates. We shall denote the weak eigenstates by the symbols α, β or simply e, μ, τ, etc., whereas the mass eigenstates will be denoted by the symbols i, j, k, etc. The mixing angles will be denoted by $U_{\alpha i}$; they relate the two sets of eigenstates as follows:

$$\begin{pmatrix} \nu_e \\ \nu_\mu \\ \nu_\tau \end{pmatrix} = U \begin{pmatrix} \nu_1 \\ \nu_2 \\ \nu_3 \end{pmatrix} . \tag{10.4}$$

Using this equation, one can derive the well-known oscillation formulae for the survival probability of a particular weak eigenstate α:

$$P_{\alpha\alpha} = 1 - 4 \sum_{i<j} |U_{\alpha i}|^2 |U_{\alpha j}|^2 \sin^2 \Delta_{ij} , \qquad (10.5)$$

where $\Delta_{ij} = (m_i^2 - m_j^2)L/4E$. The transition probability from one weak eigenstate to another is given by

$$P_{\alpha\beta} = 4 \sum_{i<j} U_{\alpha i} U_{\beta i}^* U_{\alpha j}^* U_{\beta j} \sin^2 \Delta_{ij} . \qquad (10.6)$$

Since Majorana masses violate lepton number, a very important constraint on any discussion of neutrino mass patterns arises from the negative searches for neutrinoless double beta decay [10.16].[3] The most stringent present limits are obtained from the Heidelberg–Moscow enriched-germanium-76 experiment at Gran Sasso and imply an upper limit on the following combination of masses and mixings:

$$\langle m_\nu \rangle \equiv \sum_i U_{ei}^2 m_{\nu_i} \leq 0.2 \text{ eV} . \qquad (10.7)$$

This upper limit depends on the nuclear matrix element calculated by the Heidelberg group [10.18]. There could be an uncertainty of a factor of two to three in this estimate. This would then relax the above upper bound to at most 0.6 eV. This is a very stringent limit and becomes especially relevant when one considers whether the neutrinos constitute a significant fraction of the hot dark matter of the Universe. A useful working formula is $\sum_i m_{\nu_i} \simeq 24\Omega_\nu$ eV, where Ω_ν is the neutrino fraction that contributes to the dark matter of the Universe. For instance, if the dark-matter fraction is 20%, then the sum total of neutrino masses must be 4.8 eV. The situation at the moment has become uncertain [10.20] after the results from the high-z supernova searches indicated a possible nonzero cosmological constant. Nevertheless, from structure formation data, a total neutrino mass of 2–3 eV cannot strictly be ruled out, and in fact one particular fit [10.21] prefers a cold+hot dark-matter scenario with a similar mass, as a better fit than any other (e.g. CDM+Λ). The proposed GENIUS experiment [10.19] by the Heidelberg–Moscow collaboration has the promise to bring down the upper limit on $\langle m_\nu \rangle$ by two orders of magnitude. This would profoundly affect the current ideas on neutrino masses and would help to more sharply define the theoretical directions in the field.

In view of the several levels of uncertainty that currently surround the various pieces of information on neutrino masses, we shall consider various different scenarios. It is not an unfair reflection of the present consensus in the community to say that the solar and atmospheric neutrino results are

---

[3] See [10.17] for a review and the future outlook.

considered to be the most secure pieces of evidence for neutrino masses. We shall therefore first consider the implications of taking these two sets of data seriously, supplemented by the very useful information from neutrinoless double beta decay. We shall include the LSND data subsequently.

10.3.1   Solar and Atmospheric Data and Neutrino Mass Patterns

If one wants to fit only the solar and atmospheric neutrino data, regardless of the nature of the solution to the solar neutrino puzzle (i.e. MSW with small or large angle or vacuum oscillation), it is possible to produce consistent scenarios using only the three known neutrinos. There are many mixing patterns and neutrino mass matrices that can be used for the purpose. In discussing these patterns, it is important to remember that a solution to the atmospheric neutrino puzzle requires that in the context of a three-neutrino model, ν_μ and ν_τ must mix maximally. There are two interesting mass patterns that have been widely discussed in the literature: (i) a hierarchical pattern with $m_1 \ll m_2 \ll m_3$, and (ii) an approximately degenerate pattern [10.22] with $m_1 \simeq m_2 \simeq m_3$, where the m_i are the eigenvalues of the neutrino mass matrix. In the first case, the atmospheric and the solar neutrino data give direct information on m_3 and m_2, respectively. On the other hand, in the second case, the mass differences between the first and the second eigenvalues are chosen to fit the solar neutrino data, and the differences between the second and the third to fit the atmospheric neutrino data. The limits from neutrinoless double beta decay imply very stringent constraints on the mixing pattern in the degenerate case; but before we proceed to this discussion let us focus for a while on the hierarchical mass pattern.

In proceeding with this discussion it is important to remember that the Chooz reactor data [10.23] imply that for $\Delta m^2 \geq 10^{-3}$ eV^2, the electron neutrino mixing angle has a rough upper bound $|U_{ei}| \leq 0.2$. Furthermore, it is now certain that atmospheric neutrino data cannot be fitted by ν_μ–ν_e oscillation. Together, these considerations imply that in the construction of any neutrino mass matrix, one must require that $|U_{e3}| \leq 0.2$ [10.24]. Note that U_{e2} can be larger, since the Δm^2 which corresponds to the solar neutrino puzzle is lower than that to which the Chooz experiment is sensitive. On the other hand, the solar neutrino results admit both small and large mixing angles. Combining these two inputs, one can conclude that the 3×3 neutrino mixing matrix is given by [10.25]

$$U_\nu = \begin{pmatrix} c & s & 0 \\ s/\sqrt{2} & c/\sqrt{2} & -1/\sqrt{2} \\ s/\sqrt{2} & c/\sqrt{2} & +1/\sqrt{2} \end{pmatrix}. \tag{10.8}$$

In writing the above mixing matrix, we have assumed that the atmospheric neutrino data have been fitted with $\sin^2 2\theta = 1$. On the other hand, if we take this value to be $\sin^2 2\theta = 8/9$, the above mixing matrix changes to

$$U_\nu = \begin{pmatrix} c & s & 0 \\ s/\sqrt{3} & c/\sqrt{3} & -\sqrt{2}/\sqrt{3} \\ s\sqrt{2}/\sqrt{3} & c\sqrt{2}/\sqrt{3} & +1/\sqrt{3} \end{pmatrix}. \tag{10.9}$$

There are two special cases where these mixing matrices take especially appealing forms: case A, where maximal-mixing solutions are chosen for the solar neutrino puzzle, and case B, where it is assumed that the neutrino masses are degenerate with a mass m_0 such that the effective mass deduced from neutrinoless double beta decay $\langle m_\nu \rangle \ll m_0$ [10.26]. In either case, the first matrix reduces to the so called bimaximal form [10.25–10.27], whereas the second matrix reduces to the democratic form [10.28]. The mixing matrices are obtained from the above two equations by taking $c = s = 1/\sqrt{2}$. The neutrino mass matrices in both these cases can be obtained by writing

$$\mathcal{M}_\nu = U M^{\mathrm{d}} U^{\mathrm{T}}, \tag{10.10}$$

where $M^{\mathrm{d}} = \mathrm{Diag}(m_1, m_2, m_3)$.

Degenerate case. The most compelling physical motivation for this case comes from the requirement that neutrinos constitute a significant fraction of the dark matter of the Universe. Using our previous discussion, we see that if the neutrino hot dark matter (HDM) constitutes about 20% of the critical mass of the Universe, then the total neutrino mass, i.e. $\sum_i |m_i|$, must be about 4.8 eV. If all three neutrinos are degenerate, the share of each species is 1.6 eV. Note that this is much bigger than the present upper limits on the effective mass $\langle m_\nu \rangle$ from neutrinoless double beta decay. If we ignore CP phases, then this implies a strong constraint on the mixings.

Two particularly interesting mass matrices emerge for the case where the neutrino masses are approximately degenerate. To derive them, let us choose $M^{\mathrm{d}} = \mathrm{Diag}(m_0, -m_0, m_0)$. If we further assume that $U_{e3} \simeq 0$, then we conclude that the mixing matrix elements $U_{e1} \simeq U_{e2} \simeq 1/\sqrt{2}$. It therefore follows that the small-angle MSW solution is automatically eliminated. For the bimaximal case, one obtains a neutrino mass matrix of the form [10.29]:

$$M_\nu = m_0 \begin{pmatrix} 0 & 1/\sqrt{2} & 1/\sqrt{2} \\ 1/\sqrt{2} & 1/2 & -1/2 \\ 1/\sqrt{2} & -1/2 & 1/2 \end{pmatrix}. \tag{10.11}$$

For derivation of this mass matrix in gauge models, see [10.30].

There is a corresponding mass matrix for the democratic case: it is given by

$$
M_\nu = m_0 \begin{pmatrix} 0 & 1/\sqrt{3} & \sqrt{2}/\sqrt{3} \\ 1/\sqrt{3} & 2/3 & -\sqrt{2}/3 \\ \sqrt{2}/\sqrt{3} & -\sqrt{2}/3 & 1/3 \end{pmatrix}. \tag{10.12}
$$

Finally, we want to note the form of the neutrino mass matrices that lead in the general case to mixing matrices which have either the democratic or the bimaximal form regardless of the nature of the eigenvalues. The importance of this discussion is that in trying to construct gauge models for neutrinos, the mass matrix follows directly from the Lagrangian and the mixing follows from this afterwards.

Bimaximal Case. The matrix in this case is

$$
M_\nu = \begin{pmatrix} A+D & F & F \\ F & A & D \\ F & D & A \end{pmatrix}. \tag{10.13}
$$

Note that vanishing of neutrinoless double beta decay implies that $A = -D$. A special case of this is when $A = D = 0$. Such mass matrices have been discussed in the literature for the three-neutrino case in [10.32]; for construction of gauge models for this case, see [10.33]. One can look at two special cases of (10.13); (a) $F \gg A, D$ and (b) $F \ll A, D$. In case (a), one has $\Delta m_{23}^2 > 0$ and in case (b) $\Delta m_{23}^2 < 0$. These signs can be determined in a neutrino factory.

Democratic Case. This case is more complicated. As was noted in [10.31], one must choose the charged-lepton mass matrix to be in the form given below, while keeping the neutrino mass matrix diagonal:

$$
M_\ell = \begin{pmatrix} a & 1 & 1 \\ 1 & a & 1 \\ 1 & 1 & a \end{pmatrix}. \tag{10.14}
$$

The symmetric form of the matrix in the democratic case is clear, and in fact it exhibits a permutation symmetry S_3 operating on the lepton doublets, which provides another clue for possible building of gauge models.

Hierarchical Case. One can also find some clue for model building by analyzing the case where the neutrino masses are hierarchical, i.e. $m_1 \ll m_2 \ll m_3$. To see this, let us first ignore the couplings of the first generation and consider only the 2×2 mass matrix involving the second and the third generation. It has been pointed out [10.34] that one very simple way to obtain a natural hierarchy while having a maximal (or large) mixing is to have a matrix M_ν of rank one, i.e.

$$M_\nu = \begin{pmatrix} x^2 & x \\ x & 1 \end{pmatrix}. \tag{10.15}$$

This matrix has one zero eigenvalue, and choosing $x \simeq 1$ leads to a large mixing angle. Further interest in this idea arises from the observation that such an M_ν can arise via the seesaw mechanism if one chooses

$$M_D = \begin{pmatrix} 0 & 0 \\ x & 1 \end{pmatrix}.$$

The interesting point about this form for M_D is that if we define it as $\bar{\nu}_L M_D \nu_R$ (where $\nu_{L,R}$ are column vectors), the above form for M_D mixes only the right-handed neutrinos and not the left-handed ones. In a quark–lepton unified model this would mean that only right-handed quarks mix with a large mixing angle. This, of course, is completely unconstrained by observations that confirm the Standard Model, since there are no right-handed charged currents in the Standard Model. More importantly, it leaves the left-handed mixing small. Thus this observation may provide one resolution of the apparent conundrum that the quark mixing angles are small whereas the neutrino mixng angles are large. This particular form has the disadvantege that it cannot arise in models such as left–right models, which generally lead to symmetric Dirac masses.

There are also other ways to generate maximal mixing angles. For instance, a choice for the neutrino mass matrix (for the 2–3 sector) of the form

$$M_{23} = \begin{pmatrix} a & b \\ b & a \end{pmatrix} \tag{10.16}$$

leads to maximal mixing. Such a mass matrix respects permutation symmetries (e.g. S_3) and therefore can be stable under radiative corrections. Akhmedov [10.35] has discussed the stability of neutrino mass matrices with respect to small variations of the parameters for various neutrino mass hierarchies.

10.3.2 Solar, Atmospheric and LSND Data and Scenarios with Sterile Neutrinos

In order to explain the solar, atmospheric and LSND data simultaneously using the oscillation picture, one must invoke additional neutrinos, since the different Δm^2s needed to explain each piece of data are very different from each other and could never add up to zero, as would be the case if there were only three neutrino species, i.e. $\Delta m_{12}^2 + \Delta m_{23}^2 + \Delta m_{31}^2 = 0$. This is a revolutionary result, since the LEP and SLC measurements imply precisely (within very small errors) that there are almost precisely three active neutrinos coupling to the Z boson (the actual LEP result is $N_\nu = 2.994 \pm 0.012$ [10.36]).

Furthermore, even though the conclusions from Big Bang nucleosynthesis (BBN)[4] are not as precise as the LEP–SLC data, the helium abundance is very sensitive to the number of neutrinos and a total number of neutrinos anywhere from 3.2 to 4.4 could be accomodated. Already these two results imply strong constraints on the theories that include extra neutrinos beyond those present in the Standard Model. For instance, the LEP–SLC data imply that the extra neutrinos cannot couple to the Z boson with the same strength as other neutrinos. They have therefore been designated in the literature as "sterile" neutrinos. Their "sterility" also helps them to evade the bounds from Big Bang nucleosynthesis as follows. Suppose the coupling of a sterile neutrino to known leptons is given by the four-Fermi interaction with a strength given by $G_F\epsilon$. Then, if $\epsilon \leq 10^{-3}$ [10.1], the sterile neutrinos decouple before the QCD phase transition temperature of about 100 MeV, and their effective contribution in the BBN era becomes equivalent to about 0.1 neutrino species. There are further constraints on Δm^2 and $\sin^2 2\theta$ that arise when the sterile neutrinos oscillate in the BBN era [10.39]. Typically, the constraint is

$$\Delta m_{es}^2 \sin^2 2\theta_{es} \leq 10^{-6} \text{ eV}^2 . \tag{10.17}$$

These considerations are important in constructing detailed theories of the sterile neutrinos. While we postpone the detailed discussion of theories with extra neutrinos to a separate section, here we study the possible four or more neutrino patterns and their mass matrices that correspond to the exsting neutrino data.

Case B1. The Four-Neutrino Case. It is possible to construct three possible scenarios for neutrino data using four neutrinos. Denoting the sterile neutrino by ν_s, the three scenarios are the following:

Case B1.1. The problem of the solar neutrino data is solved via the MSW small-angle oscillation between ν_e and ν_s, which then have a mass difference of $\Delta m_{es}^2 \simeq 10^{-5}$ eV^2, and the problem of the atmospheric neutrino data is solved via the ν_μ–ν_τ oscillation [10.22, 10.40]. (Of course, the solar neutrino problem could also be solved via the vacuum oscillation of ν_e and ν_s.) In either case, if we have $m_{\nu_\mu} \simeq m_{\nu_\tau} \simeq 1$ eV, then there can be a significant hot-dark-matter (HDM) component in the Universe, with enormous implications for structure formation. Note that the LSND data is then fitted by ν_μ–ν_e oscillation, with m_{ν_μ} being determined by Δm_{LSND}^2. It is interesting that the present LSND data has a considerable range where the neutrinos contribute significantly to the HDM component of the Universe.

Case B1.2. It is also possible to have a scenario where the atmospheric neutrino data are fitted by the maximal ν_μ–ν_s oscillation and the solar neutrino

---

[4] For a recent discussion, see [10.37]; see [10.38] for an earlier review.

data are fitted by the ν_e–ν_τ oscillation. As before, Δm^2_{LSND} determines the splitting between the two pairs of levels and will determine the HDM content of the Universe. In a series of papers, Bilenky et al. have proposed tests of these models [10.41].

It is possible to distinguish experimentally between these two scenarios. It will be possible to test the ν_e–ν_s oscillation solution to the solar neutrino puzzle once the SNO experiment measures the neutral-current flux of the solar neutrinos, since the ν_ss do not have any neutral-current interaction, unlike the $\nu_{\mu,\tau}$s. Similarly, for the atmospheric neutrino data, a search for neutral-current events with pion production, $\nu + N \rightarrow N + \pi^0 + N$, can discriminate between the oscillation of the atmospheric ν_μs to ν_τs and oscillation to ν_ss [10.42]. The present data appears to favor the oscillation to ν_τs at the 2σ level but cannot be taken as a conclusive proof.

Case B1.3. Another possibility is to have the three active neutrinos bunched together with very minute mass differences such that the ν_e–ν_μ oscillation explains the solar neutrino data and the ν_μ–ν_τ oscillation explains the atmospheric data. The LSND data can then be explained by including a sterile neutrino (or three of them) which is separated in mass from the three known neutrinos by an amount $\sqrt{\Delta m^2_{\text{LSND}}}$ and using indirect oscillation between the ν_e and ν_μ via the sterile neutrino state [10.44]. It has, however, been recently pointed out that this possibility may be marginally inconsistent with the observed up–down asymmetry in the atmospheric neutrino data when combined with the CDHS and Bugey limits [10.45].

Let us study the possible mass textures for four neutrinos. The general mass matrix in this case is a 4×4 symmetric matrix, which has ten nonzero entries if CP conservation is assumed. An interesting question to explore is that of the minimum number of parameters that are necessary to fit the observations. This would then isolate useful mass matrices with specific textures, which may then lead to clues to their theoretical origin. For this purpose let us do a bit of parameter counting. If the solar neutrino puzzle is assumed to be solved by the small-angle MSW mechanism, then there must be at least five parameters in the 4×4 symmetric neutrino mass matrix: three corresponding to $\Delta m^2_{\text{S,A,L}}$ and two small mixings θ_L and θ_S, respectively, (where S, A, L denote the solar, atmospheric and LSND experiments, respectively). If, on the other hand, the solar neutrino puzzle is solved via the vacuum oscillation mechanism, then one parameter is eliminated and one can make do with four parameters. Finally, if either one adopts the large-angle MSW solution (between ν_e and ν_s) for solar neutrinos or the ν_μ–ν_s alternative is used to solve the atmospheric neutrino puzzle, then we have an additional relation between the parameters, i.e.

$$\frac{\Delta m^2_S}{\Delta m^2_A} \approx \frac{\Delta m^2_A}{\Delta m^2_L} . \tag{10.18}$$

This reduces the number of parameters to three. An example of this type has recently been provided in [10.46].

Let us start with the simplest case, with three parameters [10.46]. (Here we have used the basis $(\nu_s, \nu_e, \nu_\mu, \nu_\tau)$. It could be easily reshuffled to consider the ν_μ-ν_s oscillation possibility for atmospheric neutrinos.) In this case,

$$
M = \begin{pmatrix} m_1 & \mu & m_1 - \epsilon & 0 \\ \mu & m_1 & 0 & m_1 - \epsilon \\ m_1 - \epsilon & 0 & m_1 & \mu \\ 0 & m_1 - \epsilon & \mu & m_1 \end{pmatrix}.
\tag{10.19}
$$

Here ϵ is the solar mass splitting and m_1 is the atmospheric mass splitting.

The next simplest case involves only four parameters and has the following form [10.47]:

$$
M = \begin{pmatrix} 0 & \mu_3 & 0 & 0 \\ \mu_3 & 0 & \epsilon & 0 \\ 0 & \epsilon & 0 & m \\ 0 & 0 & m & \delta \end{pmatrix},
\tag{10.20}
$$

where we choose $\delta \ll m \simeq 1$ eV. However, since in this case the sterile neutrino has a smaller mass than the ν_e (as can be seen by diagonalizing the above mass matrix), the solution to the solar neutrino puzzle must involve the vacuum oscillation of the ν_e to the ν_s. This can therefore clearly be tested once sufficient solar neutrino data has accumulated (say, for example, from SuperKamiokande and Borexino) in favor of or against the seasonal variation.

Going to one more parameter, we have a mass matrix which has one of the two following forms [10.48]:

$$
M = \begin{pmatrix} \mu_1 & \mu_3 & 0 & 0 \\ \mu_3 & 0 & 0 & \epsilon \\ 0 & 0 & \delta & m \\ 0 & \epsilon & m & \pm\delta \end{pmatrix}.
\tag{10.21}
$$

In this case, by an appropriate choice of μ_1, the sterile neutrino can be made heavier than the elctron neutrino. As a result, one can have a useful MSW transition between the ν_e and ν_s that can help to solve the solar neutrino puzzle via the small-angle MSW solution. This possibility has its characteristic predictions, and it is expected that it can be tested in future colliders such as the muon collider. For discussion of CP violation in these four-neutrino models see [10.49].

Case B2. Six-Neutrino Models.
One obvious question that arises as we consider sterile-neutrino models is "how many such neutrinos are there?" Symmetry would suggest that there are three sterile neutrinos rather than one. In fact, there are models of this type in the literature [10.50–10.52].[5] In one class of models [10.51], the additional neutrinos (i.e. the fifth and the

---

[5] See also [10.53], where it is argued that the present data are fitted slightly better by a ν_e-ν_s vacuum oscillation than by a sequential neutrino.

sixth) do not play any role in describing neutrino observations; but in the other two [10.50, 10.52], they play an essential role. The profile of the oscillation explanations in this latter case is very different and more symmetric, in the sense that both the solar and the atmospheric neutrino data are explained by maximal-mixing-angle vacuum oscillation between active and sterile neutrinos (i.e. the solar data via ν_e–ν_{es} and the atmospheric data via ν_μ–$\nu_{\mu s}$ vacuum oscillations) [10.54]; the LSND results in this case are explained via generational mixing, which is small like all other charged-fermion mixings. The physics of this scheme is different from that of the previous ones, as are the experimental tests. We shall discuss typical theories for such scenarios in subsequent sections.

Before closing this section, it is worth pointing out that the introduction of the sterile neutrino is such a far-reaching idea that attempts have been made to reconcile the LSND data without invoking oscillations, in which case the solar and atmospheric neutrino puzzles as well as the LSND data can be accommodated within the standard three-neutrino framework. The idea is to look for consistent models where a sufficiently significant amplitude for the anomalous decay mode of the muon $\mu^+ \to e^+ + \bar{\nu}_e + \bar{\nu}_\mu$ exists without conflicting with other known data, to explain the 0.3% rate for the observed $\bar{\nu}_e$s. However, it has been pointed out [10.55] that the muonium to antimuonium (M–$\bar{M}$) transition is related to the anomalous muon decay in several models (such as left–right models and supersymmetric models with R-parity breaking) in such a way that the existing experimental bounds from PSI experiment [10.56] on M–$\bar{M}$ transition suppresses the anomalous muon decay amplitude far below that required to explain the LSND data. Thus, barring some really drastic new idea to explain the LSND data as an anomalous muon decay, the confirmation of the LSND data in future work would require the existence of an additional one or more sterile neutrinos.[6] It will then be dependent on the data as to which of the scenarios with extra sterile neutrinos will win in the end.

10.4   Why Neutrino Mass Necessarily Means Physics Beyond the Standard Model

As is well known, the Standard Model is based on the gauge group $SU(3)_c \times SU(2)_L \times U(1)_Y$, under which the quarks and leptons transform as described in Table 10.1.

---

[6] There have also been attempts to explain the atmospheric neutrino data by nonoscillation mechanisms [10.57] (for a review, see [10.58]) such as flavor-changing interactions or neutrino decays, etc. In a recent paper, Bergman et al. [10.59] have argued that the first alternative runs into problem with large flavor-changing effects in τ decays, which have not been observed, and therefore that mechanism is theoretically not very plausible.

Table 10.1. The assignment of particles to the Standard Model gauge group $SU(3)_c \times SU(2)_L \times U(1)_Y$

Field	Gauge transformation
Quarks Q_L	$(3, 2, 1/3)$
Right-handed up quarks u_R	$(3, 1, 4/3)$
Right-handed down quarks d_R	$(3, 1, -2/3)$
Left-handed leptons L	$(1, 2 - 1)$
Right-handed leptons e_R	$(1, 1, -2)$
Higgs Boson H	$(1, 2, +1)$
Color gauge fields G_a	$(8, 1, 0)$
Weak gauge fields $W^{\pm}$, Z, γ	$(1, 3 + 1, 0)$

The electroweak symmetry $SU(2)_L \times U(1)_Y$ is broken by the vacuum expectation of the Higgs doublet $\langle H^0 \rangle = v_{\mathrm{wk}} \simeq 246$ GeV, which gives mass to the gauge bosons and the fermions, all fermions except the neutrino. Thus the neutrino is massless in the Standard Model, at the tree level. There are several questions that arise at this stage. What happens when one goes beyond the above simple tree-level approximation? Secondly, do nonperturbative effects change this tree-level result? Finally, how will this result be modified when quantum gravity effects are included?

The first and second questions are easily answered by using the $B - L$ symmetry of the Standard Model. The point is that since the Standard Model has no $SU(2)_L$ singlet neutrino-like field, the only possible mass terms that are allowed by Lorentz invariance are of the form $\nu_{iL}^T C^{-1} \nu_{jL}$, where i, j stand for the generation index and C is the Lorentz charge conjugation matrix. Since ν_{iL} is part of the $SU(2)_L$ doublet field and has lepton number $+1$, the above neutrino mass term transforms as an $SU(2)_L$ triplet, and furthermore it violates total lepton number (defined as $L \equiv L_e + L_\mu + L_\tau$) by two units. However, a quick look at the Standard Model Lagrangian convinces one that the model has exact lepton number symmetry both before and after symmetry breaking; therefore such terms can never arise in perturbation theory. Thus, to all orders in perturbation theory, the neutrinos are massless. As far as the nonperturbative effects go, the only known source is the weak instanton effects. Such effects could change the result if they broke the lepton number symmetry. One way to see if such breaking occurs is to look for anomalies in lepton number current conservation from triangle diagrams. Indeed, one has the equation $\partial_\mu j_\ell^\mu = cW\tilde{W} + c'B\tilde{B}$ owing to the contribution of the leptons to the triangle involving the lepton number current and Ws or Bs. Luckily, it turns out that the contribution of the anomaly to the nonconservation of baryon number current also has an identical form, so that the $B - L$ current j_{B-L}^μ is conserved to all orders in the gauge couplings. As a consequence, nonperturbative effects from the gauge sector cannot induce $B -$

L violation. Since the neutrino mass operator described above also violates $B - L$, this proves that neutrino masses remain zero even in the presence of nonperturbative effects.

Let us now turn to the effect of gravity. Clearly, as long as we treat gravity in perturbation theory, the above symmetry arguments hold, since all gravitational coupling respect $B - L$ symmetry. However, once nonperturbative gravitational effects, e.g. black holes and wormholes are included [10.60], there is no guarantee that global symmetries will be respected in the low-energy theory. The intuitive way to appreciate the argument is to note that throwing baryons into a black hole does not lead to any detectable consequence except through a net change in the baryon number of the Universe. Since one can throw an arbitrary number of baryons into the black hole, an arbitrary information loss about the net number of missing baryons would prevent us from defining a baryon number for the visible Universe – thus baryon number in the presence of a black hole cannot be an exact symmetry. Similar arguments can be given for any global charge, such as lepton number, in the Standard Model. A field-theoretic parameterization of this statement is that the effective low-energy Lagrangian for the Standard Model in the presence of black holes and wormholes, etc. must contain baryon- and lepton-number-violating terms. In the context of the Standard Model, the only such terms that one can construct are nonrenormalizable terms of the form $LHLH/M_{\mathrm{P}\ell}$ where $M_{\mathrm{P}\ell}$ is the Planck mass. After gauge symmetry breaking, they lead to neutrino masses. However, it becomes immediately clear that these masses are not enough to understand present observations, since they are at most of order $v_{\mathrm{wk}}^2/M_{\mathrm{P}\ell} \simeq 10^{-5}$ eV [10.61], and, as we discussed in the previous section, in order to solve the atmospheric neutrino problem, one needs masses at least three orders of magnitude higher.

Thus one must seek physics beyond the Standard Model to explain the observed evidence for neutrino masses.

10.5   Scenarios for Small Neutrino Mass Without Right-Handed Neutrinos

There exist a number of extensions of the Standard Model that lead to neutrino masses. Even restricting ourselves to the cases where the neutrinos are Majorana particles, there come to mind at least three different mechanisms for neutrino masses. All these models have lepton number violation built into them and therefore lead to a plethora of new phenomenological tests. The existence or nonexistence of right-handed neutrinos divides these models into two broad classes: one class that uses right-handed neutrinos and another that does not. In this section we consider models that do not introduce right-handed neutrinos to understand small neutrino masses.

As discussed in Chap. 2, the electrical neutrality of the neutrino allows for the existence of two types of mass terms consistent with Lorentz invari-

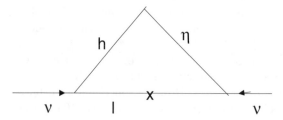

Fig. 10.1. One-loop diagram for neutrino mass in the Zee model

ance: the Majorana mass, which violates lepton number but does not require the inclusion of a right-handed neutrino, and the Dirac (or combined Dirac–Majorana) mass term, which requires the existence of a right-handed neutrino. In this section we consider models where the neutrino Majorana mass arises without the right-handed neutrino. The argument of Sect. 10.4 then implies that there must be violation of the $B - L$ quantum number either by interactions or by the vacuum state. This leads to two classes of models, which we discuss below.

10.5.1   Radiative Generation of Neutrino Masses

There are two classes of models where the introduction of lepton-number-violating interactions leads to radiative generation of small neutrino masses. One is the Zee model [10.62] and the other is the Babu model [10.63]. Let us first discuss the Zee model. In this case, one adds a charged $SU(2)_{\mathrm{L}}$ singlet field η^{+} to the Standard Model along with a second Higgs doublet H'. This allows the following additional Yukawa couplings beyond those present in the Standard Model:

$$L_Y(\eta) = \Sigma_{\alpha\beta} f_{\alpha\beta} L_\alpha^{\mathrm{T}} C^{-1} L_\beta \eta^{+} + \text{ H.c.} \tag{10.22}$$

Note that the coupling f is antisymmetric in the family indices α and β. Owing to the presence of the extra Higgs field, there are also additional terms in the Higgs potential, but the one of interest in connection with the neutrino masses is $\eta^* H^{\mathrm{T}} \tau_2 H'$. It is then easy to see that, while the neutrino masses vanish at the tree level, they arise from one-loop diagrams owing to the exchange of η fields in combination with the second Higgs (Fig. 10.1). The typical strength of these diagrams is of order $m_\ell^2 f/16\pi^2 M$ [10.64]. The dominant contribution to the $\nu_{e,\mu}$ mass in these models comes from the τ lepton in the loop and can be estimated to be of order 1–10 keV for $f \simeq 10^{-2}$, and the $\eta H H'$ coupling can also be estimated to be of order 10^{-2}–10^{-1}. These values for the masses are much bigger than being contemplated in the context of the neutrino puzzles. Thus, in the opinion of this author, the Zee model cannot account for the present observations in a natural manner without modification.

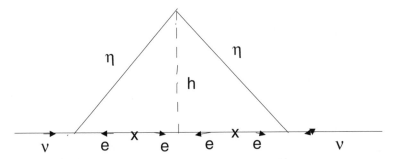

Fig. 10.2. Two-loop diagram for neutrino mass in the Babu model

Let us now pass to the second class of models [10.63]. In this case one adds to the Standard Model the fields η^+ as before, and a doubly charged field h^{++} but not a second Higgs doublet. The new Yukawa couplings of the model then are given by (repeated indices are summed)

$$\mathcal{L}_Y(\eta, h) = f_{\alpha\beta}\eta L_\alpha L_\beta + f'_{\alpha\beta}h^{++}e^T_{\alpha,\mathrm{R}}C^{-1}e_{\beta,\mathrm{R}} + \text{ H.c.} \tag{10.23}$$

The Higgs potential has one term that is of interest in connection with neutrino masses, i.e. $\lambda''v_{\mathrm{wk}}h^*\eta\eta$. The model leads to nonzero contribution to neutrino masses at the two-loop level via the diagrams of Fig. 10.2.

A typical estimate for the neutrino mass in this case is of order $f^2 f'm_\ell^2\lambda''/(16\pi^2)^2 M$. For $f \sim 10^{-1.5} \sim f' \sim \lambda''$, the two-loop contribution to the neutrino mass is of order 0.01 eV, which is of the right order of magnitude for solving the present neutrino puzzles. No excessive fine tuning of couplings is needed. In building realistic models, however, one has to pay attention to possible flavor-changing neutral-current effects such as the $\mu \to e\gamma$ decay, for which there exist rather stringent constraints from the recent MEGA experiment [10.65]: $B(\mu \to e + \gamma) \leq 10^{-11}$. This puts constraints on the parameters as follows: $f_{13}f_{23} \leq 10^{-5}$ and $(f'_{13}f'_{23} + f'_{12}f'_{11,22}) \leq 10^{-5}$. We shall not discuss detailed model building for this case.

One process that distinguishes the second class of models from the first is the muonium to antimuonium $(M–\bar{M})$ oscillation, which is mediated at the tree level by doubly charged bosons [10.66]. The present limit on the strength of the effective four-Fermi interaction describing the $M–\bar{M}$ transition is given by $G_{M–\bar{M}} \leq 3 \times 10^{-3}G_{\mathrm{F}}$ [10.56]. For a 100 GeV h^{++} boson, this implies that $h_{ee}h_{\mu\mu} \leq 3 \times 10^{-4}$, which is a nontrivial constraint on the model. Since, for the neutrino masses to be in the interesting range, the Higgs masses should not be more than 1 TeV, further improvement of the $M–\bar{M}$ transition limit would provide important information on this model.

10.5.2   High-Mass Higgs Triplet and Induced Neutrino Masses

Another way to generate nonzero neutrino masses without using the right-handed neutrino is to include in the Standard Model an $SU(2)_{\mathrm{L}}$ triplet Higgs field with $Y = 2$ so that the electric-charge profile of the members of the multiplet is given as follows: $(\Delta^{++}, \Delta^{+}, \Delta^{0})$. This allows an additional Yukawa coupling of the form $f_L L^T \tau_2 \tau L . \Delta$, where the Δ^{0} couples to the neutrinos. Clearly the Δ field has $L = 2$. When the Δ^{0} field has a nonzero vacuum expectation value (vev), it breaks lepton number by two units and leads to a Majorana mass for the neutrinos. There are two questions that arise now: how does the vev arise in the model, and how does one understand the smallness of the neutrino masses in this scheme? There are two answers to the first question. One can maintain exact lepton number symmetry in the model and generate the vev of the triplet field via the usual "Mexican hat" potential. There are two problems with this case. It leads to a massless triplet Majoron [10.67], which has been ruled out by LEP data on the Z width. Also, though it is now redundant, it may be worth pointing out that in this model the smallness of the neutrino mass is not naturally understood.

There is, however, another way to generate the induced vev, by keeping a large but positive mass (M_Δ) for the triplet Higgs boson and allowing for a lepton-number-violating coupling $M \Delta^* H H$. In this case, minimization of the potential induces a vev for the Δ^{0} field when the doublet field acquires a vev:

$$ v_{\mathrm{T}} \equiv \langle \Delta^{0} \rangle = \frac{M v_{\mathrm{wk}}^2}{M_\Delta^2} . \tag{10.24} $$

Since the mass of the Δ field is invariant under $SU(2)_{\mathrm{L}} \times U(1)_{\mathrm{Y}}$, it can be very large, connected perhaps with some new scale of physics. If we assume that $M_\Delta \sim M \sim 10^{13}$ GeV or so, we get $v_{\mathrm{T}} \sim 1$ eV. Now, in the Yukawa coupling $f_L L^T \tau_2 \tau L . \Delta$, since the Δ^{0} couples to the neutrinos, its vev leads to a neutrino mass in the eV range or less, depending on the values of the Yukawa couplings [10.68]. We shall see later, when we discuss the seesaw models, that unlike the case in those models, the neutrino mass in this case is not hierarchically dependent on the charged-fermion masses. Another point worth emphasizing is that, unlike the previous radiative scenarios, this model is more in the spirit of grand unification and can in fact be implemented in models [10.1] such as those based on the $SU(5)$ group, where there is no natural place for the right-handed neutrino.

10.5.3   The Baryogenesis Problem in Models Without Right-Handed Neutrinos

While, strictly, the models just discussed provide a way to understand the small neutrino masses without fine tuning, they may face problems in explaining the origin of matter in the Universe. Let us first consider the triplet vev models. It was shown by Ma and Sarkar [10.68] that the decay of the triplet Higgs to leptons provides a way to generate enough baryons in those models. However, for that to happen, one must satisfy one of Sakharov's three conditions that the decaying particle which leads to baryon or lepton asymmetry must be out of equilibrium. This requires that when the temperature of the Universe equals the decaying particle's mass, its decay rate must be smaller than the expansion rate of the Universe, i.e.

$$\frac{f_{\mathrm{L}}^2 M_\Delta}{12\pi} < \sqrt{g^*}\, \frac{M_\Delta^2}{M_{\mathrm{P}\ell}} \ . \tag{10.25}$$

This implies a lower limit on the mass $M_\Delta \geq f_{\mathrm{L}}^2 M_{\mathrm{P}\ell}/12\pi \sqrt{g^*}$. For $f_{\mathrm{L}} \sim 10^{-1}$, as would be required by the atmospheric neutrino data, one obtains, conservatively, $M_\Delta \geq 10^{13}$ GeV. The problem with such a large mass arises from the fact that in an inflationary model of the Universe, the typical reheating temperature dictated by the gravitino problem of supergravity is at most 10^9 GeV. Thus there is an inherent conflict [10.69] between the standard inflationary picture of the Universe and baryogenesis in the simple triplet model for neutrino masses.

As far as the radiative models go, they have explicit lepton-number-violating terms in the Lagrangian which are significant enough to keep these interactions in equilibrium for all temperatures low enough to be of interest ($T < 10^{12}$ GeV or so). So it is not clear how one would ever generate any baryon number in this class of models.

10.6 The Seesaw Mechanism and Left–Right Symmetric Unification Models for Small Neutrino Masses

A very natural and elegant way to generate neutrino masses is to include right-handed neutrinos in the Standard Model. However, the inclusion of right-handed neutrinos transforms the dynamics of gauge models in a profound way. To clarify what we mean, note that in the Standard Model (if it does not contain a ν_{R}) the $B - L$ symmetry is only linearly anomaly-free, i.e. $\mathrm{Tr}[(B - L)Q_a^2] = 0$, where the Q_a are the gauge generators of the Standard Model, but $\mathrm{Tr}(B - L)^3 \neq 0$. This means that $B - L$ is only a global symmetry and cannot be gauged. However, as soon as the ν_{R} is added to the Standard Model, one obtains $\mathrm{Tr}[(B - L)^3] = 0$, implying that the $B - L$ symmetry is now gaugeable, and one could choose the gauge group of nature to be either $SU(2)_{\mathrm{L}} \times U(1)_{I_{3\mathrm{R}}} \times U(1)_{B-L}$ or $SU(2)_{\mathrm{L}} \times SU(2)_{\mathrm{R}} \times U(1)_{B-L}$, the latter being the gauge group of the left–right symmetric models [10.70]. Furthermore, the presence of the ν_{R} makes the model quark–lepton symmetric and leads to a Gell-Mann–Nishijima-like formula for the electric charges [10.71], i.e.

$$Q = I_{3\mathrm{L}} + I_{3\mathrm{R}} + \frac{B - L}{2} \ . \tag{10.26}$$

The advantage of this formula over the charge formula in the Standard Model is that in this case all entries have a physical meaning. It also provides a natural understanding of the Majorana nature of neutrinos, as can be seen

Table 10.2. Assignment of the fermion and Higgs fields to the representation of the left–right symmetry group

Field	$SU(2)_L \times SU(2)_R \times U(1)_{B-L}$ representation
Q_L	(2,1,+1/3)
Q_R	(1,2,1/3)
L_L	(2,1,−1)
L_R	(1,2,−1)
ϕ	(2,2,0)
Δ_L	(3,1,+2)
Δ_R	(1,3,+2)

by looking at the distance scale where the $SU(2)_L \times U(1)_Y$ symmetry is valid but the left–right gauge group is broken. In this case,

$$\Delta Q = 0 = \Delta I_{3L} , \tag{10.27}$$

$$\Delta I_{3R} = -\Delta \left(\frac{B-L}{2} \right) .$$

We see that if the Higgs fields that break the left–right gauge group carry right-handed isospin of one, one must have $|\Delta L| = 2$, which means that the neutrino mass must be of the Majorana type, and the theory breaks lepton number by two units.

Let us now proceed to discuss the left–right symmetric model and demonstrate how small neutrino masses are understood in this model.

The gauge group of the theory is $SU(2)_L \times SU(2)_R \times U(1)_{B-L}$, with quarks and the leptons transforming as doublets under $SU(2)_{L,R}$. In Table 10.2, we present the transformation properties of the quark, lepton and Higgs fields of the model under the gauge group.

The first task is to specify how the left–right symmetry group breaks to the Standard Model, i.e. how one breaks the $SU(2)_R \times U(1)_{B-L}$ symmetry so that the successes of the Standard Model, including the observed predominant V–A structure of weak interactions at low energies, is reproduced. Another question of naturalness that also arises simultaneously is that since the charged fermions and the neutrinos are treated completely symmetrically (quark–lepton symmetry) in this model, how does one understand the smallness of the neutrino masses compared with those of the other fermions?

It turns out that both of the above problems of the left–right model have a common solution. The process of spontaneous breaking of the $SU(2)_R$ symmetry that suppresses the V+A currents at low energies also solves the problem of ultralight neutrino masses. To see this, let us write the Higgs fields in terms of their components:

$$\Delta = \begin{pmatrix} \Delta^+/\sqrt{2} & \Delta^{++} \\ \Delta^0 & -\Delta^+/\sqrt{2} \end{pmatrix}, \quad \phi = \begin{pmatrix} \phi_1^0 & \phi_2^+ \\ \phi_1^- & \phi_2^0 \end{pmatrix}. \tag{10.28}$$

All these Higgs fields have Yukawa couplings to the fermions given symbolically as below:

$$\mathcal{L}_Y = h_1 \bar{L}_L \phi L_R + h_2 \bar{L}_L \tilde{\phi} L_R + h_1' \bar{Q}_L \phi Q_R + h_2' \bar{Q}_L \tilde{\phi} Q_R$$
$$+ f(L_L L_L \Delta_L + L_R L_R \Delta_R) + \text{H.c.} \tag{10.29}$$

The $SU(2)_R \times U(1)_{B-L}$ symmetry is broken down to the Standard Model hypercharge $U(1)_Y$ by choosing $\langle \Delta_R^0 \rangle = v_R \neq 0$, since this carries both $SU(2)_R$ and $U(1)_{B-L}$ quantum numbers. It gives mass to the charged and neutral right-handed gauge bosons, i.e. $M_{W_R} = g v_R$ and $M_{Z'} = \sqrt{2} g v_R \cos \theta_W / \sqrt{\cos 2\theta_W}$. Thus, by adjusting the value of v_R one can suppress the right-handed current effects in both neutral- and charged-current interactions arbitrarily, leading to an effective near-maximal left-handed form for the charged-current weak interactions at low energies.

The fact that at the same time the neutrino masses also become small can be seen by looking at the form of the Yukawa couplings. Note that the f term leads to a mass for the right-handed neutrinos only at the scale v_R. Next, as we break the Standard Model symmetry by turning on the vev's for the ϕ fields as $\text{Diag}\langle \phi \rangle = (\kappa, \kappa')$, we give masses not only to the W_L and the Z bosons but also to the quarks and the leptons. In the neutrino sector the above Yukawa couplings after $SU(2)_L$ breaking by $\langle \phi \rangle \neq 0$ lead to Dirac masses for the neutrino connecting the left- and right-handed neutrinos. In the two-component neutrino language, this leads to the following mass matrix for ν, N (where we have denoted the left-handed neutrino by ν and the right-handed component by N):

$$M = \begin{pmatrix} 0 & h\kappa \\ h\kappa & f v_R \end{pmatrix}. \tag{10.30}$$

By diagonalizing this 2×2 matrix, we find the light-neutrino eigenvalue to be $m_\nu \simeq (h\kappa)^2 / f v_R$ and the heavy one to be $f v_R$. Note that typical charged-fermion masses are given by $h'\kappa$, etc. So, since $v_R \gg \kappa, \kappa'$, the light-neutrino mass is automatically suppressed. This way of suppressing the neutrino masses is called the seesaw mechanism [10.72]. Thus, in one stroke, one explains the smallness of the neutrino mass as well as the suppression of the V+A currents.[7]

---

[7] There is an alternative class of left–right symmetric models where small neutrino masses can arise from radiative corrections, if one chooses only doublet Higgses to break the gauge symmetry and heavy vector-like charged fermions to understand fermion mass hierarchies. In these models the neutrinos are Dirac particles [10.73]. For another idea, which uses composite Higgs bosons in a left–right symmetric framework to explain small Dirac neutrino masses, see [10.74].

In deriving the above seesaw formula for neutrino masses, it has been assumed that the vev of the left-handed triplet is zero so that the $\nu_L\nu_L$ entry of the neutrino mass matrix is zero. However, in most explicit models, such as the left–right model, which provide an explicit derivation of this formula, there is an induced vev for Δ_L^0 of order $\langle\Delta_L^0\rangle = v_T \simeq v_{wk}^2/v_R$. In the presence of this term the seesaw formula undergoes a fundamental change. Let us therefore distinguish between two types of seesaw formulae:

- *Type I seesaw formula.* This formula is of the form

$$M_\nu \simeq -M_D^T M_{N_R}^{-1} M_D ,$$ (10.31)

where M_D is the Dirac neutrino mass matrix and $M_{N_R} \equiv f v_R$ is the right-handed neutrino mass matrix in terms of the Δ Yukawa coupling matrix f.
- *Type II seesaw formula.* This formula is of the form

$$M_\nu \simeq f\frac{v_{wk}^2}{v_R} - M_D^T M_{N_R}^{-1} M_D .$$ (10.32)

Note that in the type I seesaw formula, what appears is the square of the Dirac neutrino mass matrix, which in general is expected to have the same hierarchical structure as the corresponding charged-fermion mass matrix. In fact, in some specific GUT models such as $SO(10)$, $M_D = M_u$. This is the origin of the common statement that neutrino masses given by the seesaw formula are hierarchical, i.e. $m_{\nu_e} \ll m_{\nu_\mu} \ll m_{\nu_\tau}$, and even a more model-dependent statement that $m_{\nu_e} : m_{\nu_\mu} : m_{\nu_\tau} = m_u^2 : m_c^2 : m_t^2$.

On the other hand, if one uses the type II seesaw formula, there is no reason to expect a hierarchy, and in fact, if the neutrino masses turn out to be degenerate as discussed before as one possibility, one way to understand this may be to use the type II seesaw formula.

While the above seesaw formulae are extremely helpful in building models with ultralight neutrino masses, there are circumstances where they are not helpful. An obvious example is one where the right-handed neutrino mass matrix is singular owing to additional symmetries (e.g. $L_e - L_\mu - L_\tau$, say), in which case its inverse does not exist and alternative ways to understand the lightness of neutrinos must be explored. There is one such mechanism in the literature, which we shall call the *type III seesaw formula*, which involves a 3×3 seesaw pattern [10.75] rather than the 2×2 one just described. For one generation, it involves three fermions (ν, N, S), where ν and N are the usual left- and right-handed neutrinos and S is a singlet neutrino (singlet under the left–right group). The relevant seesaw matrix is given by

$$\mathcal{M}_\nu = \begin{pmatrix} 0 & m_D & 0 \\ m_D & 0 & M \\ 0 & M & \mu \end{pmatrix}.$$ (10.33)

We assume that $\mu, m_D \ll M$, in which case the three eigenvalues are given by

$$m_\nu \sim \frac{m_D^2 \mu}{M^2} . \qquad (10.34)$$

Its generalization to the case where each element is a matrix is obvious. The important point here is to note that the matrix μ is in the numerator. As a result, if we want the light-neutrino mass matrix to be singular (say $L_e - L_\mu - L_\tau$-invariant), then this can be built into the matrix μ and one can then use the seesaw formula given here to understand the smallness of the neutrino masses. In the context of left–right models this kind of structure for neutrino masses arises if the triplets above are replaced by $B - L = 1$ doublets and we supplement the model with one singlet fermion per generation. (*Problem: show this by explicit construction and work out the example with $L_e - L_\mu - L_\tau$ symmetry*). Another feature of the type III seesaw formula is that one could choose the $B - L$ breaking scale to be much lower than in the case of the type I and II formulae.

Let us now address the following question: to what extent one can understand the details of the neutrino masses and mixings using the seesaw formulae? The answer to this question is quite model-dependent. While one can say that there exist many models which fit the observations, none (except a few) are completely predictive. The problem in general is that the seesaw formula of type I has 12 parameters, which is why its predictive power is so limited. One number that is predicted in a particular class of seesaw models based on the $SO(10)$ group that embodies the left–right symmetric unification model or the $SU(4)$ color is the tau neutrino mass. In this class of models, one maintains the quark–lepton symmetry to leading order so that one has the relation $m_t(M_U) = m_{\nu_\tau}^D$. The tau neutrino mass is then given by the seesaw formula as $m_{\nu_\tau} \simeq m_t^2/M_{N3R}$. In one class of string-inspired models, where the $SU(2)_R$ symmetry and the GUT symmetry break at the same scale, the right-handed neutrino masses are generated by nonrenormalizable operators, and they are given in the simplest approximation by $f v_R^2/M_{P\ell} \simeq f \times 10^{14}$ GeV. If $f = 1$, then one obtains $m_{\nu_\tau} \simeq 0.03$ eV, which is the right value to fit the atmospheric neutrino data. The rest of the data, such as the maximal mixing with the muon neutrino, etc., are not predicted without further assumptions. Even the prediction of the ν_τ mass requires an assumption that the Yukawa coupling f must be unity.

10.6.1 $SO(10)$ Realization of the Seesaw Mechanism

The most natural grand unified theory for the seesaw mechanism is the $SO(10)$ model. In this section, some of the salient features of this realization are summarized. From the previous paragraph, we learn that the simplest left–right model with $B - L = 2$ triplets provides provides a direct realization of the seesaw mechanism. In the context of the $SO(10)$ model, the first point to note is that the **16**-dimensional spinor representation contains all the fermions of each generation in the Standard Model plus the right-handed neutrino. Thus the right-handed neutrino is automatic in the $SO(10)$ model. Secondly, in order to break the $B - L$ symmetry present in the $SO(10)$ group, one may use the Higgs multiplets in either the **16**- or **126**-dimensional representation. Under the left–right symmetric group $SU(2)_L \times SU(2)_R \times U(1)_{B-L} \times SU(3)_c$, these fields decompose as follows:

$$\mathbf{16}_H = (2, 1, 1/3, 3) \oplus (1, 2, -1/3, 3^*) \oplus (2, 1, -1, 1) \oplus (1, 2, +1, 1)$$
$$\mathbf{126} = (1, 1, -2/3, 3) \oplus (1, 1, 2/3, 3^*) \oplus (3, 1, -2, 1) \oplus (1, 3, +2, 1) + \dots$$

$$(10.35)$$

where the ellipsis denotes other multiplets that have no role in the discussion of neutrino mass. In order to break the $B - L$ symmetry it is the last entry in each of the multiplets whose neutral elements need to pick up a large vev. Note, however, that the first multiplet (i.e. the $\mathbf{16}_H$) does not have any renormalizable coupling with the **16** spinors which contain the ν_R, whereas there is a renormalizable $SO(10)$ invariant coupling of the form $\mathbf{16 \, 16 \, \overline{126}}$ for the second multiplet. Therefore, if we decided to stay with the renormalizable model, then one would need a **126**-dimensional representation to implement the seesaw mechanism, whereas if we used the **16** to break the $B - L$ symmtery, we would require nonrenormalizable couplings of the form $\mathbf{16}^2 \overline{\mathbf{16}}_H^2/M_{P\ell}$. This has important implications for the $B - L$ scale. In the former case, the $B - L$ breaking scale is at an intermediate level such as $\sim 10^{13}$ GeV or so, whereas in the latter case we can have a $B - L$ scale that coincides with the GUT scale of 2×10^{16} GeV[8] as in the typical SUSY GUT models.

Another advantage of $SO(10)$ models in understanding neutrino masses is that if one uses only the **10**-dimensional representation for giving masses to the quarks and leptons, one has the mass matrix of the up quark M_u being equal to the Dirac mass matrix of the neutrinos, which goes into the seesaw formula. As a result, if we work in a basis where the up quark masses are diagonal so that all CKM mixings come from the down mass matrix, then the number of arbitrary parameters in the seesaw formula goes down from 12 to 6. Thus, even though one cannot predict the neutrino masses and mixings, the parameters of the theory are fixed by the values of the masses and mixings used as inputs. This may then be testable through the other predictions of the theory. In this model, however, there are tree-level mass relations in the down sector such as $m_d/m_s = m_e/m_\mu$, which are renormalization-group-invariant and are in disagreement with observations. It may be possible in supersymmetric models to generate enough one-loop corrections out of the supersymmetry-breaking terms (nonuniversal) to save the situation.

There is one very special cases where all 12 of the parameters are predicted by the quark and lepton masses [10.76]. The Higgs content of this model is

---

[8] For a review of SUSY GUTs, see [10.78].

only one **10** and one **126**. The Yukawa couplings involve only nine parameters, all of which, along with the input vevs of the doublets in both these multiplets, are fixed by the quark masses and mixings and the three charged-lepton masses. The important point is that the same **126** responsible for the fermion masses also has a vev along the $\nu_R\nu_R$ directions so that it generates the right-handed neutrino mass matrix. Thus, in the light-neutrino sector, it is a completely parameter-free model. But this minimal $SO(10)$ model cannot fit both the solar and the atmospheric neutrino data and is therefore ruled out. Recently other $SO(10)$ models have been considered where, under different assumptions, the atmospheric and solar neutrino data can be explained together [10.77]. These models have many interesting features, which we do not go into here for lack of space.

In the minimal supersymmetric left–right model, an analogous situation happens where the neutrino Dirac masses are found to be equal to the charged-lepton masses [10.79]. Thus, in this model too, one has only six parameters to describe the neutrino sector, and once the neutrino data have been fitted all parameters in the model are fixed, so that one has predictions that can be tested. For instance, it has been emphasized in [10.79] that there is a prediction for $B(\tau \to \mu + \gamma)$, in this model that is about two orders of magnitude below the present limits [10.80] and could therefore be used to test the model.

There is yet another class of models, where by assigning $U(1)$ charges to the fermions of the Standard Model as well as to the ν_R fields [10.81], one restricts both the Dirac and the Majorana mass matrices (for the ν_R). One then assumes that the $U(1)$ charges originate from strings or some high-scale physics.

Finally, let us comment that in models where the light neutrino mass is understood via a seesaw mechanism that uses heavy right-handed neutrinos, there is a very simple mechanism for the generation of baryon asymmetry of the Universe. Since the right-handed neutrino has a high mass, it decays at a high temperature to generate a lepton asymmetry [10.82],[9] and this lepton asymmetry is converted to baryon asymmetry via sphaleron effects [10.84], at lower temperature. It also turns out that one of the necessary conditions for sufficient leptogenesis is that the right-handed neutrinos must be heavy, as is required by the seesaw mechanism. To see this, note that one of the Sakharov conditions for leptogenesis is that the decay of the right-handed neutrino must be slower than the expansion rate of the Universe at a temperature $T \sim M_{N_R}$. The corresponding condition is

$$\frac{h_\ell^2 M_{N_R}}{16\pi} \leq \sqrt{g^*}\frac{M_{N_R}^2}{M_{P\ell}} \ . \tag{10.36}$$

---

[9] For recent discussions, see [10.83]. See these papers also for references to earlier work.

This implies that $M_{N_R} \geq h_\ell^2 M_{P\ell}/16\pi\sqrt{g^*}$. For the second generation, it implies that $M_{N_{2R}} \geq 10^{13}$ GeV and for the third generation a value even higher. Note that these are above the upper bound for reheating during inflation alluded to before. However, for the first generation, the mass limit is about 10^8 GeV, so that there is no conflict with the gravitino bound on the reheating temperature. Incidentally, the leptogenesis condition also imposes limits on the matrix elements of the right-handed neutrino mass, thereby reducing the arbitrariness of the seesaw predictions slightly.

10.7   Naturalness of Degenerate Neutrinos

In this section we would like to discuss some issues related to the degenerate-neutrino hypothesis. Recall that this is the only scenario that fits all observations if one does not include the LSND results (for instance if they are not confirmed by MiniBooNE) and the Universe has 10% to 20% of its matter in the form of neutrinos. Thus it is appropriate to discuss how such models can arise in theoretical schemes and how stable they are under radiative corrections.

The first point, already alluded to before and first made in [10.22], is that degenerate neutrinos arise naturally in models that employ the type II seesaw since the first term in the mass formula is not connected to the charged-fermion masses. One way that has been discussed is to consider schemes where one uses symmetries such as $SO(3)$ or $SU(2)$ or the permutation symmetry S_4 [10.85] so that the Majorana Yukawa couplings f_i are all equal. This then leads to the dominant contribution to all neutrinos being equal. This symmetry, however, must be broken in the charged-fermion sector in order to explain the observed quark and lepton masses. Models of this kind consistent with known data have been constructed on the basis of $SO(10)$ as well as other groups. The interesting point about the model based on $SO(10)$ is that the dominant contributions to the Δm^2s in this model come from the second term in the type II seesaw formula, which in simple models is hierarchical. It is of course known that if the MSW solution to the solar neutrino puzzle is the right solution (or an energy-independent solution), then we have $\Delta m^2_{\text{solar}} \ll \Delta m^2_{\text{atmos}}$. In fact, if we use the feature of $SO(10)$ models that $M_u = M_D$, then we have $\Delta m^2_{\text{atmos}} \simeq m_0 m_t^2/f v_R$ and $\Delta m^2_{\text{solar}} \simeq m_0 m_c^2/f v_R$ where m_0 is the common mass for the three neutrinos. It is interesting that for m_0 of the order of a few eV and $f v_R \approx 10^{15}$ GeV, both the Δm^2s are close to the required values.

An interesting theoretical issue about these models has been raised in several recent papers [10.86]. It has been noted that even though one may have tree-level models with a degenerate neutrino spectrum, it is not clear that this mass degeneracy will survive the radiative corrections. In fact, it has been convincingly argued that in models with maximal mixings the departure from degeneracy may be significant. This provides a further challenge to model building, and one must ensure that should maximal-mixing models win the day, the degeneracy is preserved at least up to two loops.

10.8   Theoretical Understanding of the Sterile Neutrino

If the existence of the sterile neutrino becomes confirmed, say by a confirmation of the LSND observation of ν_μ–ν_e oscillation or directly by SNO neutral-current data in the early part of the 21st century, a key theoretical challenge will be to construct an underlying theory that embeds the sterile neutrino along with the others with an appropriate mixing pattern, while naturally explaining its ultralightness.

If a sterile neutrino was introduced into the Standard Model, the gauge symmetry would not forbid its bare mass, implying that there would be no reason for its mass to be small. It is a common experience in physics that if a particle has a mass lighter than that normally expected on the basis of known symmetries, then this is an indication of the existence of new symmetries. This line of reasoning has been pursued in recent literature to understand the ultralightness of the sterile neutrino by using new symmetries beyond the Standard Model.

There are several suggestions for this new symmetry that might help us to accommodate an ultralight ν_s and in the process lead us to new classes of extensions of physics beyond the Standard Model. We shall consider two examples: (i) one based on the E_6 grand-unification model, and (ii) another based on the possible existence of mirror matter in the Universe. There are also other theoretical models for the sterile neutrino that involve different assumptions [10.87], but we do not discuss them here.

A completely different way to understand the ultralightness of the sterile neutrino is to introduce large extra dimensions in a Kaluza–Klein framework [10.88] and not rely on any symmetries. In these models, the sterile neutrino is supposed to live in the bulk; therefore, if its Lagrangian mass is zero, then the mass of the first Kaluza–Klein mode is inversely proportional to the size of the extra dimension. Thus a size of the extra dimension of the order of a millimeter would lead to the sterile-neutrino masses in the range of meV which are of interest for the problem of solar neutrino oscillations. There have been recent attempts [10.89] to build realistic models using a slight variation of this idea [10.90]. Models of this class are distinguished from others by their prediction of an infinite tower of sterile neutrinos, which at the moment seems perfectly consistent with observations. Eventually, this feature may provide a way to test such theories. We shall not elaborate on these models any further.

10.9   The E_6 Model for the Sterile Neutrino

E_6 [10.91] is an interesting and viable unification group beyond the $SO(10)$ group and as such has been extensively discussed in the literature [10.92].

In this model, matter belongs to the **27**-dimensional representation of the E_6 group, which, under its $SO(10) \times U(1)$ subgroup, decomposes to $\mathbf{16}_{+1} \oplus \mathbf{10}_{-2} \oplus \mathbf{1}_{+4}$ (the subscripts represent the $U(1)$ charges). The **16** is well known to contain the left- and right-handed neutrinos (to be denoted here by ν_i and ν_i^c, i being the family index). The **10** contains two neutral, colorless fermions which behave like neutrinos but are $SU(2)_L$ doublets, and the last neutral, colorless fermion in the **27**, which we identify as the sterile neutrino, is the one contained in the $SO(10)$ singlet multiplet (denoted by ν_{is}). This has all the properties desired of a sterile neutrino. Thus, in this model, there are three sterile neutrinos, which will be denoted by the corresponding flavor labels, and we shall show first how the small mass for both the active and the sterile neutrino results from a generalization of the the seesaw mechanism as a consequence of the symmetries of the group [10.93]. Furthermore, we shall see how, as a consequence of the smallness of the Yukawa couplings of the Standard Model, we can obtain not only maximal mixing between the active and sterile neutrinos of each generation, but also the necessary ultrasmall Δm^2 needed in the vacuum oscillation solution to the solar neutrino puzzle without fine tuning of parameters [10.52]. Thus, in this model, both the solar and the atmospheric neutrino puzzles are solved by the maximal vacuum oscillation of active to sterile neutrinos.

In general, in this model, we shall have for each generation a 5×5 "neutrino" mass matrix. To give the essence of the basic steps that lead to the seesaw mechanism, it is necessary to describe the symmetry breaking of E_6. We work with a supersymmetric E_6, and use three pairs of $\mathbf{27} + \overline{\mathbf{27}}$ representations and one **78**-dimensional field for symmetry breaking. The pattern of symmetry breaking is as follows:

1. $\langle 27_1 \rangle$ and $\langle \overline{27}_1 \rangle$ have GUT-scale vevs in the $SO(10)$ singlet direction.
2. $\langle 27_{16} \rangle$ and $\langle \overline{27}_{16} \rangle$ have GUT-scale vevs in the ν and ν^c directions, respectively. They break $SO(10)$ down to SU(5).
3. The $\langle 78_{[1,45]} \rangle$ completes the breaking of SU(5) down to the Standard Model gauge group at the GUT scale. We assume the vevs reside both in the adjoint and in the singlet of $SO(10)$.
4. $\langle 27_{10} \rangle$ and $\langle \overline{27}_{10} \rangle$ contain the Higgs doublets of the minimal supersymmetric model. It is assumed that H_u and H_d are both linear combinations, arising partially from the $\langle 27_{10} \rangle$ and partially from the $\langle \overline{27}_{10} \rangle$.

In addition to the above, there is another field, labelled by $27'$, whose ν^c component mixes with a singlet S, and one linear combination of this pair remains light below the GUT scale. As a consequence of radiative symmetry breaking, this picks up a vev at the electroweak scale. This was shown in [10.52]. The remaining components of $27'$ have GUT-scale masses. Let us now write down the relevant terms in the superpotential that lead to a 5×5 "neutrino" mass matrix of the form we desire. To keep matters simple, let us ignore generation mixings, which can be incorporated very trivially. The terms are

$$W = \lambda_i \psi_i \psi_i 27_{10} + f_i \psi_i \psi_i 27' + \frac{\alpha_i}{M_{P\ell}} \psi_i \psi_i 27_1 78_{[1,45]} + \frac{\gamma_i}{M_{P\ell}} \psi_i \psi_i \overline{27}_{16} \overline{27}_{16} \,.$$

(10.37)

We have shown only a subset of allowed terms in the theory that play a role in the neutrino mass physics, and we believe that it is reasonable to assume a discrete symmetry (perhaps in the context of a string model) that would allow only this subset. In any case, since we are dealing with a supersymmetric theory, radiative corrections will not generate any new terms in the superpotential.

Note that in (10.37), since the first term leads to lepton and quark masses of various generations, this term carries a generation label and obeys a hierarchical pattern, whereas the f_is not being connected to known fermion masses, need not obey a hierarchical pattern. We shall from now on assume that each $f_i \approx 1$.

After substituting the vevs for the Higgs fields in the above equation, we find a 5×5 mass matrix of the following form for the neutral-lepton fields of each generation in the basis $(\nu, \nu_s, \nu^c, E_u^0, E_d^0)$:

$$M = \begin{pmatrix} 0 & 0 & \lambda_i v_u & f_i v' & 0 \\ 0 & 0 & 0 & \lambda_i v_d & \lambda_i v_u \\ \lambda_i v_u & 0 & M_{\nu^c,i} & 0 & 0 \\ f_i v' & \lambda_i v_d & 0 & 0 & M_{10,i} \\ 0 & \lambda_i v_u & 0 & M_{10,i} & 0 \end{pmatrix} \,.$$

(10.38)

Here $M_{\nu^c,i}$ is the mass of the right-handed neutrino and $M_{10,i}$ is the mass of the entire **10**-plet in the **27** matter multiplet. Since **10** contains two full $SU(5)$ multiplets, gauge coupling unification will not be affected even though we choose its mass to be below the GUT scale.

Note that the 3×3 mass matrix involving ν^c, E_u^0, E_d^0 have superheavy entries and will therefore decouple at low energies. Their effects on the spectrum of the light neutrinos will be dictated by the seesaw mechanism [10.72]. The light-neutrino mass matrix involving ν_i, ν_{is} can be written down as

$$M_{\text{light}} \simeq \frac{1}{M_{\nu^c,i}} \begin{pmatrix} \lambda_i v_u & f_i v' & 0 \\ 0 & \lambda_i v_d & \lambda_i v_u \end{pmatrix} \begin{pmatrix} 1 & 0 & 0 \\ 0 & 0 & \epsilon \\ 0 & \epsilon & 0 \end{pmatrix} \begin{pmatrix} \lambda_i v_u & 0 \\ f_i v' & \lambda_i v_d \\ 0 & \lambda_i v_u \end{pmatrix} \,,$$

(10.39)

where $\epsilon_i = M_{10,i}/M_{\nu^c,i}$. Note that ϵ_i is expected to be of order one. This leads to a 2×2 mass matrix for the (ν, ν_c) fields of each generation of the form

$$M_i = m_{0i} \begin{pmatrix} \lambda_i^2 & \lambda_i \bar{f}_i \\ \lambda_i \bar{f}_i & \lambda_i^2 \bar{\epsilon}_i \end{pmatrix} \,.$$

(10.40)

Here $m_{0i} = v_u^2/M_{\nu^c,i}$, $\bar{f}_i = f_i \epsilon_i v'/v_u$ and $\bar{\epsilon}_i = 2\epsilon_i \cot \beta$. Taking $M_{\text{Pl}} \sim 10^{19}$ GeV, $M_{\text{GUT}} \sim 10^{16}$ GeV and reasonable values of the unknown parameters, e.g. $\alpha_i \approx 0.1$, $\gamma_i \approx 0.1$, $f_i \approx 1$ and $v' \approx v_u$, we obtain $m_{0i} \simeq 20$ eV and

$\epsilon \approx 1$, which leads us to the desired pattern of masses and mass differences, where active neutrinos of each generation mix with the corresponding sterile neutrino maximally and the Δm^2s scale as the cube of the corresponding charged-fermion mass. Thus, if we fix $\Delta m^2_{\text{atmos}} \sim 10^{-3}$ eV^2, then we obtain the vacuum oscillation solution to the solar neutrino puzzle.

This model also provides an explanation of the LSND results, and the smallness of the mixing angle observed in the experiment is now similar to that observed in the quark sector and therefore easily understood by the same mechanisms that provide an explanation of the small quark mixing angles.

10.10   The Mirror Universe Model of the Sterile Neutrino

The second suggestion that explains the ultralightness of the ν_{s} is the mirror matter model [10.50, 10.51] where the basic idea is that there is a complete duplication of matter and forces in the Universe (i.e. two sectors in the Universe, with matter and gauge forces identical prior to symmetry breaking). The mirror sector of the model then has three light neutrinos which do not couple to the Z boson and would not therefore have been seen at LEP. We denote the fields in the mirror sector by primes on the Standard Model fields. We shall call the ν_i' the sterile neutrinos, of which we now have three. The lightness of ν_i' is dictated by the mirror $B' - L'$ symmetry in a manner parallel to what happens in the Standard Model. Thus the ultralightness of the sterile neutrinos is understood in the most "painless" way.

The two "universes" are assumed to communicate only via gravity [10.94, 10.95] or other forces that are equally weak. This leads to a mixing between the neutrinos of the two universes and can cause oscillations between the ν_e of our universe and the ν_e' of the parallel universe in order to explain, for example, the solar neutrino deficit.

At an overall level, such a picture emerges quite naturally in superstring theories, which lead to $E_8 \times E_8'$ gauge theories below the Planck scale, with the two E_8s connected by gravity. For instance, one may assume the sub-Planck GUT group to be a subgroup of $E_8 \times E_8'$ in anticipation of possible future string embedding. One may also imagine the visible sector and the mirror sector as being in two different D-branes, which are then necessarily connected very weakly owing to exchange of massive bulk Kaluza–Klein excitations.

In the mirror model, either both sectors can remain identical after symmetry breaking or there can be asymmetry. We shall consider the second scenario. This was suggested in [10.51].

As suggested in [10.51], we shall assume that the process of spontaneous symmetry breaking introduces asymmetry between the two universes, e.g. the weak scale v_{wk}' in the mirror universe is larger than the weak scale $v_{\text{wk}} = 246$ GeV in our universe. The ratio of the two weak scales $v_{\text{wk}}'/v_{\text{wk}} \equiv \zeta$ is the only parameter that enters the fit to the solar neutrino data. It was shown in [10.51] that with $\zeta \simeq 20$–30, the gravitationally generated neutrino masses [10.61] can provide a resolution of the solar neutrino puzzle (i.e. one parameter generates both the required $\Delta m^2_{e\text{-s}}$ and the mixing angle $\sin^2 2\theta_{e\text{-s}} \simeq 10^{-2}$). There can also be a large-angle MSW fit with a reasonable choice of parameters, e.g. a smaller ζ but with the coefficients of higher-dimensional operators allowed to vary between 0.3 and 3.

There are other ways to connect the visible sector with the mirror sector, using, for instance, a bilinear term involving the right-handed neutrinos from the mirror and visible sectors. An $SO(10) \times SO(10)$ realization of this idea was studied in detail in a recent paper [10.96], where a complete realistic model for known particles and forces, including a fit to the fermion masses and mixings, was worked out and the resulting predictions for the masses and mixings for the normal and mirror neutrinos were presented. In this model, the fermions of each generation are assigned to the $(\mathbf{16}, \mathbf{1}) \oplus (\mathbf{1}, \mathbf{16'})$ representation of the gauge group. The $SO(10)$ symmetry is broken down to the left–right symmetric model by the combination of $\mathbf{45} \oplus \mathbf{54}$ representations in each sector. The $SU(2)_\mathrm{R} \times U(1)_{B-L}$ gauge symmetry in turn is broken by the $\mathbf{126} \oplus \overline{\mathbf{126}}$ representations. These latter fields serve two purposes: first, they guarantee automatic R-parity conservation, and second, they lead to the seesaw suppression of the neutrino masses. We shall not go into detailed discussion of the masses and mixings of neutrinos in such a model here. The mixing between the two sectors is caused by a multiplet belonging to the representation $(\mathbf{16}, \mathbf{16'}) + (\overline{\mathbf{16}}, \overline{\mathbf{16}}^{\,'})$. If the mass of this last multiplet is kept at the GUT scale, the mixing between the two right-handed neutrinos caused by this is large and the ensuing effects on the light-neutrino mixings are small.

Since the mirror matter model has many ultralight particles, the issue of consistency with Big Bang nucleosynthesis must be addressed. Recall that present observations of helium and deuterium abundance can allow for at most 4.53 neutrino species [10.37] if the baryon-to-photon ratio is chosen appropriately. Since the model has three extra neutrinos, an extra photon and an extra $e^+ e^-$ pair, a priori, the effective neutrino count in the model could be as large as 8.2. However, in the asymmetric mirror model, since the neutrinos decouple above 200 MeV or so owing to weakness of the mirror Fermi coupling, their contribution at the time of nucleosynthesis is negligible (i.e. they contribute about 0.3 to δN_ν.) On the other hand, the mirror photon could be completely in equilibrium at $T = 1$ MeV, so that it will contribute $\delta N_\nu = 1.11$. There is also a contribution from the mirror electron–positron pair of $N_\nu = 2$. Altogether, the total contribution to N_ν is less than 6.4. So, to solve this problem, we invoke an idea called asymmetric inflation, whereby the reheating temperature in the mirror sector is lower than the reheating temperature in the familiar sector. Models with this kind of possibility are discussed in [10.97].[10]

---

[10] For an alternative model of asymmetric inflation, see [10.98].

There may be another potentially very interesting application of the idea of the mirror universe. It appears that there may be a crisis in understanding the microlensing observations [10.99].[11] This has to do with the fact that the best-fit mass for the 14 microlensing events observed by the MACHO and EROS groups is around $0.5M_\odot$ and it appears difficult to use normal baryonic objects of similar mass such as white dwarfs to explain these events, since they lead to a number of cosmological and astrophysical problems [10.101]. Speculations have been advanced that this crisis may be resolved by the postulate that the MACHOs (massive compact halo objects) with $0.5M_\odot$ may be mirror stars [10.102, 10.103], which would then have none of the difficulties that arise from interpretations in terms of conventional baryonic matter. In this model, mirror baryons can form the dark matter of the universe so that there is no need for neutralino dark matter.

10.11   Conclusions and Outlook

In this review, we have tried to provide a brief look at the new physics implied by the discovery of neutrino oscillations. We started with a summary of the possible textures for neutrino masses implied by the present data with and without the inclusion of the LSND data. Several three-, four- and six-neutrino scenarios have been outlined. We then briefly described the various ideas primarily aimed at understanding the smallness of the neutrino masses, in the context of both grand and simple unified models. Finally, two theories that can naturally incorporate an ultralight sterile neutrino were discussed. Clearly, there is considerable subjective judgment used in the selection of models, and an apology is due to all authors whose models are not discussed here. Limitation of space is certainly one simple excuse that can be given. In any case, the final story in this field is yet to be written and it could very easily be that the models described here do not eventually win out. Even that will be a very valuable piece of information, since the ideas touched on here are certainly some of the more salient ones around now. But it is the fervent hope of this author (presumably shared by many workers in the field) that some of the ideas described in the literature (and reviewed here) will survive the test of time. The ball is right now in the court of the experimentalists and we pin our hopes on the large number of continuing (SNO, Borexino, K2K, MINOS) as well as planned (KAMLAND, ICARUS, LENS, GENIUS, CUORE, ORLAND, etc.) experiments in this field. On the theoretical side, true progress can be said to have been made only in understanding the smallness of the neutrino masses in different scenarios, but the complete picture of mixing angles is far from being at hand, although there are many models that under various assumptions lead to realistic fits. One of the hardest theoretical problems is to understand the ultrasmall mass difference that would be required if the vacuum oscillation solution to the solar neutrino problem

---

[11] For a recent review, see [10.100].

were eventually to win the race (although there exist a handful of interesting suggestions [10.50, 10.52, 10.104]). One will then have to check whether the radiative corrections in these models destabilize the result.

The bottom line is that the field of neutrino mass has become one of the most vibrant and exciting fields in particle physics as we move into the new millennium and is at the moment the only beacon of new physics beyond the Standard Model.

Acknowledgment

This work was supported by the National Science Foundation grant number PHY-9802551. I would like to acknowledge many useful conversations on the subject of this review with K. S. Babu, D. Caldwell, Z. Berezhiani, B. Kayser, S. Nussinov and V. L. Teplitz.

References

10.1 R. N. Mohapatra and P. B. Pal, *Massive Neutrinos in Physics and Astro-physics*, 2nd edn (World Scientific, Singapore, 1998).

10.2 P. Fisher, B. Kayser and K. Macfarland, Annu. Rev. Nucl and Part. Sci., to appear (1999), hep-ph/9906244; S. Bilenky, C. Giunti and W. Grimus, Prog. Part. Nucl. Phys. (to appear) (1999), hep-ph/9812360.

10.3 C. W. Kim and M. Pevsner, "Neutrinos in Physics and Astrophysics" (Harwood Publishers, 1993)

10.4 Y. Fukuda et al., Phys. Rev. Lett. **81**, 1562 (1998).

10.5 Y. Fukuda et al. (Kamiokande collaboration), Phys. Lett. B **335**, 237 (1994).

10.6 R. Becker-Szendy et al. (IMB collaboration), Nucl. Phys. B, Proc. Suppl. **38**, 331 (1995).

10.7 W. W. M. Allison et al. (Soudan collaboration), Phys. Lett. B **391**, 491 (1997).

10.8 M. Ambrosio et al. Phys. Lett. **B 434**, 451 (1998).

10.9 J. Bahcall, P. Krastev and A. Y. Smirnov, hep-ph/9807216.

10.10 B. T. Cleveland et al., Astrophys. J. **496**, 505 (1998); K. S. Hirata et al. (Kamiokande collaboration), Phys. Rev. Lett. **77**, 1683 (1996); W. Hampel (GALLEX collaboration), Phys. Lett. B **388**, 384 (1996); J. N. Abdurashitov et al. (SAGE collaboration), Phys. Rev. Lett. **77**, 4708 (1996); Y. Suzuki (SuperKamiokande collaboration), talk presented at *Neutrino '98*, Takayama, Japan.

10.11 A. Guth, L. Randall and M. Serna, hep-ph/9903464; R. M. Crocker, R. Foot and R. Volkas, hep-ph/9905461.

10.12 C. Athanassopoulos, Phys. Rev. Lett. **77**, 3082 (1996); nucl-ex/9709006.

10.13 W. Louis, talk at the Institute of Nuclear Theory workshop, Seattle, July 1999.

10.14 B. Armbruster et al., Phys. Rev. C **57**, 3414 (1998); G. Drexlin, talk at Beyond99 workshop, Ringberg Castle, June 1999.

10.15 B. Pontecorvo, Sov. Phys. JETP **6**, 429 (1958); Z. Maki, M. Nakagawa and S. Sakata, Prog. Theor. Phys. **28**, 870 (1962).

10.16 L. Baudis et al. Phys. Rev. Lett. **83**, 41 (1999).

10.17 H. Klapdor-Kleingrothaus, hep-ex/9907040.

10.18 A. Staudt, K. Muto and H. Klapdor-Kleingrothaus, Europhys. Lett. **13**, 31 (1990).

10.19 H. Klapdor-Kleingrothaus et al., GENIUS proposal, hep-ex/9910205.

10.20 J. Primack and M. Gross, astro-ph/9810204.

10.21 E. Gawiser and J. Silk, Nature **280**, 1405 (1998).

10.22 D. Caldwell and R. N. Mohapatra, Phys. Rev. D **48**, 3259 (1993).

10.23 M. Appolonio et al., Phys. Lett. B **420**, 397 (1998).

10.24 G. Fogli, E. Lisi, A. Marrone and G. Scioscia, hep-ph/9808205; M. C. Gonzales-Garcia, H. Nunokawa, O. Peres and J. W. F. Valle, hep-ph/9807305; S. Bilenky, C. Giunti and W. Grimus, hep-ph/9809368.

10.25 V. Barger, S. Pakvasa, T. Weiler and K. Whisnant, hep-ph/9806387; S. Bilenky and C. Giunti, hep-ph/9802201; A. Baltz, A. Goldhaber and M. Goldhaber, hep-ph/9806540.

10.26 F. Vissani, hep-ph/9708483.

10.27 G. Altarelli and F. Feruglio, Phys. Lett. B **439**, 112 (1998); M. Jezabek and A. Sumino, hep-ph/9807310.

10.28 H. Fritzsch and Z. Z. Xing, hep-ph/9808272; Y. Koide, Phys. Rev. D **39**, 1391 (1989).

10.29 H. Georgi and S. L. Glashow, hep-ph/9808293.

10.30 R. N. Mohapatra and S. Nussinov, hep-ph/9809415; Phys. Rev. D **60**, 013002 (1999).

10.31 R. N. Mohapatra and S. Nussinov, hep-ph/9808301; Phys. Lett. B **441**, 299 (1998).

10.32 R. Barbieri et al., hep-ph/9807235; R. Barbieri, L. Hall and A. Strumia, hep-ph/9808333.

10.33 A. Joshipura and S. Rindani, hep-ph/9907390.

10.34 G. Altarelli and F. Feruglio, Phys. Lett. B **451**, 388 (1998).

10.35 E. Akhmedov, hep-ph/9909217.

10.36 Particle Data Group, Eur. Phys. J. C **3**, 1 (1998).

10.37 K. Olive, astro-ph/9903309; D. Olive and D. Thomas, hep-ph/9811444.

10.38 S. Sarkar, Rep. Prog. Phys. **59**, 1493 (1996).

10.39 R. Barbieri and A. Dolgov, Phys. Lett. B **237**, 440 (1990); K. Enquist, K. Kainulainen and J. Maalampi, Phys. Lett. B **249**, 531 (1992); D. P. Kirilova and M. Chizov, hep-ph/9707282.

10.40 J. Peltoniemi and J. W. F. Valle, Nucl. Phys. B **406**, 409 (1993).

10.41 S. Bilenky, C. Giunti and W. Grimus, Eur. Phys. J. C **1**, 247 (1998); Phys. Rev. D **57**, 1920 (1998); Phys. Rev. D **58**, 033001 (1998).

10.42 A. Smirnov and F. Vissani, Phys. Lett. B **432**, 376 (1998); J. Learned, S. Pakvasa and J. Stone, Phys. Lett. **B435**, 131 (1998); L. Hall and H. Murayama, Phys. Lett. B **436**, 323 (1998).

10.43 K. Takita, invited talk at the COSMO99 workshop, Trieste, September 1999.

10.44 B. Balatenkin, G. Fuller, A. Fetter and G. Mclaughlin, Phys. Rev. C **59**, 2873 (1999).

10.45 S. Bilenky, C. Giunti, W. Grimus and T. Schwetz, hep-ph/9904316.

10.46 J. Gelb and S. P. Rosen, hep-ph/9909293.

10.47 S. Mohanty, D. P. Roy and U. Sarkar, hep-ph/9805429; K. S. Babu, E. Ma and S. Nandi, talk by K. S. Babu, Pheno 99, Madison, Wisconsin (1999).

10.48 V. Barger, T. Weiler and K. Whisnant, Phys. Lett. B **427**, 97 (1998); S. Gibbons, R. N. Mohapatra, S. Nandi and A. Raichoudhuri, Phys. Lett. **430**, 296 (1998).

10.49 S. Goswami, Phys. Rev. D **55**, 2931 (1997); N. Okada and O. Yasuda, Int. J. Mod. Phys. A **12**, 3691 (1997); E. Lipmanov, hep-ph/9909457; A. Kalliomaki, J. Maalampi and M. Tanimoto, hep-ph/9909301.

10.50 R. Foot and R. Volkas, Phys. Rev. D **52**, 6595 (1995).

10.51 Z. Berezhiani and R. N. Mohapatra, Phys. Rev. D **52**, 6607 (1995).

10.52 Z. Chacko and R. N. Mohapatra, hep-ph/9905388.

10.53 S. Goswami, D. Majumdar and A. Raichoudhuri, hep-ph/9909453.

10.54 J. Bowes and R. Volkas, J. Phys. G **24**, 1249 (1998); A. Geiser, CERN-EP-98-056; M. Kobayashi, C. S. Lim and M. Nojiri, Phys. Rev. Lett. **67**, 1685 (1991).

10.55 P. Herczeg, in *Proceedings of Beyond 97 Workshop* (1997), p. 124, ed. by H. Klapdor-Kleingrothaas; Y. Grossman and S. Bergman, Phys. Rev. **59**, 093005 (1999); L. Johnson and D. Mckay, Phys. Lett. B **433**, 355 (1998).

10.56 K. Jungman et al., Phys. Rev. Lett. **82**, 49 (1999).

10.57 E. Ma and P. Roy, Phys. Rev. Lett. **80**, 4637 (1998); G. Brooijmans, hep-ph/9808498; M. Gonzales-Garcia et al., Phys. Rev. Lett. **82**, 3202 (1999); V. Barger, J. G. Learned, S. Pakvasa and T. Weiler, Phys. Rev. Lett. **82**, 2640 (1999).

10.58 S. Pakvasa, hep-ph/9910246.

10.59 S. Bergmann, Y. Grossman and D. Pierce, hep-ph/9909390.

10.60 S. Giddings and A. Strominger, Nucl. Phys. B **306**, 890 (1987).

10.61 R. Barbieri, J. Ellis and M. K. Gaillard, Phys. Lett. B **90**, 249 (1980); E. Akhmedov, Z. Berezhiani and G. Senjanović, Phys. Rev. Lett. **69**, 3013 (1992).

10.62 A. Zee, Phys. Lett. B **93**, 389 (1980).

10.63 K. S. Babu, Phys. Lett. B **203**, 132 (1988).

10.64 L. Wolfenstein, Nucl. Phys. B **175**, 93 (1980).

10.65 M. L. Brooks et al. (MEGA collaboration), hep-ex/9905013; Phys. Rev. Lett. (1999) (to appear).

10.66 D. Chang and W. Keung, Phys. Rev. Lett. **62**, 2583 (1989); R. N. Mohapatra, in *Proceedings of the 8th Workshop on Grand Unification*, Syracuse, ed. by K. C. Wali (1987).

10.67 G. Gelmini and M. Roncadelli, Phys. Lett. B **99**, 411 (1981).

10.68 R. N. Mohapatra and G. Senjanović, Phys. Rev. D **23**, 165 (1981); C. Wetterich, Nuc. Phys. B **187**, 343 (1981); E. Ma and U. Sarkar, Phys. Rev. Lett. **80** , 5716 (1998).

10.69 D. Delepine and U. Sarkar, Phys. Rev. D **60**, 055005 (1999).

10.70 J. C. Pati and A. Salam, Phys. Rev. D **10**, 275 (1974); R. N. Mohapatra and J. C. Pati, Phys. Rev. D **11**, 566, 2558 (1975); G. Senjanović and R. N. Mohapatra, Phys. Rev. D **12**, 1502 (1975).

10.71 R. N. Mohapatra and R. E. Marshak, Phys. Lett. B **91**, 222 (1980); A. Davidson, Phys. Rev. D **20**, 776 (1979).

10.72 M. Gell-Mann, P. Ramond and R. Slansky, in *Supergravity*, ed. by P. van Niewenhuizen and D. Z. Freedman (North-Holland, Amsterdam, 1979); T. Yanagida, in *Proceedings of the Workshop on Unified Theory and Baryon Number in the Universe*, ed. by O. Sawada and A. Sugamoto (KEK, Tsukuba, 1979); R. N. Mohapatra and G. Senjanović, Phys. Rev. Lett. **44**, 912 (1980).

10.73 D. Chang and R. N. Mohapatra, Phys. Rev. Lett. **58**, 1600 (1987); G. Barenboim and F. Scheck, hep-ph/9904331.

10.74 P. Divakaran and G. Rajasekaran, Mod. Phys. Lett. A **14**, 913 (1999).

10.75 R. N. Mohapatra and J. W. F. Valle, Phys. Rev. D **34**, 1642 (1986).

10.76 K. S. Babu and R. N. Mohapatra, Phys. Rev. Lett. **70**, 2845 (1993); D. G. Lee and R. N. Mohapatra, Phys. Rev. D **51**, 1353 (1995).

10.77 S. Bludman, N. Hata, D. Kennedy and P. Langacker, Nucl. Phys., Proc. Suppl. **31**, 156 (1993); Y. Achiman and T. Greiner, Nucl. Phys. B **443**, 3 (1995); B. Brahmachari and R. N. Mohapatra, Phys. Rev. D **58**, 015001 (1998); C. Albright, K. S. Babu and S. Barr, Phys. Rev. Lett. **81**, 1167 (1998); K. S. Babu, J. C. Pati and F. Wilczek, hep-ph/9812538; Y. Achiman, hep-ph/9812389; M. Bando, T. Kugo and K. Yoshioka, Phys. Rev. Lett. **80**, 3004 (1998); K. Oda, E. Takasugi, M. Tanaka and M. Yoshimura, Phys. Rev. D **59**, 055001 (1999); T. Blazek, S. Raby and K. Tobe, hep-ph/9903340; S. Raby, hep-ph/9909279; Q. Shafi and Z. Tavartklidze, hep-ph/9910314; M. Parida and N. Singh, Phys. Rev. D **59**, 032002 (1999).

10.78 R. N. Mohapatra, TASI lectures (1997) in *Supersymmetry, Supergravity, Supercolliders*, ed. by J. Bagger (World Scientific, Singapore, 1998).

10.79 K. S. Babu, B. Dutta and R. N. Mohapatra, Phys. Lett. B **458**, 93 (1999).

10.80 A. Anastassov et al. (CLEO collaboration), hep-ex/9908025.

10.81 N. Igres, S. Lavignac and P. Ramond, Phys. Rev. D **58**, 035003 (1998); B. Stech, hep-ph/9905440; S. Lola, hep-ph/9903203; C. D. Froggatt, H. Nielsen and M. Gibsen, Phys. Lett. B **446**, 254 (1999); C. Wetterich, Phys. Lett. B **451**, 388 (1999); R. Barbieri, L. J. Hall, G. Kane and G. Ross, hep-ph/9901228.

10.82 M. Fukugita and T. Yanagida, Phys. Lett. B **74**, 45 (1986).

10.83 P. Langacker, R. D. Peccei and T. Yanagida, Mod. Phys. Lett. A **1**, 541 (1986); M. Flanz, E. Paschos and U. Sarkar, Phys. Lett. B **345**, 248 (1995); L. Covi, E. Roulet and F. Vissani, Phys. Lett. B **384**, 169 (1996); E. Ma, S. Sarkar and U. Sarkar, hep-ph/9812276.

10.84 V. Kuzmin, V. Rubakov and M. Shaposnikov. Phys. Lett. B **185**, 36 (1985).

10.85 D. G. Lee and R. N. Mohapatra, Phys. Lett. B **329**, 463 (1994); P. Bamert and C. Burgess, Phys. Lett. B **329**, 109 (1994); A. Ionissian and J. W. F. Valle, Phys. Lett. B **332**, 93 (1994); A. Joshipura, Z. Phys. C **64**, 31 (1994); Y. L. Wu, hep-ph/9810491; hep-ph/9901320.

10.86 J. A. Casas, J. R. Espinosa, A. Ibarra and I. Navarro, hep-ph/9904395; J. Ellis and S. Lola, hep-ph/9904279; N. Haba and N. Okamura, hep-ph/9906481; N. Haba, N. Okamura and M. Sugiura, hep-ph/9810471; K. R. S. Balajc, A. Dighe, R. N. Mohapatra and M. K. Parida, Phys. Rev. Lett. **84**, 5034 (2000).

10.87 K. Benakli and A. Smirnov, Phys. Rev. Lett. **79**, 4314 (1997); P. Langacker, hep-ph/9805281; D. Suematsu, Phys. Lett. B **392**, 413 (1997); E.

J. Chun, A. Joshipura and A. Smirnov, Phys. Rev. D **54**, 4654 (1996); E. Ma and P. Roy, Phys. Rev. D **52**, 4342 (1995); M. Bando and K. Yoshioka, hep-ph/9806400; N. Arkani-Hamed and Y. Grossman, hep-ph/9806223.

10.88 K. Dienes, E. Dudas and T. Gherghetta, hep-ph/9811428; N. Arkani-Hamed, S. Dimopoulos, G. Dvali and J. March-Russell, hep-ph/9811448; G. Dvali and A. Smirnov, hep-ph/9904211.

10.89 A. Perez-Lorenzana and R. N. Mohapatra, hep-ph/9910474.

10.90 R. N. Mohapatra, S. Nandi and A. Perez-Lorenzana, hep-ph/9907520.

10.91 F. Gursey and P. Sikivie, Phys. Rev. Lett. **36**, 775 (1976); P. Ramond, Nucl. Phys. B **210**, 214 (1976); Y. Achiman and B. Stech, Phys. Lett. B **77**, 389 (1978); Q. Shafi, Phys. Lett. B **79**, 301 (1979).

10.92 J. Rosner, Comments Nucl. Part. Phys. **15**, 195 (1986); hep-ph/9907438.

10.93 E. Ma, Phys. Lett. B **30**, 286 (1996).

10.94 I. Kobzarev, L. Okun and I. Ya Pomeranchuk, Sov. J. Nucl. Phys. **3**, 837 (1966).

10.95 S. I. Blinikov and M. Yu. Khlopov, Sov. Astron. **27**, 371 (1983); Z. Silagadze, hep-ph/9503481.

10.96 B. Brahmachari and R. N. Mohapatra, hep-ph/9805429.

10.97 Z. Berezhiani, A. Dolgov and R. N. Mohapatra, Phys. Lett. B **375**, 26 (1996).

10.98 V. Berezinsky and A. Vilenkin, hep-ph/9908257.

10.99 C. Alcock et al. Astrophys. J. **486**, 697 (1997); R. Ansari et al. Astron. Astrophys **314**, 94 (1996).

10.100 W. Sutherland, astro-ph/9811185; Rev. Mod. Phys. (1999) (to appear).

10.101 K. Freese, B. Fields and D. Graff, astro-ph/9901178.

10.102 Z. Berezhiani, A. Dolgov and R. N. Mohapatra, Phys. Lett. B **375**, 26 (1996); S. Blinnikov, astro-ph/9801015.

10.103 R. N. Mohapatra and V. L. Teplitz, astro-ph/9902085; Phys. Lett. B (to appear).

10.104 A. Joshipura and S. Rindani, hep-ph/9811252; hep-ph/9907390; E. Ma, Phys. Rev. Lett (1999) (to appear).

11 Neutrino Flavor Transformation in Supernovae and the Early Universe

George M. Fuller

11.1 Introduction

In this chapter we shall examine the physics of matter-enhanced neutrino flavor conversion in the Big Bang and in supernovae. We are in the midst of a revolution in experimental neutrino physics, which has confronted us with compelling evidence for nonzero neutrino rest masses and flavor mixings. Concurrently, we are undergoing a fantastic increase in the capabilities of observational astronomy, and this promises to provide new insights into the evolution of the Universe. Tantalizingly, these two great trends in science are synergistically linked, at least when it comes to neutrinos!

Here is the central fact: the dynamics and associated nucleosynthesis of the environments of both the supernova and early Universe could be altered markedly by neutrino flavor transformation. Therefore, if we can understand, for example, the synthesis of the heaviest elements (rapid neutron capture, or r-process nucleosynthesis) in supernovae, the synthesis of the lightest elements in Big Bang nucleosynthesis (BBN) and the contribution of neutrino rest masses to dark matter, then perhaps we can learn something about neutrino flavor conversion (or the absence thereof) in these venues.

Of course, this is a "two-way street". New laboratory findings on neutrino masses and mixings could translate directly into deeper and perhaps radically different understandings of supernova physics and the evolution of the Universe. Even without direct detection of the neutrinos from supernovae or direct measurement of the cosmic neutrino background, with luck nucleosynthesis and dynamical effects could represent a "fossil record" of what neutrinos have done.

The recent success of new experiments and detectors has made neutrino physics and astrophysics an experimental science again. The suggestions from these experiments that there is a rich and unexpected phenomenology associated with neutrino rest masses are exciting and are loaded with significance for physics beyond the Standard Model. Ultimately, the astrophysical considerations discussed here could provide complementary and perhaps unique insights into the vacuum neutrino mass and mixing spectrum. That is the seductive promise of neutrino astrophysics.

But there are "speed bumps" on the road to employing astrophysics to probe neutrinos. Chief among these is the difficulty of treating the nonlinear evolution of neutrino flavor transformation in supernovae and the early Universe.

The Mikeyev–Smirnov–Wolfenstein (MSW) mechanism [11.1, 11.2] for matter-enhanced conversion of one kind of neutrino to another has sparked tremendous excitement in the solar neutrino problem. (See Chap. 4 in this book.) This is as it should be, as solar neutrinos are eminently detectable. However, the neutrinos produced in the sun play no direct role in solar dynamics or nucleosynthesis; they are merely ghostly witnesses of nuclear reactions.

By contrast, neutrinos and their interactions with matter determine nearly every aspect of the physics of supernova core collapse and the early Universe, and they may play an important role in the dark-matter problem as well. So, it is not very surprising that the implications of MSW-driven neutrino flavor conversion for these sites can be profound. One problem is that, unlike the case of the sun, the weak potential which governs the interconversion of neutrino flavor labels can be itself dependent on the flavor states of the neutrinos!

Another problem is that this nonlinear neutrino flavor evolution can feed back on the environment and alter the ratio of neutrons to protons, n/p, or, equivalently, the electron fraction, or *net* number of electrons per baryon, $Y_e = 1/(1 + n/p)$. (In terms of the proper number densities of electrons and positrons n_{e^-} and n_{e^+}, respectively, and the proper number density of baryons n_b, the net electron number per baryon is $Y_e \equiv (n_{e^-} - n_{e^+})/n_b$.) This is because, in many astrophysical environments in which intense neutrino fluxes are present, Y_e is set by the competition between the lepton capture reactions on neutrons and protons [11.3, 11.4]:

$$\nu_e + n \rightleftharpoons p + e^- , \tag{11.1}$$
$$\bar{\nu}_e + p \rightleftharpoons n + e^+ . \tag{11.2}$$

In turn, the neutron-to-proton ratio is a critical determinant of the freeze-out nucleosynthesis which characterizes BBN and the r-process in neutrino-heated supernova ejecta. It is obvious from these reactions that active–active or active–sterile neutrino flavor conversion in the channel(s)

$$\nu_\alpha \rightleftharpoons \nu_\beta \quad \text{and/or} \quad \bar{\nu}_\alpha \rightleftharpoons \bar{\nu}_\beta \tag{11.3}$$

(where $\alpha = e$, and where likewise $\beta = \mu, \tau$ or a sterile species s) could affect the n/p ratio. This is obviously the case if the energy spectra of the various neutrino flavors are not the same, since the rates for the reactions in (11.1) and (11.2) are energy-dependent. Likewise, such alterations in the $\bar{\nu}_e$ or ν_e energy spectra resulting from matter-enhanced conversion can also affect the dynamics of, for example, supernovae, since the heating of the material to be ejected can come from the forward processes in (11.1) and (11.2).

The laboratory neutrino oscillation data, if taken at face value, do not seem to be amenable to a three-neutrino fit. Introducing a fourth light neutrino would have profound implications for particle physics and astrophysics because such a neutrino would have to be "sterile" (no Standard Model weak interactions) in order to be consistent with the Z^0 width measurements. These issues are discussed in Chaps. 7, 9 and 10.

At the present time, two four-neutrino mass/mixing schemes crafted to fit the laboratory data are under discussion: (1) a lower-mass "triplet" of neutrino mass states with mixing angles that make them closely associated with the active neutrinos $(\nu_e, \nu_\mu, \nu_\tau)$, and a considerably heavier sterile neutrino; and (2) a scheme with two doublets, corresponding closely to a lower-mass doublet of ν_e and ν_s, and a higher-mass doublet corresponding closely to a maximally mixed ν_μ and ν_τ.

In what follows we shall outline the special aspects of neutrino flavor conversion in the nonlinear environments of supernovae and the early Universe. We go on to discuss supernovae in particular, giving the basics of freeze-out nucleosynthesis in the neutrino-driven wind, and touching on the consequences of large-scale neutrino flavor or type (e.g. active-to-sterile) conversion. We shall then examine many of the same issues, but in the context of the early stages of the Big Bang. Finally, we shall speculate on the broader implications of these results for particle physics and for cosmology.

11.2   Matter-Enhanced Neutrino Conversion in Nonlinear Environments

The basics of the MSW mechanism are set out in the discussion by W. Haxton in Chap. 4. Here we shall extend this discussion to environments where there is nonlinear feedback on the weak potential which governs neutrino flavor conversion.

Both supernovae and the early Universe can be characterized by very intense neutrino fluxes and, perhaps, high electron/positron and baryon densities. These facts have two consequences for the simple picture of matter-enhanced neutrino flavor transformation presented earlier. First, the density of weak-charge-carrying targets may be so high that nonforward neutrino scattering processes become important. Second, neutrino–neutrino forward-scattering contributions to the weak potential can become important, even dominant in these regimes.

For convenience and for illustrative puposes we shall consider the two-by-two neutrino flavor evolution problem. The results obtained in this case are easily generalizable to problems of higher dimensionality involving more neutrino flavors. With this simplification, we can represent the state vector for a neutrino as a coherent sum of amplitudes in the "flavor basis",

$$|\Psi_\nu(t)\rangle = a_\alpha(t)|\nu_\alpha\rangle + a_\beta(t)|\nu_\beta\rangle , \tag{11.4}$$

where $a_\alpha(t)$ and $a_\beta(t)$ are the time-dependent amplitudes for the neutrino to be in the flavor eigenstates $|\nu_\alpha\rangle$ and $|\nu_\beta\rangle$, respectively. Here α and β are flavor labels and can take the appropriate values: for active neutrinos, e, μ, τ; or for a sterile neutrino, s. If neutrino scattering is important, and it *is* generally in the supernova core and in the very early Universe, then one might question the use of the coherent sum in the above equation. Indeed, every real non-forward scattering event will effectively reset the phase and the amplitudes in the decomposition of $|\Psi_\nu\rangle$. In fact, this phase resetting process can inhibit neutrino flavor transformation by truncating the development of phase in the neutrino oscillation cycle. This is Quantum Damping, or the quantum Zeno effect [11.46, 11.47]. However, in cases where the neutrino mean free path is larger than the MSW resonance width, coherent evolution of the neutrino state through the resonance is frequently still a good approximation. Even when it is not, the form of the MSW equations can still describe flavor conversion. Every time the neutrino scatters, there will be a probability (proportional to $\sin^2 2\theta_M(t)$ – see below) that it scatters into another flavor state. At MSW resonances this probability is maximal.

Let us assume for the present that the following condition obtains: coherence on the scale of the resonance width. The salient features of neutrino flavor evolution in this case can be summarized with a Schrödinger-like equation, expressed in the flavor basis as

$$
i\frac{d}{dt}\begin{bmatrix} a_\alpha(t) \\ a_\beta(t) \end{bmatrix} = H_{\text{TOTAL}} \begin{bmatrix} a_\alpha(t) \\ a_\beta(t) \end{bmatrix}. \tag{11.5}
$$

Here t is any neutrino evolution variable, for example the time since the neutrino was created, the radius above the neutrino sphere of a supernova, or even the Friedmann–Robertson–Walker (FRW) time coordinate in a homogeneous and isotropic universe. For our particular environments of interest, the propagation Hamiltonian in (11.5) can be approximated as the sum of four terms,

$$
H_{\text{TOTAL}} = H_{\text{VAC}} + H_{\text{BARYON}} + H_{\nu\nu} + H_{\text{THERMAL}}, \tag{11.6}
$$

which give the contributions related to the vacuum masses, the "baryonic" term from forward scattering on electrons/positrons and nucleons, forward scattering on other neutrinos (the neutrino "background"), and the "thermal" terms, respectively. These latter contributions represent mass renormalization terms from forward scattering on loops, and second order (in G_F) forward scattering involving charged leptons (see [11.4–11.7]).

In the two-by-two case it is simple to separate the neutrino propagation Hamiltonian into a traceless part and a piece proportional to the identity. The operator which is proportional to the identity merely provides a common phase to all components of the neutrino state and so can be ignored for the purposes of following neutrino flavor conversion. The traceless component of the neutrino propagation Hamiltonian is

$$
H_{\text{TOTAL}} = \frac{1}{2}\begin{pmatrix} -\Delta\cos 2\theta + A + B + C & \Delta\sin 2\theta + B_{\text{off}} \\ \Delta\sin 2\theta + B_{\text{off}}^* & \Delta\cos 2\theta - A - B - C \end{pmatrix}, \tag{11.7}
$$

where $\Delta \equiv \delta m^2/(2E_\nu)$, E_ν is the neutrino energy and $\delta m^2 \equiv m_2^2 - m_1^2$ is the difference of the squares of the vacuum neutrino mass eigenvalues, in this case m_1 and m_2. The baryonic contribution to the neutrino propagation Hamiltonian arises from A, while B and B_{off} represent the effect of the neutrino–neutrino forward-scattering contributions. (B_{off}^* is the complex conjugate of B_{off}.) Here the thermal terms are denoted by C.

In the above expressions we have assumed that there exists *in vacuum* a simple unitary relationship between the flavor basis and the underlying energy eigenstate, or "mass", basis, for example,

$$|\nu_\alpha\rangle = \quad \cos\theta|\nu_1\rangle + \sin\theta|\nu_2\rangle \, , \tag{11.8}$$
$$|\nu_\beta\rangle = -\sin\theta|\nu_1\rangle + \cos\theta|\nu_2\rangle \, , \tag{11.9}$$

where $|\nu_1\rangle$ and $|\nu_2\rangle$ represent the energy eigenstates corresponding to the vacuum mass eigenvalues m_1 and m_2, respectively. The vacuum mixing angle is θ.

In matter, and where neutrino fluxes are nonnegligible, it *may* be possible to provide a similar unitary transformation between the flavor basis and, in this case, the *instantaneous* mass (energy) eigenstate basis,

$$|\nu_\alpha\rangle = \quad \cos\theta_{\text{M}}(t)|\nu_1(t)\rangle + \sin\theta_{\text{M}}(t)|\nu_2(t)\rangle \, , \tag{11.10}$$
$$|\nu_\beta\rangle = -\sin\theta_{\text{M}}(t)|\nu_1(t)\rangle + \cos\theta_{\text{M}}(t)|\nu_2(t)\rangle \, , \tag{11.11}$$

where $|\nu_1(t)\rangle$ and $|\nu_2(t)\rangle$ are the instantaneous mass eigenstates, and $\theta_{\text{M}}(t)$ is the instantaneous matter mixing angle. Such a unitary decomposition is possible only in the limit where the neutrino mean free path is much larger than the MSW resonance width and only where the off-diagonal terms in the neutrino propagation Hamiltonian are real (e.g. $B_{\text{off}}^* = B_{\text{off}}$ in (11.7)). This latter requirement places a severe constraint on the form of the neutrino density operator. In the general case involving more than two neutrinos and where the requirement on the off-diagonal terms cannot be met, then neutrino flavor evolution must be followed numerically in the flavor basis. Nevertheless, it is useful to proceed with the discussion of the two-by-two case, referred to the mass basis, as a means of highlighting the key physics issues.

The single-neutrino density operator for a neutrino which is intially in flavor state $|\nu_\alpha\rangle$ is

$$|\Psi_{\nu_\alpha}(t)\rangle\langle\Psi_{\nu_\alpha}(t)| = |a_1(t)|^2|\nu_1(t)\rangle\langle\nu_1(t)| + |a_2(t)|^2|\nu_2(t)\rangle\langle\nu_2(t)| + \mathcal{X} \, , \tag{11.12}$$

where the "cross terms" in the single neutrino density operator are

$$\mathcal{X} = a_1(t)a_2^*(t)|\nu_1(t)\rangle\langle\nu_2(t)| + a_1^*(t)a_2|\nu_2(t)\rangle\langle\nu_1(t)| \, . \tag{11.13}$$

Here the amplitudes to be in the instantaneous mass eigenstates are $a_1(t) = \langle \nu_1(t)|\Psi_{\nu_\alpha}(t)\rangle$ and $a_2(t) = \langle \nu_2(t)|\Psi_{\nu_\alpha}(t)\rangle$. Clearly, the requirement that the single-neutrino density operator be real is satisfied if the cross terms vanish.[1]

The density operator for neutrinos propagating in a pencil of directions $\mathrm{d}\Omega_\nu = \sin\theta_\nu\,\mathrm{d}\theta_\nu\,\mathrm{d}\phi_\nu$ and with momentum $\boldsymbol{p}$ in an interval d^3p is

$$\rho_{\boldsymbol{p}}\,\mathrm{d}^3p = \sum_\alpha \mathrm{d}n_{\nu_\alpha}|\Psi_{\nu_\alpha}(t)\rangle\langle\Psi_{\nu_\alpha}(t)| \,, \tag{11.14}$$

where the sum is over all relevant neutrino flavors in the problem, and where, in the case of a thermal Fermi–Dirac energy distribution, the number of neutrinos in the given pencil of momenta and directions is

$$\mathrm{d}n_{\nu_\alpha} = \frac{1}{2\pi^2}\left(\frac{\mathrm{d}\Omega_{\nu_\alpha}}{4\pi}\right)T_{\nu_\alpha}^3\,F_2\left(\eta_{\nu_\alpha}\right)f_{\nu_\alpha}\left(E_\nu\right)\mathrm{d}E_\nu \,, \tag{11.15}$$

where we employ natural units, and where the temperature of the distribution for the neutrinos ν_α is T_{ν_α} and the degeneracy parameter (chemical potential divided by temperature) for this distribution is η_{ν_α}. Here the Fermi integral of order k is defined to be

$$F_k\left(\eta\right) \equiv \int_0^\infty \frac{x^k}{e^{x-\eta}+1}\,\mathrm{d}x \tag{11.16}$$

and the normalized distribution function for the neutrino species ν_α is

$$f\left(E_\nu\right) \approx \frac{1}{T_{\nu_\alpha}^3\,F_2\left(\eta_{\nu_\alpha}\right)}\frac{E_\nu^2}{\exp\left(E_\nu/T_{\nu_\alpha} - \eta_{\nu_\alpha}\right)+1}\mathrm{d}E_\nu \,. \tag{11.17}$$

It is important to note that although the neutrino and antineutrino energy spectra in both supernovae and the early Universe may start out as thermal and Fermi–Dirac in character, this shape could change drastically once large-scale neutrino flavor or type transformation begins.

However, for a neutrino energy spectrum which *is* thermal, a useful way to characterize the spectrum is in terms of an average neutrino energy. For a neutrino species ν_α, this is

$$\langle E_{\nu_\alpha}\rangle = \int_0^\infty E_\nu f_{\nu_\alpha}\left(E_\nu\right)\mathrm{d}E_\nu = T_{\nu_\alpha}\frac{F_3\left(\eta_{\nu_\alpha}\right)}{F_2\left(\eta_{\nu_\alpha}\right)} \,. \tag{11.18}$$

It is sometimes useful to note that $F_2\left(0\right) = (3/2)\zeta\left(3\right)$, where the Riemann zeta function of argument 3 is ≈ 1.20206, and $F_3\left(0\right)/F_2\left(0\right) \approx 3.151$, and $F_3\left(3\right)/F_2\left(3\right) \approx 3.992$. For the early Universe the initial neutrino energy spectra are generally taken to be thermal and Fermi–Dirac, in which case it is convenient to represent the number of neutrinos in the interval $\mathrm{d}\epsilon$ (where $\epsilon \equiv$

---

[1] J. Pantaleone was the first to recognize the potentially pernicious effects of the cross terms; however, the reduction of these terms stemming from phase averaging is quite efficient in most astrophysically interesting situations [11.4].

E_ν/T_ν) in terms of the local equilibrium proper number density of photons $n_\gamma = [2\zeta(3)/\pi^2]T_\gamma^3$ as

$$\mathrm{d}n_{\nu_\alpha} \approx \frac{n_{\nu_\alpha}}{F_2(\eta_{\nu_\alpha})} \frac{\epsilon^2 \mathrm{d}\epsilon}{e^{\epsilon - \eta_{\nu_\alpha}} + 1} \approx \left(\frac{n_\gamma}{4\zeta(3)}\right) \frac{\epsilon^2 \mathrm{d}\epsilon}{e^\epsilon + 1} . \tag{11.19}$$

In the last approximation in this equation we have assumed that the neutrino degeneracy parameter is zero ($\eta_{\nu_\alpha} = 0$). As is appropriate in a homogeneous and isotropic Universe, we have integrated over all directions in the early Universe to find the differential number of neutrinos ν_α in the interval $\mathrm{d}\epsilon$, i.e. we have taken $\int \mathrm{d}\Omega_{\nu_\alpha}/4\pi = 1$ in going from (11.15) to (11.19). The integrations over angles and energies inherent in many manipulations of the density operator can in some cases (the early Universe and the cores and hot bubble regions of supernovae) average out the cross terms in the single-neutrino density operator [11.4].

With this prescription for representing the neutrino density operator we can write the neutrino–neutrino forward scattering contribution to the propagation Hamiltonian,

$$H_{\nu\nu} = \sqrt{2}G_\mathrm{F} \int (1 - \cos\theta_q)\{\rho_q - \bar\rho_q\}\mathrm{d}^3q , \tag{11.20}$$

where we integrate over neutrino three-momenta $\boldsymbol{q}$, and where $\bar\rho$ is the density operator for antineutrinos, which is constructed in obvious analogy to that in (11.14). The factor of $(1 - \cos\theta_q)$ in (11.20), involving the angle θ_q between the test neutrino whose effective mass we shall want to calculate and a background neutrino with momentum $\boldsymbol{q}$ in the direction pencil $\mathrm{d}\Omega_q$, comes from the structure of the weak current [11.5]. (Specifically, it originates from the left-handed projection aspect of the weak current–current Hamiltonian; that is, the $\gamma^\mu(1 - \gamma_5)$ term therein.)

We can decompose this neutrino–neutrino forward scattering contribution into a piece proportional to the identity, $\mathcal{Z}I$, and a traceless part:

$$H_{\nu\nu} = \mathcal{Z}I + \frac{1}{2}\begin{pmatrix} B & B_\mathrm{off} \\ B_\mathrm{off} & -B \end{pmatrix}, \tag{11.21}$$

where, from the above definitions, we can write the "flavor-diagonal" contribution as

$$B = \sqrt{2}G_\mathrm{F} \int (1 - \cos\theta_q) \left(\{\rho_q - \bar\rho_q\}_{\alpha\alpha} - \{\rho_q - \bar\rho_q\}_{\beta\beta}\right) \mathrm{d}^3q , \tag{11.22}$$

and the off-diagonal term as

$$B_\mathrm{off} = 2\sqrt{2}G_\mathrm{F} \int (1 - \cos\theta_q) \left(\{\rho_q - \bar\rho_q\}_{\alpha\beta}\right) \mathrm{d}^3q . \tag{11.23}$$

In (11.22) and (11.23) the flavor basis matrix elements of the density operators are defined as

$$\{\rho_q - \bar\rho_q\}_{\alpha\alpha} \equiv \langle\nu_\alpha|\rho_q|\nu_\alpha\rangle - \langle\bar\nu_\alpha|\bar\rho_q|\bar\nu_\alpha\rangle , \tag{11.24}$$

$$\{\rho_q - \bar\rho_q\}_{\beta\beta} \equiv \langle\nu_\beta|\rho_q|\nu_\beta\rangle - \langle\bar\nu_\beta|\bar\rho_q|\bar\nu_\beta\rangle \,, \tag{11.25}$$

$$\{\rho_q - \bar\rho_q\}_{\alpha\beta} \equiv \langle\nu_\alpha|\rho_q|\nu_\beta\rangle - \langle\bar\nu_\alpha|\bar\rho_q|\bar\nu_\beta\rangle \,, \tag{11.26}$$

with α and β taking on any of the flavor/type labels e, μ, τ and s, but note that the off-diagonal term B_{off} vanishes if one of the neutrinos is sterile. (There is no off-diagonal flavor basis contribution to the neutrino propagation Hamiltonian for active–sterile neutrino mixing $\nu_\alpha \rightleftharpoons \nu_s$ [11.7].)

An MSW resonance, or level crossing, occurs when

$$\Delta\cos 2\theta = A + B + C \,. \tag{11.27}$$

If we can transform to the instantaneous mass basis, then it is straightforward to show that the effective matter mixing angles in this case are given by

$$\sin 2\theta(t)_{\mathrm{M}} = \frac{\Delta\sin 2\theta + B_{\mathrm{off}}}{\sqrt{(\Delta\cos 2\theta - A - B - C)^2 + (\Delta\sin 2\theta + B_{\mathrm{off}})^2}} \,, \tag{11.28}$$

$$\cos 2\theta(t)_{\mathrm{M}} = \frac{\Delta\cos 2\theta - A - B - C}{\sqrt{(\Delta\cos 2\theta - A - B - C)^2 + (\Delta\sin 2\theta + B_{\mathrm{off}})^2}} \,. \tag{11.29}$$

In a similar fashion, we can obtain the effective matter mixing angles for antineutrinos:

$$\sin 2\bar\theta(t)_{\mathrm{M}} = \frac{\Delta\sin 2\theta - B_{\mathrm{off}}}{\sqrt{(\Delta\cos 2\theta + A + B - C)^2 + (\Delta\sin 2\theta - B_{\mathrm{off}})^2}} \,, \tag{11.30}$$

$$\cos 2\bar\theta(t)_{\mathrm{M}} = \frac{\Delta\cos 2\theta + A + B + C}{\sqrt{(\Delta\cos 2\theta + A + B - C)^2 + (\Delta\sin 2\theta - B_{\mathrm{off}})^2}} \,. \tag{11.31}$$

In the latter two equations we have used the fact that the amplitudes for forward scattering of antineutrinos have the opposite sign to those for neutrino forward scattering. (The thermal term does not change sign.)

As is evident from these expressions, the behavior of the effective matter mixing angle is analogous to that of the pure electron-driven MSW case discussed in Chap. 4. As a neutrino propagates from regions where the density is well above that of the resonance, through the resonance (at t_{res}) and on into low-density regions, the effective matter mixing angle will evolve from $\theta_{\mathrm{M}}(t < t_{\mathrm{res}}) \approx \pi/2$, through $\theta_{\mathrm{M}}(t = t_{\mathrm{res}}) = \pi/4$ and on to $\theta_{\mathrm{M}}(t > t_{\mathrm{res}}) \approx \theta$.

On propagating through resonance, the probability that the neutrino changes from one flavor state to another depends on the adiabaticity of the transition. An adiabatic case is one where neutrino flavor transformation is complete and corresponds to the neutrino remaining on one (either the higher or the lower) instantaneous neutrino mass track, and not making a jump from one track to the other. Physically, this is equivalent to the local neutrino oscillation frequency being large compared with the inverse of the time required for the neutrino to transit the "width" of the resonance.

In analogy to what was done in Chap. 4, we can define the resonance width δt to be that increment in time (or density) at resonance over which the effective neutrino weak potential $V = A + B + C$ changes enough to cause $\sin^2 2\theta_M$ to drop by a factor of two from its resonance value of unity:

$$\delta t = \frac{\delta t}{\delta V} \, \delta V \approx \mathcal{H} \tan 2\theta \,, \tag{11.32}$$

where the effective weak-charge scale height is

$$\mathcal{H} = \left| \frac{1}{V} \frac{dV}{dt} \right|_{\text{res}}^{-1} \tan 2\theta \,. \tag{11.33}$$

The neutrino oscillation length at resonance is $L_{\text{res}}^{\text{osc}} = (4\pi E_\nu^{\text{res}})/(\delta m^2 \sin 2\theta) = 2\pi/\delta V$, and the adiabaticity parameter is proportional to the ratio $\delta t / L_{\text{res}}^{\text{osc}}$, and is defined as

$$\gamma \equiv 2\pi \, \frac{\delta t}{L_{\text{res}}^{\text{osc}}} \approx (\delta V)^2 \left. \frac{d\epsilon}{dV} \middle/ \frac{d\epsilon}{dt} \right|_{\text{res}} , \tag{11.34}$$

where the last approximate expression is the form relevant for a homogeneous and isotropic FRW universe. Adiabatic neutrino flavor evolution corresponds to $\gamma \gg 1$. Unlike the occasionally complicated situation for solar neutrinos, the Landau–Zener jump probability [11.8] with the form $P_{\text{LZ}} \approx \exp\left(-\pi\gamma/2\right)$ nearly always suffices as a good estimate of the probability $(1 - P_{\text{LZ}})$ of $\nu_\alpha \rightarrow \nu_\beta$ conversion at resonance in supernovae and the early Universe. The exception to this is where density fluctuations result in depolarization at resonance, rather than conversion ([11.9] and references to R. Sawyer therein).

The physical interpretation of (11.34) is straightforward. The adiabaticity parameter is proportional to the energy width of the resonance divided by the "sweep rate", $d\epsilon/dt$, of the resonance energy through the neutrino distribution function. The resonance sweep rate is determined in the case of the supernova r-process environment by the outflow rate of a given mass element and by the inherent density scale height of the envelope above the neutron star. In the early Universe, the resonance will sweep through the neutrino distribution function from low to high neutrino energy at a rate given by the local Hubble expansion rate of the Universe. In both supernovae and the early Universe, the vacuum mixing angles required for adiabaticity for a given E_ν and δm^2 can be surprisingly small.

Because neutrino–neutrino forward scattering can be important, even dominant in setting the weak potential V, there can be more channels available for matter-enhanced neutrino flavor transformation $\nu_\alpha \rightleftharpoons \nu_\beta$ in supernovae and the Big Bang than in the sun. Table 11.1 gives A and B for all possible channels of 2×2 active–sterile and active–active neutrino flavor mixing. The neutrino–neutrino forward-scattering contribution to the weak potential B depends on the neutrino numbers/fluxes and the geometry of the

Table 11.1. Weak-potential terms A and B appropriate for various matter-enhanced neutrino transformation channels; n_b is the proper total baryon number density. The notation for the net neutrino numbers/fluxes N_{ν_α} is discussed in the text

Channel	A	B
$\nu_e \rightleftharpoons \nu_s$	$(3\sqrt{2}/2)G_F n_b\,(Y_e - 1/3)$	$\sqrt{2}G_F\left(2N_{\nu_e} + N_{\nu_\mu} + N_{\nu_\tau}\right)$
$\nu_\mu \rightleftharpoons \nu_s$	$(\sqrt{2}/2)G_F n_b\,(Y_e - 1)$	$\sqrt{2}G_F\left(N_{\nu_e} + 2N_{\nu_\mu} + N_{\nu_\tau}\right)$
$\nu_\tau \rightleftharpoons \nu_s$	$(\sqrt{2}/2)G_F n_b\,(Y_e - 1)$	$\sqrt{2}G_F\left(N_{\nu_e} + N_{\nu_\mu} + 2N_{\nu_\tau}\right)$
$\nu_\mu \rightleftharpoons \nu_e$	$(\sqrt{2}/2)G_F n_b Y_e$	$\sqrt{2}G_F\left(N_{\nu_e} - N_{\nu_\mu}\right)$
$\nu_\tau \rightleftharpoons \nu_e$	$(\sqrt{2}/2)G_F n_b Y_e$	$\sqrt{2}G_F\left(N_{\nu_e} - N_{\nu_\tau}\right)$
$\nu_\mu \rightleftharpoons \nu_\tau$	0	$\sqrt{2}G_F\left(N_{\nu_\mu} - N_{\nu_\tau}\right)$

astrophysical environment through (11.22). In Table 11.1, N_{ν_α} denotes the net neutrino number/flux contribution to B from neutrino species ν_α and $\bar{\nu}_\alpha$. For the isotropic conditions relevant for neutrino tranformation in an FRW universe, $N_{\nu_\alpha} = n_{\nu_\alpha} - n_{\bar{\nu}_\alpha}$ is just the local proper *net* neutrino number density. This is also the case for the isotropic conditions deep in the core of a supernova, where the transport mean free path for neutrinos is small compared with the radius of the core. By contrast, above the surface of a neutron star, where the neutrino-driven wind originates, neutrinos are (or are close to being) optically thin, and so the N_{ν_α} terms represent average net fluxes which are weighted by $(1 - \cos\theta_q)$ in (11.22). Obviously, as the fluxes approach isotropy, the $\cos\theta_q$ term will tend to average to zero, and the N_{ν_α} will approach the net number density $n_{\nu_\alpha} - n_{\bar{\nu}_\alpha}$.

It is evident from Table 11.1 that supernovae and the early Universe could each possess a very rich (and complicated!) phenomenology of matter- or neutrino-driven neutrino flavor transformation. That is, *if* the neutrino mass and mixing spectrum in vacuum accommodates such a situation.

For example, the A term for the matter-enhanced $\nu_\mu \rightleftharpoons \nu_s$ channel is proportional to $Y_e - 1$, and is therefore negative. When A dominates the weak potential V, this implies that a neutrino mass level crossing (resonance) can be obtained in this channel for a sufficiently high baryon density n_b only when the neutrino mass eigenvalue most closely associated with the ν_μ in vacuum is larger than the corresponding mass eigenvalue most closely associated with the sterile neutrino in vacuum. Were this not the case, and the muon neutrino corresponded to the lighter species in vacuum, then the $\nu_\mu \rightleftharpoons \nu_s$ channel could not be matter-enhanced under these conditions, though the $\bar{\nu}_\mu \rightleftharpoons \bar{\nu}_s$ could be!

This is a consequence of the antineutrino forward-scattering amplitudes having opposite sign from those associated with neutrino forward-scattering. In the first case, the $\nu_\mu \rightleftharpoons \nu_s$ channel, the matter terms bring the effective mass of ν_μ *down* with increasing baryon density, to eventually guarantee a mass level crossing. In the second case, where the muon neutrino is the light one in vacuum, the matter terms bring the effective mass of the $\bar{\nu}_\mu$ *up* to produce a mass level crossing with $\bar{\nu}_s$ at a high enough value of n_b. This example is not as complicated as it might be, however, because at least it has the feature that $Y_e \leq 1$, so that A can never change sign for a given mixing channel (it is negative for neutrinos, positive for antineutrinos).

Consider the $\nu_e \rightleftharpoons \nu_s$ channel when the appropriate A term dominates the weak potential. In this case we have the same features as in the previous example, but now with the added twist that, for a given vacuum neutrino mass spectrum, *either* the $\nu_e \rightleftharpoons \nu_s$ channel *or* the $\bar{\nu}_e \rightleftharpoons \bar{\nu}_s$ channel can be matter-enhanced, depending on whether $Y_e > 1/3$ or $Y_e < 1/3$.

To make this situation even more complicated and nonlinear, the flavor transformations of ν_e and $\bar{\nu}_e$ can feed back on the electron fraction Y_e and, hence, A, through the neutrino capture reactions in (11.1) and (11.2). As we shall see, this also produces a nonlinear sensitivity to the material expansion rate in both supernovae and the early Universe.

Finally, we should face the specter of neutrino scattering. Everything we have presented so far assumes that neutrino flavor evolution through resonances is coherent. As mentioned above, this is roughly tantamount to an assumption that the scattering mean free path for neutrinos is large compared with the resonance width. We shall term this fortuitous situation the "coherence condition".

At each scattering event, the neutrino has an amplitude to become another flavor that is proportional to $\sin 2\theta_M(t_{\text{scatt}})$. Obviously this will be appreciable near resonance, where $\theta_M \approx \pi/4$. We can see that even though the coherence condition may not be met, significant neutrino transformation at resonance is still a possibility, although in this case we will in general have to include quantum damping and solve for the full Boltzmann evolution of the system of neutrinos and matter.

Quantifying the interplay of both coherent and incoherent transport in the presence of MSW resonances is a daunting problem. Reference [11.10] describes the general problem in the context of the early Universe. Fortunately, in both the early epochs of the Big Bang and in some of the events of interest in the phenomenon of the core collapse supernova, the conditions can sometimes either favor treatment of neutrino flavor evolution in the coherent-condition limit, or else allow the estimation of neutrino flavor conversion by the formalism presented in this section.

11.3 Core Collapse Supernovae

Neutrinos completely dominate the energetics and dynamics of Type II, Type Ib and Type Ic supernovae. These are the core collapse supernovae. In these objects, $\sim 10\%$ of the rest mass of the $\sim 1.4\,M_\odot$ remnant neutron star will be radiated away as neutrinos on a timescale of seconds!

It is fairly obvious from these figures that if matter-enhanced neutrino flavor transformation occurs, there may be dramatic consequences. Let us first outline the standard picture of how these events transpire, emphasizing the role of the neutrinos. We shall then discuss how this paradigm could be altered were Nature so devious as to arrange neutrino masses and mixings in such a way that efficient MSW neutrino flavor conversion could occur at key locations/times.

The progenitors of core collapse supernova events are massive stars ($M > 10\,M_\odot$) which evolve on timescales of millions of years to form near-Chandrasekhar-mass ($\sim 1.4\,M_\odot$), low-entropy (in units of Boltzmann's constant per baryon, $s \sim 1$) "iron" cores. The nuclear evolution of these stars, from hydrogen burning to helium burning and then carbon/oxygen burning and so on, and eventually to silicon burning, is an approach to nuclear statistical equilibrium (NSE).

This is a condition in which all of the strong and electromagnetic reactions that build up nuclei balance those reactions which disintegrate them. Symbolically, we can represent this balance of nuclear-reaction rates in NSE as

$$Zp + Nn \rightleftharpoons A\,(Z, N) + \gamma\,, \tag{11.35}$$

where $A\,(Z, N)$ represents a nucleus of mass number $A = Z + N$, containing Z protons and N neutrons. The result is a steady-state abundance distribution given by a Saha equation, derived by equating the chemical potentials of the reactants in (11.35) appropriately. In NSE the mass fraction of any nucleus is given by a Saha equation

$$X\,(Z, N) \sim s^{1-A} X_p^Z X_n^N \exp\left\{\mathrm{BE}[A\,(Z, N)]/T\right\}\,, \tag{11.36}$$

where X_p and X_n are the mass fractions of free protons and neutrons, respectively, and $\mathrm{BE}[A\,(Z, N)]$ is the binding energy of the nucleus $A\,(Z, N)$. Note the dramatic sensitivity of the abundance of heavy nuclei to the entropy in NSE.

From this argument it is easy to believe that entropy is a key determinant of many aspects of supernova physics, including the explosion dynamics and the nucleosynthesis of neutrino-heated ejecta. And, not surprisingly, neutrino emission and interactions determine the entropy.

For example, in the evolution of the supernova progenitor towards iron-core formation, neutrinos dominate the energy/entropy loss from the core. Indeed, from core oxygen burning on, the thermal and nuclear neutrino luminosity exceeds the photon luminosity.

This has two effects: the iron core has a low entropy $s \approx 1$ (the temperature in billions of kelvin will be $T_9 \approx 7$, but the electron Fermi energy will be high, $\mu_e \sim 10\,\mathrm{MeV}$); and the electron fraction is low, $Y_e \approx 0.42$. In these conditions, NSE obtains and the Saha equation predicts that the baryons in these cores will reside in nuclei with masses $A \sim 60$. In other words, the dominant nuclear species are iron-peak nuclei, which are at the top of the binding-energy-per-nucleon curve. Essentially all of the pressure support for the core comes from the relativistically degenerate electrons.

The iron core of the presupernova star eventually will go dynamically unstable. This comes about through a combination of general relativistic effects, NSE rearrangement involving alpha particles, and electron capture on free protons and the protons which reside in large nuclei:

$$e^- + \text{``}p\text{''} \rightarrow \text{``}n\text{''} + \nu_e \, . \tag{11.37}$$

The subsequent collapse and explosion event (if there is one!), and the post-explosion epoch divide up into three categories for our purposes.

These are (i) the infall epoch, (ii) the explosion/shock reheating epoch and (iii) the neutrino-driven-wind epoch. Let us consider the role of neutrinos and the effects of neutrino transformation on each of these regimes in turn.

The collapse from the initial "iron white dwarf" with a radius of some 10^8 cm to a hot, proto-neutron star with a radius $\sim 4 \times 10^6$ cm takes about one second. As the core collapses its entropy remains low ($s \sim 1$). The collapse is characterized by near free fall until the large nuclei merge as nuclear saturation density is approached. At this point, the pressure goes over from being dominated by relativistic electrons (adiabatic index $\Gamma_1 \approx 4/3$) to being dominated by nonrelativistic nucleons ($\Gamma_1 \approx 5/3$).

This violently halts the collapse of the inner, so-called homologous core, which contains about $\sim 0.7 \, M_\odot$. The homologous core is a kind of "instantaneous" Chandrasekhar mass, reflecting the local value of Y_e. By this time, the electron fraction has been reduced by the electron capture process in (11.37) to $Y_e \approx 0.37$, which is why the homologous core has only about half of the initial iron core's mass. In fact, the electron capture process and its reverse, ν_e capture, will come into equilibrium shortly before the collapse is halted. This "beta equilibrium" can be established once the neutrinos produced via the electron capture reaction (and those few thermal neutrino–antineutrino pairs which are produced) become trapped in the core as the material density becomes so large that the neutrino mean free path becomes smaller than the radius of the core. The ν_es will eventually form a degenerate distribution, and the chemical potentials of the reactants in (11.37) in beta equilibrium will be related by

$$\mu_e - \mu_{\nu_e} = \mu_n - \mu_p + \delta m_{np} \, , \tag{11.38}$$

where $\delta m_{np} \approx 1.29 \, \text{MeV}$ is the neutron–proton mass difference.

A shock will form at the edge of the inner core, with an initial energy $\approx 10^{51}$ ergs, reflecting the kinetic energy of the infalling outer-core material. The shock will begin to move out, traversing the remainder of the material that was in the initial iron core. This is fatal for the shock.

The entropy jump across the shock is roughly a factor of ten, and since NSE obtains, the result is that the equilibrium shifts from heavy nuclei to free baryons and alpha particles. However, since nucleons are bound in nuclei by $\sim 8\,\mathrm{MeV}$, this represents a huge endothermic loss of the shock's energy, corresponding to some 10^{51} ergs per $0.1\,M_\odot$ of material traversed! The shock will evolve quickly to become a standing accretion shock, incapable at this point of causing a supernova explosion. (Type II supernovae are observed to have some 10^{51} ergs of optical and ejecta kinetic energy.)

This brings us to the explosion/shock reheating epoch. Note that the energy in the initial shock is only of order 1% of the gravitational binding energy ($\approx 10^{53}$ ergs) of the cold neutron star which probably will result eventually from the supernova process, and only of order 10% of the gravitational binding energy ($\approx 10^{52}$ ergs) of the initial hot proto-neutron star.

Where did the rest of the gravitational binding energy go? It goes into the electron-capture-produced and thermal pair-produced neutrinos, almost all of which are trapped in the core. Inelastic, nonconservative neutrino scattering on leptons and nuclear structures will result in a kind of equipartition of energy among all six neutrino species, though there will be initially a significant net neutrino number in ν_e, corresponding to an abundance relative to baryons $Y_{\nu_e} \approx 0.04$. All neutrinos will diffuse to the edge of the neutron star, where the steep falloff in density will guarantee their transport and energy decoupling. This decoupling region, or "neutrino sphere", is more or less coincident with the edge of the proto-neutron star, and the neutrinos mostly stream freely above this point.

The neutrino luminosities at the neutrino sphere are generally similar for all six neutrino species, though, especially during the shock reheating epoch, there is a larger number-flux of electron-type neutrinos, reflecting the initial net electron lepton number of the core. During the shock reheating epoch, the neutrino luminosities are each about $L_{\nu_{e,\mu,\tau}} \sim 10^{52}\,\mathrm{erg\,s^{-1}}$ out to a time post-core-bounce $t_{\mathrm{pb}} \sim 0.2\,\mathrm{s}$, whereas, in the later ($t_{\mathrm{pb}} \approx 3\,\mathrm{s}$ to $\sim 15\,\mathrm{s}$) neutrino-driven-wind, r-process epoch, the neutrino luminosities are $L_{\nu_{e,\mu,\tau}} \sim 10^{51}\,\mathrm{erg\,s^{-1}}$ to $\sim 10^{50}\,\mathrm{erg\,s^{-1}}$.

The neutrino energy distribution functions at the neutrino sphere are roughly black-body Fermi–Dirac in character (as in (11.17)) with degeneracy parameters spanning the range $\eta_\nu \approx 0$ to 3 for the various neutrino species. It is important to note, however, that there is significant distortion from a black-body distribution on the (depleted) high-energy tail of the ν_e energy spectrum.

The neutrino opacities are not the same for all the neutrino species, and the result will be a generic average-energy hierarchy. The mu and tau neutrinos and their antiparticles have no significant charged-current opacities, owing to the relatively low temperature scale here, whereas the ν_e and $\bar{\nu}_e$ neutrinos have significant charged-current opacities arising from the reactions in (11.1) and (11.2), respectively. Hence, the mu and tau neutrinos decouple deeper in the core, where it is hotter, and, since there are more neutrons than protons in the neutron star, the ν_es have the largest opacities and so decouple furthest out, and therefore have the least energetic spectrum.

The average energies of the neutrino energy distribution functions will be in accord with the hierarchy

$$\langle E_{\nu_\tau} \rangle \approx \langle E_{\bar{\nu}_\tau} \rangle \approx \langle E_{\nu_\mu} \rangle \approx \langle E_{\bar{\nu}_\mu} \rangle \gtrsim \langle E_{\bar{\nu}_e} \rangle \gtrsim \langle E_{\nu_e} \rangle . \tag{11.39}$$

As time goes on, and the proto-neutron star becomes more neutronized, the $\bar{\nu}_e$ and mu and tau average energies will approach each other. During the shock reheating epoch and the early phase of the neutrino-driven-wind epoch, the average neutrino energies are $\langle E_{\nu_\tau} \rangle \approx \langle E_{\bar{\nu}_\tau} \rangle \approx \langle E_{\nu_\mu} \rangle \approx \langle E_{\bar{\nu}_\mu} \rangle \approx 26\,\mathrm{MeV}$, while $\langle E_{\bar{\nu}_e} \rangle \approx 16\,\mathrm{MeV}$ and $\langle E_{\nu_e} \rangle \approx 11\,\mathrm{MeV}$. Various calculations give widely different results for these quantities, though the energy hierarchy in (11.39) is respected. (Modern equation of state and opacity estimates suggest that these average energies move closer together, especially at late times.)

Above the neutron star the neutrinos maintain their near-black-body energy spectra, with the same temperatures as they had in the decoupling region. (The neutrino number densities are "diluted" in obvious fashion as a result of the $\int d\Omega_\nu \sim 1/r^2$ dependence in (11.17).) Since all material transport velocities below the shock are subsonic, hydrostatic equilibrium obtains above the neutron star. Numerical simulations show that the entropy above the neutron star, but below the shock, is nearly constant ($s \sim 10$). As a result, the temperature of the matter in this region declines with increasing distance from the neutron star center, r. At a sufficiently large radius the temperature of the local matter (nuclei, neutrons/protons, photons, and electrons and positrons, T_m) will be lower than the temperatures T_{ν_α} characterizing the neutrino energy distributions.

Since the neutrino fluxes at this epoch are huge, the forward ν_e and $\bar{\nu}_e$ capture reaction processes in (11.1) and (11.2) can lead to a net heating under the shock. This will, in turn, lead to a higher pressure differential across the shock and, hence, a higher shock Mach number. The shock will begin to move out again. In other words, neutrinos from the core freely stream outward, and a few of them are captured and transfer their energy to the shock. The details of this process are far from well understood, and the one-dimensional picture presented here is likely to be found inadequate in detail, but the basic picture cannot be that wrong. After all, all the energy resides in the neutrino seas in the core. When all is said and done, a supernova explosion results by about $t_\mathrm{pb} \approx 1\,\mathrm{s}$.

Before tackling the postexplosion physics, it is instructive at this point to see how the collapse and explosion picture presented above could be modified by matter-enhanced neutrino flavor transformation. Consider that the infall epoch is governed by low entropy. This is what caused the baryons to reside in large nuclei and, therefore, led to the pressure being dominated by relativistic electrons, which guaranteed a collapse all the way to nuclear density.

However, all of this could change if the electron neutrinos were transformed to any of ν_s, ν_μ or ν_τ. Such an MSW conversion $\nu_e \rightleftharpoons \nu_x$ would be tantamount to a reduction in the ν_e number density, because during infall there are present only tiny fluxes of ν_μ and ν_τ neutrinos. Therefore, beta equilibrium would be perturbed at high density: essentially, the chemical potential for neutrinos in (11.38) could tend towards zero for the $\nu_e \rightarrow \nu_s$ case. Re-establishment of beta equilibrium would then require a significant reduction in μ_e. A more prosaic way of putting this is that, regardless of the neutrino oscillation channel, electron capture in (11.37) would be unblocked by a diminished number density of ν_e.

In turn, this would lead to two effects [11.5]: (1) a reduction in Y_e; and (2) potentially significant entropy production if the MSW channel is $\nu_e \rightarrow \nu_{\mu,\tau}$, as these particles are trapped in the core. The first effect follows obviously from the enhanced electron capture rate. The second effect results from the nuclear physics of electron capture on the heavy nuclei present in the core. In particular, electron capture tends to leave the daughter nucleus in an excited state, and this represents an increase in the entropy per baryon.

The reduction in Y_e would have the effect of reducing the homologous-core mass, therefore increasing the amount of material that the shock must traverse and photodisintegrate. This would lead to an even weaker shock and an earlier stalling point. Does this mean that reheating cannot generate an explosion in this scenario? Possibly, though there is not really anything definitive that can be said here.

The ranges of masses and mixing angles that would be needed for such an MSW transition to take place in a damaging point in core collapse are interesting. Roughly, $\delta m^2 \approx 10^4 \, \mathrm{eV}^2$ to $10^7 \, \mathrm{eV}^2$ and $\sin^2 2\theta > 10^{-8}$ would be required. From what we now know about laboratory limits on active-neutrino masses (see the other chapters in this book), the target neutrino here would have to be sterile, with a mass in the $100 \, \mathrm{eV}$ to $1 \, \mathrm{keV}$ range. In fact, for this range in masses and δm^2 values, the A term completely dominates the weak potential and, as a consequence, the difficulties associated with the neutrino background are absent. The intriguing aspect of this scenario is the amazing power of the MSW transition to amplify a vacuum mixing of one part in a hundred million to the point where the fate of a supernova explosion is in peril! In addition, sterile neutrinos in this mass range have been proposed as candidates for warm and cold dark matter.

In stark contrast to this scenario, [11.11] has pointed out how matter-enhanced neutrino flavor transformation in the channel $\nu_e \rightleftharpoons \nu_{\mu,\tau}$ could actually *help* the shock reheating problem. Here, we require masses and mixngs with parameters in the ranges $\delta m^2 \approx 100 \, \mathrm{eV}^2$ to $10^4 \, \mathrm{eV}^2$ and $\sin^2 2\theta > 10^{-7}$, depending on δm^2. The idea is to exploit the average-neutrino-energy hierarchy shown in (11.39), and to have the energetic mu and/or tau neutrinos change to ν_es above the neutrino sphere but below the position of the stalled shock.

Since the mu and tau neutrinos have considerably higher energies on average than do the ν_es, swapping flavor labels in an MSW transition will generate ν_es with a "hot" spectrum. In turn, this will accelerate the rate of the ν_e capture reaction in (11.1) and result in more energy deposition. This could result in a 60% increase in the shock energy if the δm^2 is large enough and/or if the stalled shock has made it far enough out. Recent simulations of this effect by the Mezzacappa group [11.12] find that there will be little net increase in the heating rate for mu and tau neutrino masses consistent with experiment unless the shock has stalled out and become an accretion shock at a larger radius than their calculations indicate.

Matter-enhanced neutrino transformation in the region below (with a higher density than) the shock is characterized by weak potentials dominated by the matter term A. Again, this venue is relatively immune to the problems associated with the neutrino background [11.4].

Long after the supernova explosion ($t_{pb} > 3\,\mathrm{s}$), the hydrostatic envelope above the still-hot proto-neutron star will possess a large entropy ($s \sim 100$), corresponding to relatively low baryon density and high photon and electron/positron pair densities. The thermodynamic structure of this environment looks nothing like the well-ordered, low-entropy conditions which attend the infall epoch, or even the shock reheating regime. The high entropy in the envelope ensures that the baryons are free neutrons and protons, with very few nuclei (see (11.35)).

The total baryonic mass of this hydrostatic envelope, or "hot bubble", is $< 10^{-4}\,M_\odot$. This is a tiny fraction of the neutron star mass, with the consequence that the gravitational potential in the envelope is completely dominated by the neutron star. In hydrostatic conditions, or in subsonic steady-state outflow, the enthalpy per baryon in this case is approximately the gravitational potential energy per baryon, so that the temperature falls with increasing radius as

$$T_9 \approx \frac{22.5}{S_{100}} \left(\frac{M_{NS}}{1.4\,M_\odot} \right) \frac{1}{r_6} \,, \tag{11.40}$$

where $S_{100} = s/100$, M_{NS} is the mass of the neutron star and r_6 is the radial distance from the center of the neutron star in units of $10^6\,\mathrm{cm}$. The matter temperature at the surface of the neutron star will be on the order of the electron neutrino distribution function temperature, $T_9 \sim 20$ to 40.

Although the temperature here falls with increasing radius, the entropy is roughly constant and is dominated by relativistic particles. In that limit, the proper entropy density is

$$S = \frac{2\pi^2}{45} g_s T^3 \,, \tag{11.41}$$

where the statistical weight in relativistic particles is $g_s \approx 11/2$ for $T_9 > 3$ and ≈ 2 for lower temperatures. The entropy per baryon is then $s = S/\rho N_A$, where ρ is the matter density in units of $\mathrm{g\,cm}^{-3}$ and N_A is Avogadro's number. If we assume constant entropy per baryon in the envelope, then the density will be proportional to the cube of the temperature and, hence, inversely proportional to the cube of the radius:

$$\rho_6 \approx 38 \left(\frac{g_s}{11/2}\right) \left(\frac{M_{\mathrm{NS}}}{1.4\,M_\odot}\right)^3 \left(\frac{1}{S_{100}^4}\right) \frac{1}{r_6^3} , \qquad (11.42)$$

where ρ_6 is the density in units of $10^6\,\mathrm{g\,cm^{-3}}$. Note that the larger the entropy, the smaller the effective density scale height for the baryons ($\sim r/3$) at a given radius and the larger the neutrino flux experienced at the location of a given temperature.

The ν_e and $\bar\nu_e$ capture reactions on free nucleons ((11.1) and (11.2)) heat the material in the envelope and set the electron fraction there [11.3]. Some heating comes from neutrino–antineutrino annihilation very near the neutron star surface, but most comes from neutrino and antineutrino capture. The effect of the heating is to drive a slow (i.e. subsonic) outflow of material (see [11.13]). This is the neutrino-driven wind, which can be modeled as a homologous, steady-state expansion, with the expansion rate given in terms of the dynamic expansion timescale τ_{dyn}, as

$$\lambda_{\mathrm{exp}} = \frac{1}{\tau_{\mathrm{dyn}}} \sim \frac{S_{100}^4 \dot{M}}{M_{\mathrm{NS}}^3} , \qquad (11.43)$$

where $\dot{M}$ is the mass outflow rate. Characteristic values of the dynamic expansion timescale range around $\tau_{\mathrm{dyn}} \approx 0.2\,\mathrm{s}$ in many models. The sense of (11.43) is that a higher entropy will require a faster expansion to advect outward a given mass in baryons in a unit time.

It is revealing to consider the weak-interaction and nucleosynthesis history of a representative "comoving" mass element as it is heated near the neutron star surface and then moves out with the flow of the neutrino-driven wind. As it moves out in this manner it will experience ever lower matter temperatures and larger radii and, hence, lower neutrino fluxes.

If we denote the rates for the reactions in (11.1) and (11.2) as $\lambda_{\nu_e n}$ and $\lambda_{\bar\nu_e p}$, respectively, then for values of the radius below which $\lambda_{\nu_e n}, \lambda_{\bar\nu_e p} \gg \lambda_{\mathrm{exp}}$, the neutron-to-proton ratio at radius r will be in a dynamic steady state equilibrium with the neutrino fluxes at this position, and we shall have

$$\frac{n}{p}(r) = \frac{1}{Y_e(r)} - 1 \approx \frac{\lambda_{\bar\nu_e p}(r)}{\lambda_{\nu_e n}(r)} \approx \frac{L_{\bar\nu_e}\langle E_{\bar\nu_e}\rangle}{L_{\nu_e}\langle E_{\nu_e}\rangle} . \qquad (11.44)$$

This form for the steady-state result will obtain when the forward rates of the reactions in (11.1) and (11.2) dominate over the reverse rates of electron and positron capture. The last approximation in this equation suggests that, in the absence of matter-enhanced neutrino flavor transformation, and for similar neutrino luminosities for ν_e and $\bar\nu_e$, we shall have neutron-rich conditions in the neutrino-heated wind. This follows from the average-neutrino-energy hierarchy in (11.39).

At the weak-freeze-out radius, r_{WFO} (or, through (11.40), the position where T_{WFO} occurs), the neutrino capture rates are comparable to the material expansion rate. Above this radius (at a lower temperature), the neutrino capture rates will fall below the material expansion rate; the value of Y_e at the weak-freeze-out position will be "frozen in" subsequent to this point in the progress of a comoving fluid element.

Likewise, as the fluid element moves outward, to regions of lower temperature, the NSE abundance distribution of nuclei will begin to favor the production of heavier species. However, the nuclear reactions required to do this (see (11.35)) are temperature-dependent, and their rates will begin to fall below the local material expansion rate, and we shall experience NSE freeze-out. That is, the nuclear-reaction rates for strong and electromagnetic processes will not be able to keep up with the "demands" of the Saha equation. In practice, weak freeze-out occurs near where the temperature falls to $T_9^{\text{WFO}} \approx 10$, while NSE freeze-out occurs when the fluid element encounters temperatures below $T_9^{\text{NSE}} \approx 8$.

It was realized a few years ago that the conditions of neutron excess and high entropy in the neutrino-driven wind are, at first glance, seductively perfect for the production of the r-process nuclei [11.14]. This is because as NSE freeze-out is approached, nearly every nucleon which can be incorporated into an alpha particle *is* incorporated into an alpha particle.

Since the medium is evidently neutron-rich, this means that when the alpha particles form, all of the protons are "scoured out". This will leave a sea of free neutrons, alpha particles and a very, very small number of heavier, iron-peak "seed" nuclei. The number of seed nuclei is small because of the high entropy and the relatively rapid expansion rate. Since assembly of heavier nuclei from alpha particles and neutrons requires slow three-body reactions ($\alpha + \alpha + \alpha$ and, predominantly, $\alpha + \alpha + n$), the rapid expansion also augurs for a small number of seeds.

As the fluid element flows out to regions of even lower temperature, the neutrons in it will be captured on the seed nuclei to make the r-process nuclei, including, for example, iodine, platinum and uranium isotopes. Since the seed nuclei have masses $A \sim 50$, and we want to produce nuclei with $A \sim 200$, we are going to require a large neutron-to-seed ratio. It is clear from the above discussion that the neutron-to-seed ratio can be increased by either (1) decreasing Y_e at weak freeze-out; (2) increasing the expansion rate so as to reduce the number of seeds by compressing the time available for the three-body reactions; or (3) by increasing the entropy, which will also disfavor heavy seed nuclei.

Several calculations of this neutrino-heated-ejecta r-process scenario have been carried out [11.14], with seemingly remarkable results. The amount of r-process production per supernova is about right. This is true because of the small baryon content of the high-entropy wind. The abundance pattern produced is always the solar-system pattern for nuclei with masses $A > 100$, while for lighter nuclei, especially the neutron number $N = 50$ species, the abundance yield is predicted to vary widely from supernova event to supernova event. The origin of this sensitivity is the extreme dependence of the abundance yield on Y_e, coupled with the fact that these lighter species would originate very early in the "wind", in fact, in what we have called the latter part of the shock reheating epoch [11.15]. The electron fraction should vary at this early epoch from one supernova event to another. This epoch is, after all, during the chaotic regime prior to the explosion.

Recent observations of abundances of r-process species on the surfaces of ultra-metal-poor halo stars are largely consistent with the neutrino-heated-ejecta models [11.16]. These stars have extremely low iron contents, relative to the solar abundance, but large r-process enhancements. They may well be samples of only one or a few supernova events. Furthermore, it has been argued that the bulk of the r-process material cannot come from neutron star mergers and that arguments related to the galactic chemical-evolution timescale based on meteoritic abundance anomalies favor, or at least are consistent with, the models of neutrino-heated supernova ejecta [11.17]. Finally, it has been argued that certain features seen in the solar-system r-process abundance pattern could be produced by neutral-current neutrino-induced spallation of nuclei [11.18]. This would be direct evidence that the r-process occurs in an intense neutrino flux. Therefore, all of the current data support, or are at least consistent with, the origin of some or most of the r-process material in neutrino-heated winds.

There is only one problem. Neutrino-heated supernova ejecta with the indicated (relatively slow expansion) thermodynamic parameters cannot possibly produce the r-process. Here is why.

When the alpha particles assemble and scour out all of the protons, the free neutrons are left exposed to the still-intense flux of ν_es. The capture reaction in (11.1) will then be able to turn some of these neutrons into protons. As soon as a proton is produced it finds another neutron and is promptly incorporated into an alpha particle. In short order, the neutron number and, hence, the neutron-to-seed ratio is run to unacceptably low values, precluding a viable r-process. This is the "alpha effect", which will operate wherever there is an intense ν_e flux through a region undergoing NSE freeze-out at a fair entropy [11.19].

The "knobs" of entropy, expansion rate and Y_e can be "turned" to attempt to save the r-process in this site [11.20], and it suffices to say that values of these parameters can be found that favor a high enough neutron-to-seed ratio. The question is whether the neutrino-heated-outflow models can ever produce them.

The basic problem is simple. Nucleons near the surface of the neutron star are gravitationally bound by $\sim 100\,\mathrm{MeV}$. In the neutrino-heated-ejecta models it is the *neutrinos* which supply the energy to eject baryons to interstellar space. The neutrinos have energies of order $10\,\mathrm{MeV}$, so that, on average, neutrinos must interact with the matter some 10 times per ejected baryon. This *necessitates* the existence of a large ν_e flux. But such a flux guarantees a ferocious alpha effect, at least, i.e. the expansion is slow enough that weak processes are still coupled.

The model of neutrino-heated supernova ejecta could be saved through three means: (1) cheat, and have some mechanism *other than* neutrinos do the baryon lifting; (2) invoke an extremely relativistic neutron star to greatly increase the entropy and wind expansion rate; and (3) appeal to new neutrino physics. The first of these might well work, with baryons being ejected by MHD effects, but then it is *not* a model of *neutrino-heated* ejecta. So you will have to tune your MHD or hydrodynamics to obtain the mass ejection and nucleosynthesis effects you would have obtained with neutrino heating. The second option may well be viable [11.21]. The drawback in this case is that the neutron star must be near its instability radius $\approx 3M_{NS}$ in order that the enthalpy in the wind is high enough. The neutron star is therefore on a knife-edge; it could turn into a black hole in a millisecond. Is there enough material ejected to account for the data? This fix is finely tuned.

It is interesting that neutrino flavor transformation can play a role here: it is usually bad in the active–active channel, but it could be an elegant dodge to the alpha effect in the active–sterile channel.

The active–active matter-enhancement channel $\nu_e \rightleftharpoons \nu_{\mu,\tau}$ would tend to raise the energy of the ν_e spectrum and therefore *lower* the neutron number, as is obvious from (11.44) [11.3]. This militates against the r-process, but the range of δm^2 and $\sin^2 2\theta$ requisite for a matter-enhanced transition below the weak-freeze-out position is interesting: $10^4\,\mathrm{eV}^2 > \delta m^2 > 0.2\,\mathrm{eV}^2$ and, roughly, $\sin^2 2\theta > 10^{-5}$. This is in a range that could be consistent with recent laboratory experiments. For example, if the LSND experiment is seeing $\nu_\mu \rightleftharpoons \nu_e$ vacuum oscillations with $\delta m^2 > 1\,\mathrm{eV}^2$, and there is no sterile neutrino involved in the supernova case, then there can be no r-process from conventional neutrino-heated-ejecta scenarios. Unlike the shock reheating epoch, the neutrino-driven-wind epoch is characterized by appreciable, but not dominant, neutrino background terms B and, in the active–active channels, B_{off}.

The active–sterile matter-enhanced channels are intriguing because they can be engineered to remove the ν_e flux below the alpha particle formation region, but above the region near the neutron star where ν_e capture is necessary to effect baryon lifting. This, obviously, would disable the alpha effect. There are two ways that have been proposed to do this.

The first exploits the interplay of matter-enhanced $\nu_e \rightleftharpoons \nu_s$ and $\bar\nu_e \rightleftharpoons \bar\nu_s$ processes, the material outflow rate and Y_e to effect removal of the bulk of the ν_e flux as described above [11.22]. This works beautifully and relies on a fast expansion. In a slow expansion, the electron fraction will tend to its fixed-point value $Y_e = 1/3$ as outlined in Sect. 11.2. Although some have advanced this as being good for the r-process [11.23], it is actually not, since it does not solve the alpha effect problem. In a moderate to fast expansion (i.e. at the rates suggested by the above simple wind models), the electron fraction will not evolve to its fixed point but will actually plunge below it, *and* remove the bulk of the ν_e flux below the weak-freeze-out position. This result is very promising and can be made to work in the "three of a kind" neutrino mass and mixing models, with a sterile neutrino possessing a mass less than $\sim 100\,\text{eV}$, but greater than about $\sim 1\text{eV}$, depending on the entropy of the outflow. However, neutrino background effects are very important in this scenario and their numerical modeling is still at a primitive stage. The nonlinearities inherent in B and in Y_e (and, hence, A) in this model are fierce.

In a parallel but different effort, it has been discovered [11.24] that the two-doublet neutrino mass and mixing scheme favored to fit all of the experimental data [11.25] serendipitously also engineers an elegant removal of the ν_e flux below the weak-freeze-out radius. Therefore, this also solves the alpha effect problem. The solution in this case has two parts.

First, we exploit the SuperK-derived maximal mixing of the mu and tau neutrinos (see Chap. 5). We transform to a new basis, in which one of the neutrino species decouples [11.24, 11.26]. A plausible manifestation of the unitary relationship between the mass basis and the flavor basis for the two-doublet scheme in vacuum is

$$
\begin{pmatrix} |\nu_e\rangle \\ |\nu_s\rangle \\ |\nu_\mu\rangle \\ |\nu_\tau\rangle \end{pmatrix} = \begin{pmatrix} \cos\phi & \sin\phi\cos\omega & \sin\phi\sin\omega & 0 \\ -\sin\phi & \cos\phi\cos\omega & \cos\phi\sin\omega & 0 \\ 0 & -\sin\omega/\sqrt{2} & \cos\omega/\sqrt{2} & 1/\sqrt{2} \\ 0 & \sin\omega/\sqrt{2} & -\cos\omega/\sqrt{2} & 1/\sqrt{2} \end{pmatrix} \begin{pmatrix} |\nu_1\rangle \\ |\nu_2\rangle \\ |\nu_3\rangle \\ |\nu_4\rangle \end{pmatrix}. \tag{11.45}
$$

Note that the mixing between ν_e and ν_s is mostly determined by the angle ϕ, while ω controls the level of mixing between between the doublets, $\nu_{\mu,\tau}$ and $\nu_{e,s}$.

The key step is to define two linear combinations of ν_μ and ν_τ in this representation, and then transform into the new flavor basis defined by these states. The linear combinations are

$$
|\nu_\mu^*\rangle \equiv \frac{|\nu_\mu\rangle - |\nu_\tau\rangle}{\sqrt{2}}, \tag{11.46}
$$

$$
|\nu_\tau^*\rangle \equiv \frac{|\nu_\mu\rangle + |\nu_\tau\rangle}{\sqrt{2}}. \tag{11.47}
$$

Armed with this new flavor basis, we can transform the matrix in (11.45) to produce a new unitary transformation to the mass basis in vacuum:

$$
\begin{pmatrix} |\nu_e\rangle \\ |\nu_s\rangle \\ |\nu_\mu^*\rangle \\ |\nu_\tau^*\rangle \end{pmatrix} = \begin{pmatrix} \cos\phi & \sin\phi\cos\omega & \sin\phi\sin\omega & 0 \\ -\sin\phi & \cos\phi\cos\omega & \cos\phi\sin\omega & 0 \\ 0 & -\sin\omega & \cos\omega & 0 \\ 0 & 0 & 0 & 1 \end{pmatrix} \begin{pmatrix} |\nu_1\rangle \\ |\nu_2\rangle \\ |\nu_3\rangle \\ |\nu_4\rangle \end{pmatrix}. \tag{11.48}
$$

This transformation has effectively reduced a 4×4 mixing problem to a 3×3 problem. The new state $|\nu_\tau^*\rangle$ is apparently a mass eigenstate in vacuum in this representation: it is essentially decoupled. Since ν_μ and ν_τ have, in the supernova environment, identical effective weak potentials A, B and B_{off} (and, technically, C, but the thermal terms are not significant in the hot-bubble region), the transformation to the new states will remain valid at all densities/temperatures of relevance here.

As the neutrinos leave the neutrino sphere and propagate outward into lower densities, they will encounter, in this scheme the sequential resonances

$$\nu_\mu^* \rightleftharpoons \nu_s ,\tag{11.49}$$

$$\nu_\mu^* \rightleftharpoons \nu_e .\tag{11.50}$$

The ratio of the temperatures (positions) of these resonances for any E_ν will be (crudely)

$$\frac{T_{\mu s}}{T_{\mu e}} \approx \frac{\left[Y_e + \left(Y_{\nu_e} - Y_{\nu_\mu} - Y_{\nu_\tau}\right)\right]^{1/3}\Big|_{\mu e}}{\left[(1 - Y_e)/2 - Y_{\nu_e} - 2Y_{\nu_\mu} - Y_{\nu_\tau}\right]^{1/3}\Big|_{\mu s}} ,\tag{11.51}$$

where we have assumed that the entropy is uniform throughout the hot bubble, and that the vacuum mass-squared differences and the cosines of the effective vacuum *two-neutrino* mixing angles are the same for the two channels. If the neutrino background terms are not too important and $Y_e > 1/3$, the first resonance, $\nu_\mu^* \rightleftharpoons \nu_s$, will occur before the second, $\nu_\mu^* \rightleftharpoons \nu_e$. If there are no sterile neutrinos coming from the core, then the first resonance will have the effect of converting all ν_μ^* to steriles, while the second resonance will convert the ν_e to ν_μ^*. There will be no ν_μ^*s to convert to electron neutrinos at the second resonance. As in the other active–sterile scheme, this one again relies on a fast expansion, so that Y_e cannot adjust downward to reflect the equilibrium value for this quantity in the low-ν_e-flux conditions.

11.4   The Early Universe and Cosmology

So far as we can infer from fossil nucleosynthetic evidence left over from Big Bang nucleosynthesis and measurements of the cosmic microwave background radiation (CMBR), neutrinos dominate the energetics and entropy of the early Universe. But much about the neutrino physics of the early Universe remains mysterious. Matter-enhanced neutrino flavor transformation in the presence of a sterile neutrino has the potential to overturn many cherished notions about the role of neutrinos in nucleosynthesis and dark matter.

In many ways the early Universe between the end of the QCD transition and through the BBN epoch resembles the neutrino-heated hot-bubble environment discussed in the last section. This is especially true of the nucleosynthesis epoch. We shall exploit this similarity in gauging the effects of matter-enhanced neutrino flavor transformation. One can easily facilitate this by simply mapping the radius (or progress of an outward-moving fluid element) in the hot bubble to the age of the Universe.

The overriding feature of the physics of the early universe from our perspective is the huge entropy ($s \sim 10^{10}$). From (11.41) and an assumed closure fraction from baryon rest mass Ω_b, we can estimate the entropy per baryon and the baryon-to-photon ratio, which are

$$s \approx 2.53 \times 10^8 \Omega_b^{-1} h^{-2} , \tag{11.52}$$

$$\eta \approx 2.79 \times 10^{-8} \Omega_b h^2 , \tag{11.53}$$

respectively, where the Hubble parameter in units of $100 \, \mathrm{km \, s^{-1} \, Mpc^{-1}}$ is h, and we have assumed a temperature for the CMBR at the current epoch, $T_\gamma^0 = 2.75$ K. From the Burles and Tytler deuterium measurements we know that $\Omega_b h^2 \approx 0.02$ ([11.27] and references therein). Of course, this observational inference of the baryon density is valid only if these D/H (ratio of deuterium to hydrogen abundance) values reflect the primordial value, there are no significant entropy inhomogeneities during NSE freeze-out in BBN and there are equal numbers of all six active neutrino species, all with thermal Fermi–Dirac energy spectra and zero chemical potential.

This may strike some as a disturbingly long list of caveats. For example, do we really know what the neutrino degeneracy parameters are? The best current limits on these stem from BBN considerations, as we shall discuss [11.28], and involve constraints from the observationally derived abundances of other nuclei produced in BBN, principally ^{4}He [11.29]. It will prove expeditious at this point to run through the physics of standard BBN.

At a very early epoch in the Universe all the active neutrinos will be in thermal and chemical equilibrium with the matter. The expansion rate of the Universe at an epoch with temperature T is

$$H \approx \left(\frac{8\pi^3}{90} \right)^{1/2} g^{1/2} \frac{T^2}{m_{\mathrm{Pl}}} , \tag{11.54}$$

where the statistical weight in relativistic particles is a sum of both boson and fermion contributions, $g = \sum_i (g_\mathrm{b})_i + 7/8 \sum_j (g_\mathrm{f})_j \approx g_\mathrm{s}$, and where $m_{Pl} \approx 1.22 \times 10^{22}$ MeV is the Planck mass. In these radiation-dominated conditions the age of the universe is $t \approx (1/2) H^{-1}$, and the particle horizon length is $2t$.

Neutrino inelastic scattering on relativistic particles (electron/positron pairs, other neutrinos) can maintain the energy flow and thermal contact between neutrinos and matter, so long as the temperature is above $T_{\mathrm{WD}} \approx 1$ to 3 MeV. Below this "weak decoupling" temperature, the scattering rates will fall below the expansion rate of the Universe. At this point, the matter and neutrinos can no longer exchange energy on a timescale which is short compared with the age of the Universe. In a sense, this is analogous to the energy-decoupling aspect of the neutrino-decoupling process near the neutrino sphere of the neutron star, as discussed in the previous section.

However, even though the neutrinos have energy-exchange decoupled, they are still energetic enough to effect rapid interconversion of neutrons and protons through the capture reactions in (11.1) and (11.2). This is exactly analogous to the steady weak-equilibrium case in the region below the weak-freeze-out position in the neutrino-heated wind. As in that case, the neutron-to-proton ratio will be given by (11.44), so long as we interpret the rates in the ratio as *net* forward rates for the processes in (11.1) and (11.2) (that is, they include the compensating electron and positron capture rates).

Just as in the neutrino-driven-wind case, the temperature will eventually fall low enough that the Y_e value at the last equilibrium interaction point becomes frozen in. This is the weak-freeze-out point, and occurs at $T_{\mathrm{WFO}} \approx 0.7\,\mathrm{MeV}$ for zero lepton numbers in the neutrinos, where the neutron-to-proton ratio will be $n/p \approx 1/6$. The low value of this ratio reflects the low temperature of the neutrino energy distributions assumed here. The $\bar{\nu}_e + p$ capture reaction has an appreciable threshold, and so is hindered relative to the $\nu_e + n$ process. Note that we could easily change the n/p ratio either by invoking different energy spectra, or by changing the expansion rate by altering g. The latter effect moves the point (changes T_{WFO}) where the neutrino capture rates are comparable to the expansion rate.

Just as in the neutron star neutrino-driven-wind case, the material in the early Universe will experience freeze-out of the strong and electromagnetic interactions when the nuclear-reaction rates fall below the expansion rate. The entropy in the early Universe $s \sim 10^{10}$ is comparable to that in the wind, $s \sim 10^2$, in a nuclear-physics sense! Clearly, the large entropy in the early Universe will militate against the formation of heavy nuclei, as is immediately obvious from an inspection of (11.36). This is similar to, but more pronounced than, the same tendency in the wind.

Oppositely to the neutrino-driven wind environment, however, the early Universe is proton-rich after weak freeze-out, at least if we make the usual assumption of zero net neutrino numbers. Therefore, BBN is essentially the isospin mirror of nucleosynthesis in neutrino-heated supernova ejecta. Most of the nuclear-physics events in BBN follow in direct analogy to those in the wind, but with this isospin mirror image twist.

Therefore, just as in the neutrino-driven-wind scenarios, alpha particles incorporate all available nucleons which can be so incorporated and, in this case, scour out all of the neutrons. In effect, then, the neutron number at weak freeze-out governs the helium yield in BBN. Note that the remaining free protons will have difficulty in assembling heavy nuclei, not only because of the entropy, but also because of the Coulomb barriers associated with charged-particle capture. The same argument applies to the triple alpha reaction.

It is common to parameterize the ^{4}He yield by an effective number of neutrino flavors N_ν. This is a dangerous and misleading procedure, because it leads to confusion about how it is that neutrinos determine the neutron-to-proton ratio. It is the neutron-to-proton ratio which determines the ^{4}He yield, nothing else. The energy distribution functions of the neutrinos enter the bargain by affecting the expansion rate and therefore the weak-freeze-out point, and also directly through the rates of the capture processes (11.1) and (11.2).

So, what are the neutrino energy distribution functions in the early Universe? We shall follow common practice and define the net lepton number carried by an active neutrino species ν_α as

$$L_{\nu_\alpha} \equiv \frac{n_{\nu_\alpha} - n_{\bar{\nu}_\alpha}}{n_\gamma} . \tag{11.55}$$

The BBN bounds on these quantities based on the ^{4}He yield are [11.28]

$$-4.1 \times 10^{-2} \leq L_{\nu_e} \leq 0.84 , \tag{11.56}$$

$$|L_{\nu_\mu, \nu_\tau}| \leq 27.5 . \tag{11.57}$$

These numbers leave tremendous latitude for mischief in the early Universe and invite speculation about ways to avoid these bounds!

If the individual neutrino numbers are small, then the relationship between the degeneracy parameter and the lepton number is $\eta_{\nu_\alpha} \approx 1.46 L_{\nu_\alpha}$. For the general case,

$$L_{\nu_\alpha} \approx \frac{1}{4\zeta(3)} \left(\frac{\pi^2}{3} \eta_{\nu_\alpha} + \frac{1}{3} \eta_{\nu_\alpha}^3 \right) , \tag{11.58}$$

as long as the ν_αs have relativistic kinematics.

From a neutrino transformation standpoint, the early Universe is a horrible environment to follow computationally. Not only can the weak-potential (neutrino background) term B dominate, with many of the consequent "diseases" endemic to that regime, but also the thermal term can dominate at a high enough temperature. It is useful for our purposes here to rewrite the weak potential which governs neutrino flavor transformation in the channel $\nu_\alpha \rightleftharpoons \nu_s$ (here $\alpha = e, \mu, \tau$) for temperatures above the weak-freeze-out scale (where $Y_e \approx 1/2$) as

$$V \approx \frac{2\sqrt{2}\zeta(3)}{\pi^2} G_F T^3 \left(\mathcal{L} \pm \frac{\eta}{4} \right) - \delta_T G_F^2 \epsilon T^5 , \tag{11.59}$$

where the upper sign "+" is for the $\alpha = e$ channel, and the lower sign "−" is for the $\alpha = \mu, \tau$ channels, and where the coefficients of the thermal term are

$$\delta_T \approx \begin{cases} 79.3 & \alpha = e \\ 22.2 & \alpha = \mu, \tau \end{cases} . \tag{11.60}$$

(At a high enough temperature scale $\delta_T \approx 79.3$ for all flavors on account of pair population of massive charged lepton states.) Here we have subsumed the A and B terms from Table 11.1 into the term $\mathcal{L} \pm \eta/4$. The neutrino background contribution is

$$\mathcal{L} \equiv 2L_{\nu_\alpha} + \sum_{\beta \neq \alpha} L_{\nu_\beta} \, . \tag{11.61}$$

We can provide an example expression for this potential,

$$V \approx (40.2\,\text{eV}) \left(\frac{\mathcal{L} \pm \eta/4}{10^{-2}} \right) \left(\frac{T}{\text{GeV}} \right)^3 - \mathcal{F}_\epsilon \left(\frac{T}{\text{GeV}} \right)^5 \, , \tag{11.62}$$

with

$$\mathcal{F} \approx \begin{cases} 10.8\,\text{eV} & \alpha = e \\ 3.0\,\text{eV} & \alpha = \mu, \tau \end{cases} \, . \tag{11.63}$$

We see from the forms of these potentials that active neutrinos acquire increasingly large masses as we go back in time, to higher temperature, until we reach a temperature so high that the negative thermal terms begin to bring the effective neutrino mass down with increasing temperaure. Therefore, a heavier sterile neutrino will experience two level crossings with an active neutrino: one beyond the peak of the potential at very high temperature, and a later lower-temperature resonance below the potential peak. This presupposes of course that the potential peak is above the mass of the sterile neutrino. The height of the peak depends on the magnitude of the net driving lepton number. Since the expansion rate of the Universe goes as the square of the temperature scale, the higher-temperature mass level crossing will tend to be less adiabatic than the lower-temperature one.

The phenomenology of active–sterile neutrino transformation in the early Universe has been investigated over the last several decades [11.30, 11.31]. In all of this work the most important effect against which we are battling is the tendency for sterile-neutrino production. Given the tight limits on the helium abundance, there is not much "room" for additional contributions to g and the attendant increase in the n/p ratio and the ^{4}He yield. Arguments along these lines led to important limits on the large-mixing-angle active–sterile solution $\nu_e \rightleftharpoons \nu_s$ for the solar neutrino problem [11.32].

But perhaps the most fascinating development is the idea that a net lepton number in the neutrinos could be generated in situ by the aid of active–sterile plus active–active matter-enhanced neutrino transformation [11.33, 11.34]. If this could generate an excess of ν_es ($L_{\nu_e} > 0$), then perhaps the bad effects from the concomitant increase in the population of the sterile-neutrino sea could be avoided.

In that work, we again seek BBN (primarily ^{4}He) limits on various neutrino mass and mixing schemes that have been proposed for other problems. Unfortunately, the coupled active–sterile plus active–active neutrino transformation scenarios that have been discussed so far are extremely difficult to calculate with confidence. Approximations must be made in the general

Boltzmann plus coherent transport problem discussed in Sect. 11.2. Currently, there are four different groups with at least two different answers [11.35–11.41]; and the situation really is not even that good, because two groups that agree on the final result do not agree on how they got there.

The end results range all the way from those of Dolgov et al. [11.41], who maintain that there is *no* lepton number generation possible in these schemes, through those of the Foot and Volkas collaboration [11.34, 11.35], who obtain significant lepton asymmetries, to those of the author, Shi and Abazajian, who have derived stringent constraints on these schemes on the basis of the ^{4}He yield in BBN and an interesting causality effect.

In [11.39] it was recognized that the number of sterile neutrinos generated in these scenarios, and hence the increase in the ^{4}He yield which could be the undoing of those scenarios, would be greatly enhanced if the sign of the lepton number generated in these scenarios was chaotic and random. That would mean that at the epoch of NSE freeze-out in BBN there would exist domains of net neutrino lepton number with two different signs. The size of these domains would be the comoving horizon scale at the epoch when the lepton number was generated by matter-enhanced neutrino flavor transformation. During or before BBN, active neutrinos which moved across the domain boundaries would experience a lepton number gradient, which, in turn, could lead to further matter-enhanced production of sterile neutrinos. This would lead, of course, to a faster expansion rate and, hence, even more ^{4}He. The calculation by another group [11.38] seems to confirm the chaotic nature of lepton number generation, albeit not in the same range of parameters as seen in the original calculation by Shi [11.33]. Settling this issue is a key problem for the future.

Finally, suppose there really are sterile neutrinos and (small) net lepton numbers. In this case it may be possible that matter-enhanced active–sterile neutrino transformation could convert a small excess of active neutrinos with low mass into a population of massive sterile neutrinos and form a new and attractive dark-matter candidate [11.42]. The twist is that, with MSW production of these steriles, their energy spectrum could be grossly nonthermal and skewed towards low energy. Therefore, these sterile neutrinos are born "cold", with an energy spectrum that would cause them to revert to nonrelativistic kinematics at an early epoch. As a result, they would not have the "disease" that ordinary hot dark matter, or even warm dark matter [11.43], has: too much free streaming and too much damping of large-scale structure. By constrast, the sterile neutrinos of the Shi & Fuller scenario are essentially cold or warm dark matter, but with the twist of a small-mass-scale structure damping. In fact, the existence of "cusps" and too many dwarf galaxy-sized objects in standard cold-dark-matter models may signal the need for a modification of the dark-matter candidate particle along these lines (see, for example, the discussion in [11.44]).

The favored allowed mass range for the Shi & Fuller sterile neutrinos is from $100\,\mathrm{eV}$ at the light end to about $10\,\mathrm{keV}$ at the heavy end, for a total fractional contribution to closure of about $\Omega_\nu \approx 0.2$. The required driving lepton number for these cases runs from $\mathcal{L} \approx 0.1$ at the low end to $\mathcal{L} \approx 10^{-3}$ at the high end. (We cannot go to higher masses with lower initial driving lepton numbers because there would be no MSW resonance in that case.) In fact, we can relate the sterile-neutrino mass to the Hubble parameter at the current epoch and the required initial net lepton number in neutrinos,

$$\Omega_{\nu_\mathrm{s}} \approx \left(\frac{m_{\nu_\mathrm{s}}}{6\,\mathrm{keV}}\right)\left(\frac{0.65}{h}\right)^2\left(\frac{\mathcal{L}}{10^{-2}}\right). \tag{11.64}$$

The required lepton numbers here are not large enough to conflict with any of the BBN limits discussed above, but they necessitate a reexmination of how the net baryon number in the Universe is generated. In particular, the net baryon number could not be produced in the cosmic electroweak symmetry-breaking phase transition, as baryogenesis scenarios based on nonequilibrium processes on bubble walls at this epoch necessarily have $B - L = 0$. More radical models for the origin of baryon number do not have this problem [11.45].

Matter-suppressed or vacuum scattering production of warm dark matter sterile neutrinos was proposed by Dodelson and Widrow [11.43] and revisited recently [11.48]. Abazajian, the author, and Patel have revisited this issue and considered both scattering and matter-enhanced production of sterile neutrino dark matter [11.49].

I would like to acknowledge partial support from NSF grant Phy 98-00980 at University of California San Diego, and the hospitality of the Institute for Nuclear Theory, University of Washington.

References

11.1 L. Wolfenstein, Phys. Rev. D **17**, 2369 (1978); Phys. Rev. D **20**, 2634 (1979).

11.2 S.P. Mikeyev and A.Yu. Smirnov, Sov. Phys. JETP **69**, 4 (1986).

11.3 Y.-Z. Qian, G. M. Fuller, G. J. Mathews, R. W. Mayle, J. R. Wilson and S. E. Woosley, Phys. Rev. Lett. **71**, 1965 (1993).

11.4 Y.-Z. Qian and G. M. Fuller, Phys. Rev. D **51**, 1479 (1995).

11.5 G. M. Fuller, R. W. Mayle, J. R. Wilson and D. N. Schramm, Astrophys. J. **322**, 795 (1987).

11.6 D. Nötzold and G. Raffelt, Nucl. Phys. B **307**, 924 (1988).

11.7 G. Sigl and G. Raffelt, Nucl. Phys. B **406**, 423 (1993).

11.8 W. C. Haxton, Phys. Rev. D **36**, 2283 (1987).

11.9 F. Loreti, Y.-Z. Qian, G. M. Fuller and A. B. Balantekin, Phys. Rev. D **52**, 6664 (1995).

11.10 X. Shi, Phys. Rev. D **54**, 2753 (1996).

11.11 G. M. Fuller, R. W. Mayle, B. S. Meyer and J. R. Wilson, Astrophys. J. **389**, 517 (1992).

11.12 A. Mezzacappa and S. W. Bruenn, in *Proceedings of the Second International Workshop on the Identification of Dark Matter* (World Scientific, 2000, Singapore), in press.

11.13 Y.-Z. Qian and S. E. Woosley, Astrophys. J. **471**, 331 (1996).

11.14 B. S. Meyer, G. J. Mathews, W. M. Howard, S. E. Woosley and R. D. Hoffman, Astrophys. J. **399**, 1656 (1992); S. E. Woosley, G. J. Mathews, J. R. Wilson, R. D. Hoffman and B. S. Meyer, Astrophys. J. **433**, 229 (1994); K. Takahashi, J. Witti and H.-T. Janka, Astron. Astrophys. **286**, 857 (1994).

11.15 R. D. Hoffman, S. E. Woosley, G. M. Fuller and B. S. Meyer, Astrophys. J. **460**, 478 (1996).

11.16 C. Sneden, J. J. Cowan, D. L. Burris, and J. W. Truran, Astrophys. J. 496, 235 (1998); C. Sneden, A. McWilliam, G. W. Preston, J. J. Cowan, D. L. Burris and B. J. Armosky, Astrophys. J. **467**, 819 (1996); C. Sneden, J. J. Cowan, I. I. Ivans, G. M. Fuller, S. Burles, T. C. Beers and J. E. Lawler, Astrophys. J. (Lett.), in press (2000) astro-ph/0003086.

11.17 G.J. Wasserburg and Y.-Z. Qian, Astrophys. J. **529**, L21 (2000).

11.18 W. C. Haxton, K. Langanke, Y.-Z. Qian and P. Vogel, Phys. Rev. Lett. **78**, 2694 (1997); Y.-Z. Qian, W. C. Haxton, K. Langanke and P. Vogel, Phys. Rev. C **55**, 1532 (1997).

11.19 G. M. Fuller and B. S. Meyer, Astrophys. J. **453**, 792 (1995); B. S. Meyer, G. C. McLaughlin and G. M. Fuller, Phys. Rev. C, **58**, 3696 (1998).

11.20 B. S. Meyer and J. S. Brown, Astrophys. J. Suppl. **112**, 199 (1997); R. D. Hoffman, S. E. Woosley and Y.-Z. Qian, Astrophys. J. **482**, 951 (1996).

11.21 C. Y. Cardall and G. M. Fuller, Astrophys. J. **486**, L111 (1997); G. M. Fuller and Y.-Z. Qian, Nucl. Phys. A **606**, 167 (1996); J. Salmonson and J. R. Wilson, Astrophys. J., in press (1999).

11.22 G. C. McLaughlin, J. Fetter, A. B. Balantekin and G. M. Fuller, Phys. Rev. C **59**, 2873 (1999).

11.23 J. T. Peltoniemi, Astron. Astrophys. **254**, 121 (1992); J. T. Peltoniemi, hep-ph/9511323, (1995); H. Nunokawa, J. T. Peltoniemi, A. Rossi and J. W. Valle, Phys. Rev. D **56**, 1704 (1997).

11.24 D. O. Caldwell, G. M. Fuller and Y.-Z. Qian, Phys. Rev. D **61**, 123005 (2000), astro-ph/9910175.

11.25 D. O. Caldwell and R. N. Mohapatra, Phys. Rev. D **48**, 3259 (1993); D. O. Caldwell and R. N. Mohapatra, Phys. Lett. B **354**, 371 (1995); J. T. Peltoniemi and J. W. F. Valle, Nucl. Phys. B **406**, 409 (1993); J. M. Gelb and S. P. Rosen, hep-ph/9909293; G. M. Fuller, J. R. Primack and Y.-Z. Qian, Phys. Rev. D **52**, 1288 (1995).

11.26 A. B. Balantekin and G. M. Fuller, Phys. Lett. B, in press, hep-ph/9908465.

11.27 S. Burles and D. Tytler, in *Proceedings of the Second Oak Ridge Symposium on Atomic and Nuclear Astrophysics*, ed. by A. Mezzacappa (Institute of Physics, Bristol, 1998).

11.28 K. Kang and G. S. Steigman, Nucl. Phys. B **372**, 494 (1992).

11.29 K. A. Olive, E. Skillman and G. Steigman, Astrophys. J. **483**, 788 (1997); Y. I. Izotov and T. X. Thun, Astrophys. J. **500**, 188 (1998).

11.30 A. Dolgov, Sov. J. Nucl. Phys. **33**, 700 (1981); D. Fargion and M. Shepkin, Phys. Lett B **146**, 46 (1984); P. G. Langacker, University of Pennsylvania Preprint UPR0401T (unpublished).

11.31 R. Barbieri and A. Dolgov, Phys. Lett. B **349**, 743 (1991); K. Enqvist, K. Kainulainen and M. Thomson, Nucl. Phys. B **373**, 498 (1992); K. Enqvist, K. Kainulainen and J. Maalampi, Nucl. Phys. B **349**, 754 (1991); A. D. Dolgov, J. H. Hansen and D. Y. Semikoz, Nucl. Phys. B **503**, 426 (1997); Nucl. Phys. B **543**, 269 (1999); D. P. Kirilova and M. V. Chizhov, Phys. Lett B **393**, 375 (1997); Phys. Rev. D **58**, 073004 (1998); Nucl. Phys. B **534**, 447 (1998).

11.32 X. Shi, D. N. Schramm and B. D. Fields, Phys. Rev. D **48**, 2563 (1993).

11.33 X. Shi, Phys. Rev. D **54**, 2753 (1996).

11.34 R. Foot and R. R. Volkas, Phys. Rev. Lett. **75**, 4350 (1995); R. Foot, M. J. Thomson, and R. R. Volkas, Phys. Rev. D **53**, 5349 (1996); R. Foot and R. R. Volkas, Phys. Rev. D **56**, 6653 (1997).

11.35 R. Foot and R. R. Volkas, Phys. Rev. D **55**, 5147 (1997).

11.36 P. Di Bari, P. Lipari and M. Lusignol, hep-ph/9907548.

11.37 X. Shi, G. M. Fuller and K. Abazajian, Phys. Rev. D **60**, 063002 (1999).

11.38 K. Enqvist, K. Kainulainen and A. Sorri, hep-ph/9906452.

11.39 X. Shi and G. M. Fuller, Phys. Rev. Lett. **83**, 3120 (1999), astro-ph/9904041.

11.40 K. Abazajian, G. M. Fuller and X. Shi, Phys. Rev. D **62**, 093003 (2000).

11.41 A. D. Dolgov, S. vH. Hansen, S. Pastor and D. vV. Semikoz, preprint, TAC-1999-018.

11.42 X. Shi and G. M. Fuller, Phys. Rev. Lett. **82**, 2832 (1999).

11.43 S. Dodelson and L. M. Widrow, Phys. Rev. Lett. **72**, 17 (1994).

11.44 A. Klypin, J. Holtzman, J. Primack and E. Rego, Astrophys, J. **416**, 1 (1993).

11.45 L. Affleck and M. Dine, Nucl. Phys. B **249**, 361 (1985).

11.46 L. Stodolsky, Phys. Rev. D **36**, 2273 (1987).

11.47 B. H. J. McKellar and M. J. Thomson, Phys. Rev. D **49**, 2710 (1994).

11.48 A. Dolgov and S. H. Hansen, hep-ph/0009083.

11.49 K. Abazajian, G. M. Fuller, M. Patel, astro-ph/0101524.

12 Hot Dark Matter in Cosmology

Joel R. Primack and Michael A. K. Gross

12.1 Historical Summary

Cosmological dark matter (DM) in the form of neutrinos with masses of up to a few electron volts is known as hot dark matter (HDM). In 1979–83, this appeared to be perhaps the most plausible dark-matter candidate. Such HDM models of cosmological structure formation led to a top-down formation scenario, in which superclusters of galaxies are the first objects to form, with galaxies and clusters forming through a process of fragmentation. Such models were abandoned when it was realized that if galaxies formed sufficiently early to agree with observations, their distribution would be much more inhomogeneous than it is observed to be. Since 1984, the most successful structure formation models have been those in which most of the mass in the Universe is in the form of cold dark matter (CDM). But mixed models with both cold and hot dark matter (CHDM) were also proposed in 1984 [12.1], although not investigated in detail until the early 1990s.

The recent atmospheric neutrino data from SuperKamiokande provide strong evidence of neutrino oscillations and therefore of nonzero neutrino mass. These data imply a lower limit on the HDM (i.e. light-neutrino) contribution to the cosmological density $\Omega_\nu \gtrsim 0.001$ – almost as much as that of all the stars in the centers of galaxies – and permit higher Ω_ν. The "standard" COBE-normalized critical-matter-density (i.e. $\Omega_m = 1$) CDM model has too much power on small scales. It was discovered in 1992–95 that CDM with the addition of neutrinos with a total mass of about 5 eV, corresponding to $\Omega_\nu \approx 0.2$, results in a much improved fit to the data on the distribution of nearby galaxies and clusters. Indeed, the resulting cold + hot dark-matter (CHDM) cosmological model is arguably the most successful $\Omega_m = 1$ model for structure formation [12.2–12.6].

However, other recent data have begun to make a convincing case for $0.3 \lesssim \Omega_m \lesssim 0.5$. In the light of all these new data, several authors have considered whether cosmology still provides evidence favoring a neutrino mass of a few eV in flat models with a cosmological constant $\Omega_\Lambda = 1 - \Omega_m$. The conclusion is that the possible improvement of the low-Ω_m flat (ΛCDM) cosmological models with the addition of light neutrinos appears to be rather limited, but that ΛCHDM models with $\Omega_\nu \lesssim 0.1$ may be consistent with currently available data. Data expected soon may permit detection of such a hot-dark-matter contribution, or alternatively provide stronger upper limits on Ω_ν and neutrino masses.

12.2   Hot, Warm and Cold Dark Matter

"Hot DM" refers to particles, such as neutrinos, that were moving at nearly the speed of light at redshift $z \sim 10^6$ (or time $t \sim 1$ yr), when the temperature T was of order 3×10^2 eV and the cosmic horizon first encompassed $10^{12} M_\odot$, the amount of dark matter contained in the halo of a large galaxy like the Milky Way. Hot-DM particles must also still have been in thermal equilibrium after the last phase transition in the hot early Universe, the QCD confinement transition, which presumably took place at $T_{\mathrm{QCD}} \approx 10^2$ MeV. Hot-DM particles have a cosmological number density roughly comparable to that of the microwave background photons, which (as we shall see shortly) implies an upper bound to their mass of a few tens of eV. This then implies that free streaming of these relativistic particles destroys any fluctuations smaller than supercluster size, $\sim 10^{15} M_\odot$.

The terminology "hot", "warm" and "cold" DM was introduced in 1983 [12.7, 12.8]. Warm-DM (WDM) particles interact much more weakly than neutrinos. They decouple (i.e. their mean free path first exceeds the horizon size) at $T \gg T_{\mathrm{QCD}}$, and are not heated by the subsequent annihilation of hadronic species. Consequently their number density is roughly an order of magnitude lower, and their mass an order of magnitude higher, than for hot-DM particles. Fluctuations corresponding to sufficiently large galaxy halos, $\gtrsim 10^{11} M_\odot$, could then survive free streaming. In theories of local supersymmetry broken at $\sim 10^6$ GeV, gravitinos could be DM of the warm variety [12.9–12.11]. Other WDM candidates are also possible, of course, such as right-handed neutrinos [12.12]. WDM does not fit the observations if $\Omega_{\mathrm{m}} = 1$ [12.13], but for low Ω_{m} some have suggested that it may be worth reconsidering, to avoid some possible problems of CDM [12.14, 12.15]. However, the cutoff in the power spectrum $P(k)$ at large k implied by WDM will also inhibit the formation of small dark-matter halos at high redshift. But such small halos are presumably where the first stars form, which produce metals rather uniformly throughout the early Universe, as indicated by observations of the Lyman α forest (neutral-hydrogen clouds seen in absorption in quasar spectra).

CDM consists of particles for which free streaming is of no cosmological importance. Two different sorts of CDM consisting of elementary particles have been proposed: heavy thermal remnants of annihilation such as supersymmetric neutralinos, and a cold Bose condensate such as axions. A universe where the matter is mostly CDM and there is a large cosmological constant looks very much like the one astronomers actually observe, and this low-Ω_{m} ΛCDM model [12.16] is the current favorite model for structure formation in the Universe [12.17–12.19].

12.3 Galaxy Formation with HDM

The standard HDM candidate is massive neutrinos [12.20–12.23], although other, more exotic theoretical possibilities have been suggested, such as a "majoron" of nonzero mass which is lighter than the lightest neutrino species, and into which all neutrinos decay. Neutrinos appeared to be an attractive DM candidate because of the measurement of an electron neutrino mass of about 30 eV in 1980 [12.24]. This coincided with the improving cosmic-microwave-background (CMB) limits on the primordial fluctuation amplitude, which forced Zel'dovich and other theorists to abandon the idea that all the dark matter could be made of ordinary baryonic matter. The version of HDM that they worked out in detail, with adiabatic Gaussian primordial fluctuations, became the prototype for the subsequent $\Omega_m = 1$ CDM theory.

12.3.1 Mass Constraints

Direct measurements of neutrino masses have given only upper limits (see also Chap. 3). A secure upper limit on the electron neutrino mass is roughly 15 eV. The Particle Data Group [12.25] notes that a more precise limit cannot be given, since unexplained effects have resulted in significantly negative measurements of $m(\nu_e)^2$ in tritium beta decay experiments. However, this problem is at least partially resolved, and the latest experimental upper limits on the electron neutrino mass are 2.8 eV from the Mainz [12.26] and 2.5 eV from the Troitsk [12.27] tritium beta decay experiments (both 95% C.L.). There is an upper limit on an effective Majorana neutrino mass of ~ 1 eV from neutrinoless-double-beta-decay experiments [12.28] (cf. [12.29]). The upper limits from accelerator experiments on the masses of the other neutrinos are $m(\nu_\mu) < 0.17$ MeV (90% C.L.) and $m(\nu_\tau) < 18$ MeV (95% C.L.) [12.25, 12.30],[1] but since stable neutrinos with such large masses would certainly "overclose the Universe" (i.e. contribute such a large cosmological density that the Universe could never have attained its present age), cosmology implies a much lower upper limit on these neutrino masses.

Before going further, it will be necessary to discuss the thermal history of neutrinos in the standard hot Big Bang cosmology in order to derive the corresponding constraints on their mass. Left-handed neutrinos of mass ≤ 1 MeV remain in thermal equilibrium until the temperature drops to $T_{\nu d}$, at which point their mean free path first exceeds the horizon size and they essentially cease interacting thereafter, except gravitationally [12.31]. Their mean free path is, in natural units ($\hbar = c = 1$), $\lambda_\nu \sim (\sigma_\nu n_{e^\pm})^{-1} \sim [(G_F^2 T^2)(T^3)]^{-1}$, where $G_F \approx 10^{-5}$ GeV^{-2} is the Fermi constant that measures the strength of the weak interactions. The horizon size is $\lambda_h \sim (G\rho)^{-1/2} \sim M_{P\ell} T^{-2}$, where

---

[1] Reference [12.30] gives a comprehensive summary of neutrino data and ongoing experiments.

the Planck mass $M_{P\ell} \equiv G^{-1/2} = 1.22 \times 10^{19}$ GeV. Thus $\lambda_h/\lambda_\nu \sim (T/T_{\nu d})^3$, with the neutrino decoupling temperature

$$T_{\nu d} \sim M_{P\ell}^{-1/3} G_F^{-2/3} \sim 1 \,\text{MeV} . \qquad (12.1)$$

After T drops below 1/2 MeV, $e^+ e^-$ annihilation ceases to be balanced by pair creation, and the entropy of the $e^+ e^-$ pairs heats the photons. Above 1 MeV, the number density $n_{\nu i}$ of each left-handed neutrino species and its right-handed antiparticle is equal to that of the photons, n_γ, times a factor 3/4 from Fermi versus Bose statistics. But then $e^+ e^-$ annihilation increases the photon number density relative to that of the neutrinos by a factor of 11/4.[2] As a result, the neutrino temperature $T_{\nu,0} = (4/11)^{1/3} T_{\gamma,0}$. Thus today, for each species,

$$n_{\nu,0} = \frac{3}{4} \frac{4}{11} n_{\gamma,0} = 109 \, \theta^3 \,\text{cm}^{-3} , \qquad (12.2)$$

where $\theta \equiv (T_0/2.7 \,\text{K})$. With the cosmic-background-radiation temperature $T_0 = 2.728 \pm 0.004$ K measured by the FIRAS instrument on the COBE satellite [12.32], $T_{\nu,0} = 1.947$ K and $n_{\nu,0} = 112 \,\text{cm}^{-3}$.

Since the present cosmological matter density is

$$\bar{\rho}_m = \Omega \rho_c = 10.54 \, \Omega_m h^2 \,\text{keV cm}^{-3} , \qquad (12.3)$$

it follows that

$$\sum_i m_{\nu i} < \bar{\rho}_m / n_{\nu,0} \leq 96 \, \Omega_m h^2 \theta^{-3} \,\text{eV} \approx 93 \, \Omega_m h^2 \,\text{eV} , \qquad (12.4)$$

where the sum runs over all neutrino species with $M_{\nu i} \leq 1$ MeV. (Heavier neutrinos will be discussed in the next paragraph.) Observational data imply that $\Omega_m h^2 \approx 0.1$–0.3, since $\Omega_m \approx 0.3$–0.5 and $h \approx 0.65 \pm 0.1$ [12.19]. Thus, if all the dark matter were light neutrinos, the sum of their masses would be ≈ 9–28 eV.

In deriving (12.4), we have been assuming that all the neutrino species are light enough to still be relativistic at decoupling, i.e. lighter than 1 MeV. The

---

[2] In the argument giving the 11/4 factor, the key ingredient is that the entropy in interacting particles in a comoving volume S_I is conserved during ordinary Hubble expansion, even during a process such as electron–positron annihilation, so long as it occurs in equilibrium. That is, $S_I = g_I(T) N_\gamma(T) = $ constant, where $N_\gamma = n_\gamma V$ is the number of photons in a given comoving volume V, and $g_I = [g_B + (7/8) g_F]_I$ is the effective number of helicity states in the interacting particles (with the factor of 7/8 reflecting the difference in energy density between fermions and bosons). Just above the temperature of electron–positron annihilation, $g_I = g_\gamma + 7/8 \times g_e = 2 + (7/8) \times 4 = (11/2)$; while below it, $g_I = g_\gamma = 2$. Thus, as a result of the entropy of the electrons and positrons being dumped into the photon gas at annihilation, the photon number density is thereafter increased relative to that of the neutrinos by a factor of 11/4.

bound (12.4) shows that they must then be much lighter than that. In the alternative case that a neutrino species is nonrelativistic at decoupling, it has been shown [12.33–12.37] that its mass must then exceed several GeV, which is not true of the known neutrinos (ν_e, ν_μ and ν_τ). (One might at first think that the Boltzmann factor would sufficiently suppress the number density of neutrinos weighing a few tens of MeV to allow compatibility with the present density of the Universe. It is the fact that they "freeze out" of equilibrium well before the temperature drops to their mass that leads to the higher mass limit.) We have also been assuming that the neutrino chemical potential is negligible, i.e. that $|n_\nu - n_{\bar\nu}| \ll n_\gamma$. This is very plausible, since the net baryon number density $(n_b - n_{\bar b}) \lesssim 10^{-9} n_\gamma$, and Big Bang nucleosynthesis restricts the allowed parameters [12.38] (see also Chap. 11).

12.3.2   Phase Space Constraint

We have just seen that light neutrinos must satisfy an upper bound on the sum of their masses. But now we shall discuss a lower bound on neutrino mass that arises because they must be rather massive to form the dark matter in galaxies, since their phase space density is limited by the Pauli exclusion principle. A slightly stronger bound follows from the fact that they were not degenerate in the early Universe.

The phase space constraint [12.39] follows from Jeans's theorem in classical mechanics to the effect that the maximum six-dimensional phase space density cannot increase as a system of collisionless particles evolves. At early times, before density inhomogeneities become nonlinear, the neutrino phase space density is given by the Fermi–Dirac distribution

$$n_\nu(p) = \frac{g_\nu}{h^3} \left[1 + \exp\left(\frac{pc}{kT_\nu(z)} \right) \right]^{-1}, \tag{12.5}$$

where here h is Planck's constant and $g_\nu = 2$ for each species of left-handed ν plus right-handed $\bar\nu$. Since momentum and temperature both scale as the redshift z as the Universe expands, this distribution remains valid after neutrinos drop out of thermal equilibrium at ~ 1 MeV, and even into the nonrelativistic regime $T_\nu < m_\nu$ [12.31]. The standard version of the phase space constraint follows from demanding that the central phase space density $9[2(2\pi)^{5/2} G r_c^2 \sigma m_\nu^4]^{-1}$ of the DM halo, assumed to be an isothermal sphere of core radius r_c and one-dimensional velocity dispersion σ, not exceed the maximum value of the initial phase space density $n_\nu(0) = g_\nu/2h^3$. The result is

$$m_\nu > (120\,\mathrm{eV}) \left(\frac{100\,\mathrm{km\,s^{-1}}}{\sigma} \right)^{1/4} \left(\frac{1\,\mathrm{kpc}}{r_c} \right)^{1/2} \left(\frac{g_\nu}{2} \right)^{-1/4}. \tag{12.6}$$

The strongest lower limits on m_ν follow from applying this to the smallest galaxies. Both theoretical arguments regarding the dwarf spheroidal (*dS*) satellite galaxies of the Milky Way [12.40] and data on Draco, Carina and Ursa Minor made it clear some time ago that dark matter dominates the gravitational potential of these dS galaxies, and the case has only strengthened with time [12.41]. The phase space constraint then sets a lower limit [12.42] $m_\nu > 500$ eV, which is completely incompatible with the cosmological constraint (12.4). However, this argument only excludes neutrinos as the DM in certain small galaxies; it remains possible that the DM in these galaxies is (say) baryonic, while that in larger galaxies such as our own is (at least partly) light neutrinos. A more conservative phase space constraint was obtained for the Draco and Ursa Minor dwarf spheroidals [12.43], but the authors concluded that neutrinos consistent with the cosmological upper bound on m_ν cannot be the DM in those galaxies. A similar analysis applied to the gas-rich low-rotation-velocity dwarf irregular galaxy DDO 154 [12.44] gave a limit $m_\nu > 94$ eV, again inconsistent with the cosmological upper bound.

12.3.3   Free Streaming

The most salient feature of HDM is the erasure of small fluctuations by free streaming. Thus even collisionless particles effectively exhibit a Jeans mass. It is easy to see that the minimum mass of a surviving fluctuation is of order $M_{\mathrm{Pl}}^3/m_\nu^2$ [12.45, 12.46]. Let us suppose that some process in the very early Universe – for example, thermal fluctuations subsequently vastly inflated in the inflationary scenario – gave rise to adiabatic fluctuations on all scales. In adiabatic fluctuations, all the components – radiation and matter – fluctuate together. Neutrinos of nonzero mass m_ν stream relativistically from decoupling until the temperature drops to $T \sim m_\nu$, during which time they traverse a distance $d_\nu = R_{\mathrm{H}}(T = m_\nu) \sim M_{\mathrm{Pl}}\, m_\nu^{-2}$. In order to survive this free streaming, a neutrino fluctuation must be larger in linear dimension than d_ν. Correspondingly, the minimum mass of neutrinos in a surviving fluctuation is $M_{\mathrm{J},\nu} \sim d_\nu^3 m_\nu n_\nu(T = m_\nu) \sim d_\nu^3 m_\nu^4 \sim M_{\mathrm{Pl}}^3\, m_\nu^{-2}$. By analogy with Jeans's calculation of the minimum mass of an ordinary fluid perturbation for which gravity can overcome pressure, this is referred to as the (free-streaming) Jeans mass.

A more careful calculation [12.46, 12.47] gives

$$d_\nu = 41(m_\nu/30\,\mathrm{eV})^{-1}(1+z)^{-1} \text{ Mpc} , \tag{12.7}$$

that is, $d_\nu = 41(m_\nu/30\mathrm{eV})^{-1}$ Mpc in comoving coordinates, and correspondingly

$$M_{\mathrm{J},\nu} = 1.77\, M_{\mathrm{Pl}}^3\, m_\nu^{-2} = 3.2 \times 10^{15}(m_\nu/30\,\mathrm{eV})^{-2} M_\odot , \tag{12.8}$$

which is the mass scale of superclusters. Objects of this size are the first to form in a ν-dominated universe, and smaller-scale structures such as galaxies can form only after the initial collapse of supercluster-size fluctuations.

When a fluctuation of total mass $\sim 10^{15} M_\odot$ enters the horizon at $z \sim 10^4$, the density contrast δ_{RB} of the radiation plus baryons ceases growing and instead starts oscillating as an acoustic wave, while that of the massive neutrinos δ_ν continues to grow linearly with a scale factor $R = (1 + z)^{-1}$ since the Compton drag that prevents growth of δ_{RB} does not affect the neutrinos. By recombination, at $z_r \sim 10^3$, $\delta_{RB}/\delta_\nu \lesssim 10^{-1}$, with possible additional suppression of δ_{RB} by Silk damping. Thus the HDM scheme with adiabatic primordial fluctuations predicts small-angle fluctuations in the microwave background radiation that are lower than in the adiabatic baryonic cosmology, which was one of the reasons HDM appealed to Zel'dovich and other theorists. Similar considerations apply in the warm- and cold-DM schemes. However, as we shall discuss in a moment, the HDM top-down sequence of cosmogony is wrong, and with the COBE normalization hardly any structure would have formed by the present.

In numerical simulations of dissipationless gravitational clustering starting with a fluctuation spectrum appropriately peaked at $\lambda \sim d_\nu$ (reflecting damping by free streaming below that size and less time for growth of the fluctuation amplitude above it), the regions of high density form a network of filaments, with the highest densities occurring at the intersections and with voids in between [12.48–12.51]. The similarity of these features to those seen in observations was cited as evidence in favor of HDM [12.52].

12.3.4   Problems with ν DM

A number of potential problems with the neutrino dominated universe had emerged by about 1983, however.

- From studies both of nonlinear clustering [12.51, 12.53] (comoving length scale $\lambda \lesssim 10$ Mpc) and of streaming velocities [12.54] in the linear regime ($\lambda > 10$ Mpc), it follows that supercluster collapse must have occurred recently: $z_{sc} \leq 0.5$ is indicated and in any case $z_{sc} < 2$ [12.51]. However, the best limits on galaxy ages, obtained from globular clusters and other stellar populations, indicated that galaxy formation took place before $z \approx 3$. Moreover, if quasars are associated with galaxies, as is suggested by the detection of galactic luminosity around nearby quasars and the apparent association of more distant quasars with galaxy clusters, the abundance of quasars at $z > 2$ is also inconsistent with the "top-down" neutrino-dominated scheme in which superclusters form first: $z_{sc} > z_{galaxies}$.
- Numerical simulations of the nonlinear "pancake" collapse, taking into account dissipation of the baryonic matter, showed that at least 85% of the baryons were so heated by the associated shock that they remained unable to condense, attract neutrino halos and eventually form galaxies [12.7, 12.55]. This was a problem for the HDM scheme for two reasons. With the primordial-nucleosynthesis constraint $\Omega_b \lesssim 0.1$, there would be difficulty in having enough baryonic matter condense to form the luminosity that we actually observe. And, where are the X-rays from the shock-heated pancakes [12.56]?

- The neutrino picture predicts [12.57] that there should be a factor of ~ 5 increase in M/M_b between large galaxies ($M \sim 10^{12} M_\odot$) and large clusters ($M \geq 10^{14} M_\odot$), since the larger clusters, with their higher escape velocities, are able to trap a considerably larger fraction of the neutrinos. Although there is some indication that the mass-to-light ratio M/L increases with M, the ratio of total to luminous mass M/M_{lum} is probably a better indicator of the value of M/M_b, and it is roughly the same for galaxies with large halos and for rich clusters.

These problems, while serious, would perhaps not have been fatal for the HDM scheme. But an even more serious problem for HDM arose from the low amplitude of the CMB fluctuations detected by the COBE satellite, $(\Delta T/T)_{\mathrm{rms}} = (1.1 \pm 0.2) \times 10^{-5}$ smoothed on an angular scale of about $10°$ [12.58]. Although HDM and CDM both have the Zel'dovich spectrum shape ($P(k) \propto k$) in the long-wavelength limit, because of the free-streaming cutoff the amplitude of the HDM spectrum must be considerably higher in order to form any structure by the present. With the COBE normalization, the HDM spectrum is only beginning to reach nonlinearity at the present epoch.

Thus the evidence against standard HDM is convincing. At the very least, it indicates that structure formation in a neutrino-dominated universe must be rather more complicated than in the standard inflationary picture.

The main alternative that has been considered is cosmic strings plus hot dark matter. Because the strings would continue to seed structure up until the present, and because these seeds are in the nature of rather localized fluctuations, HDM would probably work better with string seeds than CDM. However, strings and other cosmic-defect models are now essentially ruled out [12.59, 12.60] because they predict that the cosmic microwave background would have an angular power spectrum without the pronounced (Doppler/acoustic/Sakharov) peak at angular wavenumber $l \sim 220$ that now appears to be clearly indicated by the data, along with secondary peaks at higher l.

12.4   Cold plus Hot Dark Matter
and Structure Formation: $\Omega_{\mathrm{m}} = 1$

Even if most of the dark matter is of the cold variety, a little hot dark matter can have a dramatic effect on the predicted distribution of galaxies. In the early Universe, the free streaming of the fast-moving neutrinos washes out any inhomogeneities in their spatial distribution on the scales that will later become galaxies. If these neutrinos are a significant fraction of the total mass of the Universe, then although the density inhomogeneities will be preserved in the cold dark matter, their growth rates will be slowed. As a result, the amplitude of the galaxy-scale inhomogeneities today is less with a little hot dark matter than if the dark matter is only cold. (With the tilt n of the primordial spectrum $P_p(k) = Ak^n$ fixed – which, as we discuss below, is not necessarily reasonable – the fractional reduction in the power on small scales is $\Delta P/P \approx 8\Omega_\nu/\Omega_m$ [12.61]. See Fig. 12.1 for examples of how the power spectrum $P(k)$ is affected by the addition of HDM in $\Omega_m = 0.4$ flat cosmologies.) Since the main problem with $\Omega_m = 1$ cosmologies containing only CDM is that the amplitude of the galaxy-scale inhomogeneities is too large compared with those on larger scales, the presence of a little HDM appeared to be possibly just what was needed. And, as was mentioned at the outset, a CHDM model with $\Omega_m = 1$, $\Omega_\nu = 0.2$ and Hubble parameter $h = 0.5$ is perhaps the best fit to the galaxy distribution in the nearby Universe for any cosmological model. The effects of the relatively small amounts of HDM in a CHDM model on the distribution of matter compared with a purely CDM model are shown graphically in [12.62]; see also [12.63]. As expected, within galaxy halos the distributions of cold and hot particles are similar. But the hot particles are more widely distributed on larger scales, and the hot/cold ratio is significantly enhanced in low-density regions.

The first step in working out the theory of structure formation is to use linear perturbation theory, which is valid since cosmic-microwave-background measurements show that density fluctuations are small at the redshift of recombination, $z_r \sim 10^3$. The most extensive early calculations of this sort were carried out by Holtzman [12.64, 12.65], who concluded that the most promising cosmological models were CHDM and ΛCDM [12.66]. The most efficient method of computing the linear evolution of fluctuations now is that used in the CMBFAST code [12.67]. An alternative Monte Carlo treatment of the evolution of neutrino density fluctuations was given in [12.68], but the differences from the usual treatment appear to be small. Detailed analytic results have been given in [12.69, 12.70] and reviewed in [12.63]. But the key point can be understood simply: there is less structure in CHDM models on small scales because the growth rate of CDM fluctuations is reduced on the scales where free streaming has wiped out neutrino fluctuations. Let us define the fluctuation growth rate f by

$$f(k) \equiv \frac{d \log \delta(k)}{d \log a} \,, \tag{12.9}$$

where $\delta(k)$ is the amplitude of the fluctuations of wavenumber $k = 2\pi/\lambda$ in CDM, and as usual $a = 1/(1+z)$ is the scale factor. For $\Omega_m = 1$ CDM fluctuations, the growth rate $f = 1$. This is also true for fluctuations in CHDM, for k sufficiently small that free streaming has not significantly decreased the amplitude of neutrino fluctuations. However, in the opposite limit $k \longrightarrow \infty$ [12.46, 12.63],

$$f_\infty = (\sqrt{1 + 24\Omega_c} - 1)/4 \approx \Omega_c^{0.6} \,, \tag{12.10}$$

assuming that $\Omega_c + \Omega_\nu = 1$. For example, for $\Omega_\nu = 0.2$, $f_\infty = 0.87$. Even though the growth rate is only a little lower for these large-k (i.e. short-wavelength) modes, the result is that their amplitude is decreased substantially compared with longer-wavelength modes.

The next step in determining the implications for structure formation is to work out the effects on nonlinear scales using N-body simulations. This is harder for cold + hot models than for CDM because the higher velocities of the neutrinos require more particles to adequately sample the neutrino phase space. The simulations must reflect the fact that the neutrinos initially have a redshifted Fermi–Dirac phase space distribution [12.71]. Such CHDM simulations have been compared with observational data using various statistics. CHDM with $\Omega_\nu = 0.3$, the value indicated by approximate analyses [12.66, 12.72], was shown to lead to groups of galaxies having substantially lower velocity dispersions than CDM did, and in better agreement with observations [12.73]. But it also leads to a void probability function (VPF) with more intermediate-sized voids than are observed [12.74]. This theory had so little small-scale power that a quasi-linear analysis using the Press–Schechter approximation showed that there would not be enough of the high-column-density hydrogen clouds at high redshift $z \sim 3$ known as damped Lyman-α systems [12.75–12.77]. But CHDM with $\Omega_\nu = 0.2$ suppresses small-scale fluctuations less and therefore has a better chance of avoiding this problem [12.78]. Simulations [12.79] showed that this version of CHDM also has a VPF in good agreement with observations [12.80]. The group velocity dispersions also remained sufficiently small to plausibly agree with observations, but it had become clear that the N-body simulations used lacked sufficient resolution to identify galaxies so that this statistic could be measured reliably [12.81].

A resolution problem also arose regarding the high-redshift damped Lyman-α systems. Earlier research had been based on the idea that these systems are rather large disk galaxies in massive halos [12.82], but then high-resolution hydrodynamic simulations [12.83] showed that relatively small gaseous protogalaxies moving in smaller halos provided a good match to the new, detailed kinematic data [12.84]. It thus appeared possible that CHDM models with $\Omega_\nu \lesssim 0.2$ might produce enough damped Lyman-α systems. With the low Hubble parameter $h \sim 0.5$ required for such $\Omega_{\rm m} = 1$ models, the total neutrino mass would then be $\lesssim 5$ eV.

While neutrino oscillation experiments can determine the differences of squared neutrino masses, as we shall briefly review next, cosmology is sensitive to the actual values of the neutrino masses – for any that are larger than about 1 eV. In that case, cosmology can help to fill in the neutrino mass matrix.

One example of this is the fact that if the HDM mass is roughly evenly shared between two or three neutrino species, the neutrinos will be lighter than if the same mass were all in one species, so that the free-streaming length will be longer. A consequence is that, for the same total neutrino mass and corresponding Ω_ν, the power spectrum will be approximately 20% lower on the scale of galaxy clusters if the mass is shared between two neutrino species [12.2]. Since the amplitude and "tilt" n of the power spectrum in CDM-type models is usually fixed by comparison with the COBE results and cluster abundance, this has the further consequence that higher n (i.e. less tilt) is required when the neutrino mass is divided between comparable-mass neutrino species. Less tilt means that there is more power on small scales, which appears to be favorable for the CHDM model, for example because it eases the problems with damped Lyman-α systems [12.2, 12.85].

12.5   Evidence for Neutrino Mass from Oscillations

There is mounting astrophysical and laboratory data suggesting that neutrinos oscillate from one species to another [12.30], which can only happen if they have nonzero mass. Of these experimental results, the ones that are regarded as probably the most secure are those concerning atmospheric neutrino oscillations from SuperKamiokande (see Chap. 5) and solar neutrinos from several experiments (see Chap. 4). But the experimental results that are most relevant to neutrinos as HDM are from the Liquid Scintillator Neutrino Detector (LSND) experiment at Los Alamos (see Chap. 7).

Older Kamiokande data [12.86] showed that, for events attributable to atmospheric neutrinos with visible energy $E > 1.3$ GeV, the deficit of ν_μ increases with zenith angle. The SuperKamiokande detector has confirmed and extended the results of its smaller predecessor [12.87]. These data imply that $\nu_\mu \rightarrow \nu_\tau$ oscillations occur with a large mixing angle, $\sin^2 2\theta > 0.82$, and an oscillation length several times the height of the atmosphere, which implies that $5 \times 10^{-4} < \Delta m_{\tau\mu}^2 < 6 \times 10^{-3}$ eV^2 (90% C.L.). (Neutrino oscillation experiments measure not the masses, but rather the difference of the squared masses, of the oscillating species; here $\Delta m_{\tau\mu}^2 \equiv |m(\nu_\tau)^2 - m(\nu_\mu)^2|$.) This in turn implies that if other data require either ν_μ or ν_τ to have a large enough mass ($\gtrsim 1$ eV) to be an HDM particle, then these two neutrinos must be nearly equal in mass, i.e. the HDM mass would be shared between these two neutrino species. Both the new SuperKamiokande atmospheric ν_e data and the lack of a deficit of $\bar{\nu}_e$ in the Chooz reactor experiment [12.88] make it quite unlikely that the atmospheric neutrino oscillation is $\nu_\mu \rightarrow \nu_e$. If the oscillation were instead to a sterile neutrino, the large mixing angle implies that this sterile species would become populated in the early Universe and lead to too much ^{4}He production during the Big Bang nucleosynthesis epoch [12.89]. (Sterile neutrinos are discussed further below.) It may be possible to verify that $\nu_\mu \rightarrow \nu_\tau$ oscillations occur via a long-baseline neutrino oscillation experiment. The K2K experiment is looking for missing ν_μ due to $\nu_\mu \rightarrow \nu_\tau$ oscillations with a beam of ν_μ from the Japanese KEK accelerator directed at the SuperKamiokande detector, with more powerful Fermilab–Soudan and CERN–Gran Sasso long-baseline experiments in preparation, the latter of which will look for τ appearance.

The observation by LSND of events that appear to represent $\bar{\nu}_\mu \to \bar{\nu}_e$ oscillations followed by $\bar{\nu}_e + p \to n + e^+$, $n + p \to D + \gamma$, with coincident detection of e^+ and the 2.2 MeV neutron-capture γ ray, suggests that $\Delta m^2_{\mu e} > 0$ [12.90]. The independent LSND data [12.91] suggesting that $\nu_\mu \to \nu_e$ oscillations are also occurring are consistent with, but have less statistical weight than, the LSND signal for $\bar{\nu}_\mu \to \bar{\nu}_e$ oscillations. Comparison of the latter with exclusion plots from other experiments allows two discrete values of $\Delta m^2_{\mu e}$, around 10.5 and 5.5 eV^2, or a range 2 eV$^2 \gtrsim \Delta m^2_{\mu e} \gtrsim 0.2$ eV^2. The lower limit in turn implies a lower limit $m_\nu \gtrsim 0.5$ eV, or $\Omega_\nu \gtrsim 0.01(0.65/h)^2$. This would imply that the contribution of HDM to the cosmological density is at least as great as that of all the visible stars $\Omega_* \approx 0.0045(0.65/h)$ [12.92]. Such an important conclusion requires independent confirmation. The KArlsruhe Rutherford Medium Energy Neutrino (KARMEN) experiment has added shielding to decrease its background so that it can probe a similar region of $\Delta m^2_{\mu e}$ and neutrino mixing angle; the KARMEN results exclude a significant portion of the LSND parameter space, and the numbers quoted above take into account the current KARMEN limits. The Booster Neutrino Experiment (BooNE) at Fermilab should attain greater sensitivity.

The observed deficit of solar electron neutrinos in three different types of experiments suggests that some of the ν_e undergo Mikheyev–Smirnov–Wolfenstein matter-enhanced oscillations $\nu_e \to \nu_x$ to another species of neutrino ν_x with $\Delta m^2_{ex} \approx 10^{-5}$ eV^2 as they travel through the sun (e.g. [12.93]), or possibly "just so" vacuum oscillations with even smaller Δm^2_{ex} [12.94] (cf. [12.95]). The LSND $\nu_\mu \to \nu_e$ signal, with a much larger $\Delta m^2_{e\mu}$, is inconsistent with $x = \mu$, and the SuperKamiokande atmospheric neutrino oscillation data are inconsistent with $x = \tau$. Thus a fourth neutrino species ν_s is required if all these neutrino oscillations are actually occurring. Since the neutral weak boson Z^0 decays to only three species of neutrinos, any additional neutrino species ν_s could not couple to the Z^0, and is called "sterile". This is perhaps distasteful, although many modern theories of particle physics beyond the Standard Model include the possibility of such sterile neutrinos. The resulting pattern of neutrino masses would have ν_e and ν_s very light, and $m(\nu_\mu) \approx m(\nu_\tau) \approx (\Delta m^2_{e\mu})^{1/2}$, with the ν_μ and ν_τ playing the role of the HDM particles if their masses are high enough [12.96]. This neutrino spectrum might also explain how heavy elements are synthesized in core-collapse supernova explosions [12.97] (cf. [12.98]). Note that the required solar neutrino mixing angle is very small, unlike that required to explain the atmospheric ν_μ deficit, so a sterile neutrino species would not be populated in the early Universe and would not lead to too much ^{4}He production.

Of course, if one or more of the indications of neutrino oscillations are wrong, then a sterile neutrino is not needed and other patterns of neutrino masses are possible. But in any case the possibility remains of neutrinos having a large enough mass to be HDM. Assuming that the SuperKamiokande data on atmospheric neutrinos are really telling us that ν_μ oscillates to ν_τ, the two simplest possibilities regarding neutrino masses are as follows:

a) Neutrino masses are hierarchical like all the other fermion masses, increasing with generation, as in seesaw models. Then the SuperKamiokande $\Delta m^2 \approx 0.003$ implies $m(\nu_\tau) \approx 0.05$ eV, corresponding to

$$\Omega_\nu = 0.0013(m_\nu/0.05\text{eV})(0.65/h)^2 \ . \tag{12.11}$$

This is not big enough to affect galaxy formation significantly, but it is another puzzling cosmic coincidence that it is close to the contribution to the cosmic density from stars.

b) The strong mixing between the mu and tau neutrinos implied by the SuperKamiokande data suggests that these neutrinos are also nearly equal in mass, as in the Zee model [12.99] (cf. [12.100]) and many modern models [12.94, 12.96] (although such strong mixing can also be explained in the context of hierarchical models based on the $SO(10)$ grand unified theory [12.101]). Then the above Ω_ν is just a lower limit. An upper limit is given by cosmological structure formation. In cold + hot dark matter (CHDM) models with $\Omega_m = 1$, we saw in the previous section that if Ω_ν is greater than about 0.2 the voids are too big and there is not enough early structure. In the next section we consider the upper limit on Ω_ν if $\Omega_m \approx 0.4$, which is favored by a great deal of data.

12.6   Cold plus Hot Dark Matter and Structure Formation: $\Omega_m \approx 0.4$

We have already mentioned that the $\Omega_m = 1$ CHDM model with $\Omega_\nu = 0.2$ was found to be the best fit to nearby galaxy data for all cosmological models [12.4]. But this did not take into account the new high-z supernova data and analyses [12.102] leading to the conclusion that $\Omega_\Lambda - \Omega_m \approx 0.2$, nor the new high-redshift galaxy data. Concerning the latter, Somerville, Primack and Faber [12.103] found that none of the $\Omega_m = 1$ models with a realistic power spectrum (e.g. CHDM, tilted CDM or τCDM) makes anywhere near enough bright $z \sim 3$ galaxies. But these authors found that ΛCDM with $\Omega_m \approx 0.4$ makes about as many high-redshift galaxies as are observed [12.103]. This Ω_m value is also implied if clusters have the same baryon fraction as the Universe as a whole: $\Omega_m \approx \Omega_b/f_b \approx 0.4$, using for the cosmological density of ordinary matter $\Omega_b = 0.019h^{-2}$ [12.104], and for the cluster baryon fraction $f_b = 0.06h^{-3/2}$ [12.105] from X-ray data or $f_b = 0.077h^{-1}$ from Sunyaev–Zel'dovich data [12.106]. An analysis of the cluster abundance as a function of redshift based on X-ray temperature data also implies that $\Omega_m \approx 0.44 \pm 0.12$ [12.107, 12.108]. Thus, most probably, Ω_m is ~ 0.4 and there is a cosmological constant $\Omega_\Lambda \sim 0.6$. In the 1984 paper that helped launch CDM [12.16], we actually considered two models in parallel, CDM

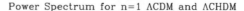

Power Spectrum for n=1 ΛCDM and ΛCHDM

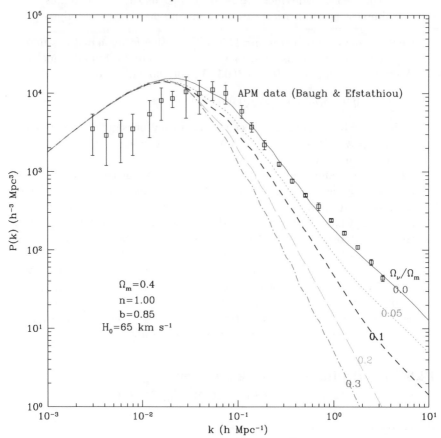

Fig. 12.1. Nonlinear dark-matter power spectrum versus wavenumber for ΛCDM and ΛCHDM models with $\Omega_\nu/\Omega_m = 0.05, 0.1, 0.2, 0.3$. Here $\Omega_m = 0.4$, the Hubble parameter $h = 0.65$, there is no tilt (i.e. $n = 1$) and the bias $b = 0.85$. Note that in this figure and in Fig. 12.2 we have "nonlinearized" all the model power spectra [12.109], to allow them all to be compared with the APM data (the small "wiggles" in the high-Ω_ν power spectra are an artifact of the nonlinearization procedure)

with $\Omega_m = 1$ and ΛCDM with $\Omega_m = 0.2$ and $\Omega_\Lambda = 0.8$, which we thought would bracket the possibilities. It looks like a ΛCDM intermediate between these extremes may turn out to be the right mix.

The success of $\Omega_m = 1$ CHDM in fitting the CMB and galaxy distribution data suggests that flat low-Ω_m cosmologies with a little HDM should be investigated in more detail. We have used CMBFAST [12.67] to examine ΛCHDM models with various h, Ω_m and Ω_ν, assuming $\Omega_b = 0.019h^{-2}$. Figure 12.1 shows the power spectrum $P(k)$ for ΛCDM and a sequence of ΛCHDM models with increasing amounts of HDM, compared with the power

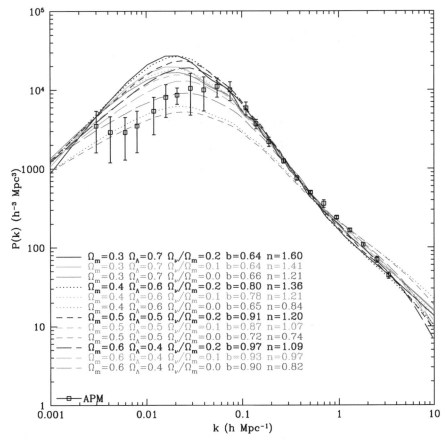

Fig. 12.2. Nonlinear dark-matter power spectrum versus wavenumber for 12 ΛCHDM models with $N_\nu = 2$ massive neutrino species and the Hubble parameter $h = 0.65$, with tilt and σ_8 determined by COBE and the ENACS cluster abundance. The bias chosen for these models is that which minimizes χ^2 over the entire range of available APM data

spectrum from APM (Automatic Plate Measuring System) [12.114]. Here we have fixed $\Omega_m = 0.4$ and the Hubble parameter $h = 0.65$. All of these models have no tilt and the same bias parameter, to make it easier to compare them with each other. As expected, the large-scale power spectrum is the same for all these models, but the amount of small-scale power decreases as the amount of HDM increases.

In Figs. 12.2 and 12.3 we consider a sequence of twelve ΛCDM and ΛCHDM models with $h = 0.65$, $\Omega_m = 0.3$, 0.4, 0.5 and 0.6, and $\Omega_\nu/\Omega_m = 0$, 0.1 and 0.2. We have adjusted the amplitude and tilt n of the primordial power spectrum for each model in order to match the four-year COBE amplitude and the ENACS differential mass function of clusters [12.115] (cf. [12.116]).

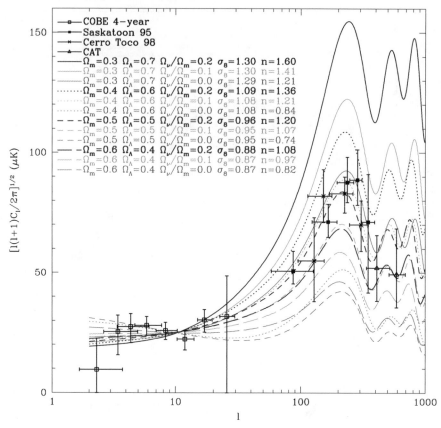

Fig. 12.3. CMB anisotropy power spectrum versus angular wavenumber for the same models as in Fig. 12.2. The data plotted are from COBE and three recent small-angle experiments [12.110–12.113] (in the case of [12.113], we have used the D1 and D2 data)

(We checked the CMBFAST calculation of ΛCHDM models against Holtzman's code, used in our earlier investigation of ΛCHDM models [12.2]. Our results are also compatible with those of recent studies [12.117, 12.118] in which $n = 1$ models were considered. But we find that some ΛCDM and ΛCHDM models require $n > 1$, called "antitilt", and it is easy to create cosmic inflation models that give $n > 1$ – cf. [12.119].) In all the ΛCHDM models the neutrino mass is shared between $N_\nu = 2$ equal-mass species – as explained above, this is required by the atmospheric neutrino oscillation data if neutrinos are massive enough to provide cosmologically significant HDM. (This results in slightly more small-scale power compared with the case of $N_\nu = 1$ massive species, as explained above, but the $N_\nu = 1$ curves are very similar to those shown.) In [12.120] we have shown similar results for a Hubble parameter $h = 0.6$, and also plotted the best CHDM and ΛCDM models

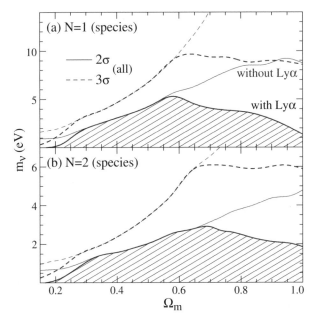

Fig. 12.4. Constraints on the neutrino mass assuming (**a**) $N_\nu = 1$ massive neutrino species and (**b**) $N_\nu = 2$ equal-mass neutrino species. The *heavier curves* show the effect of including the Lyman-α forest constraint. (From [12.122]; used by permission)

from [12.4]. Note that all these figures are easier to read in color; see the version of this chapter on the Los Alamos archive [12.121].

Of the ΛCHDM models shown, for $\Omega_\mathrm{m} = 0.4$–0.6 the best simultaneous fits to the small-angle CMB and the APM galaxy power spectrum data [12.114] are obtained for the model with $\Omega_\nu/\Omega_\mathrm{m} = 0.1$, and, correspondingly, $m(\nu_\mu) \approx m(\nu_\tau) \approx 0.8$–$1.2$ eV for $h = 0.65$. For $\Omega_\mathrm{m} < 0.4$, smaller or vanishing neutrino mass appears to be favored. Note that the antitilt permits some of the ΛCHDM models to give a reasonably good fit to the COBE plus small-angle CMB data. Thus, adding a little HDM to the moderate-Ω_m ΛCDM models may perhaps improve somewhat their simultaneous fit to the CMB and galaxy data, but the improvement is not nearly as dramatic as was the case for $\Omega_\mathrm{m} = 1$.

It is apparent that the ΛCDM models with $\Omega_\mathrm{m} = 0.4$ and 0.5 have too much power at small scales ($k \gtrsim 1h^{-1}$ Mpc), as is well known [12.123, 12.124] – although recent work [12.125] suggests that the distribution of *dark-matter halos* in the $\Omega_\mathrm{m} = 0.3$, $h = 0.7$ ΛCDM model may agree well with the APM data. On the other hand, the ΛCHDM models may have too little power on small scales – high-resolution ΛCHDM simulations and semianalytic models of early galaxy formation may be able to clarify this. Such simulations should also be compared with data from the massive new galaxy redshift surveys 2dF and SDSS using shape statistics, which have been shown to be able to discriminate between CDM and CHDM models [12.126].

Note that all the ΛCDM and ΛCHDM models that are normalized to COBE and have tilt compatible with the cluster abundance are a poor fit to the APM power spectrum near the peak. The ΛCHDM models all have the peak in their linear power spectrum $P(k)$ higher and at lower k than have the currently available data (e.g. from APM). Thus the viability of ΛCDM or ΛCHDM models with a power-law primordial fluctuation spectrum (i.e. just the tilt n) depends on this data/analysis being wrong. In fact, it has recently been argued [12.127] that because of correlations, the error bars underestimate the true errors in $P(k)$ for small k by at least a factor of two. The new large-scale surveys 2dF and SDSS will be crucial in giving the first really reliable data on this, perhaps as early as 2001.

The best published constraint on Ω_ν in ΛCHDM models is in [12.122]. Figure 12.4 shows the result of the analysis in [12.122], which uses the COBE and cluster data much as we did above, the $P(k)$ data only for $0.025(h/\mathrm{Mpc}) < k < 0.25(h/\mathrm{Mpc})$, the constraint that the age of the Universe is at least 13.2 ± 2.9 Gyr (95% C.L.) from globular clusters [12.128] and also the power spectrum at high redshift $z \sim 2.5$ determined from Lyman-α forest data. The conclusion is that the total neutrino mass m_ν is less than about 5.5 eV for all values of Ω_m, and $m_\nu \lesssim 2.4(\Omega_\mathrm{m}/0.17 - 1)$ eV for the observationally favored range $0.2 \leq \Omega_\mathrm{m} \leq 0.5$ (both at 95% C.L.). Analysis of additional Lyman-α forest data can allow detection of the signature of massive neutrinos even if m_ν is only a fraction of an eV. Useful constraints on Ω_ν will also come from large-scale weak gravitational-lensing data [12.129] combined with cosmic-microwave-background anisotropy data.

JRP acknowledges support from NASA and NSF grants at University of California Santa Cruz.

References

12.1 S. A. Bonometto and R. Valdarnini, Phys. Lett. A **103A**, 369 (1984); L.-Z. Fang, S. X. Li and S. P. Xiang, Astron. Astrophys. **140**, 77 (1984); Q. Shafi and F. W. Stecker, Phys. Rev. Lett. **53**, 1292 (1984).

12.2 J. R. Primack, J. Holtzman, A. A. Klypin and D. O. Caldwell, Phys. Rev. Lett. **74**, 2160 (1995).

12.3 J.R. Primack, "The best theory of cosmic structure formation is cold + hot dark matter", in *Critical Dialogues in Cosmology*, ed. by N. Turok (World Scientific, Singapore, 1997), p. 535; updated in astro-ph/9707285.

12.4 E. Gawiser and J. Silk, Science **280**, 1405 (1998).

12.5 J. R. Primack, Science **280**, 1398 (1998).

12.6 M. A. K. Gross, R. Somerville, J. R. Primack, J. Holtzman and A. A. Klypin, Mon. Not. R. Astron. Soc. **301**, 81 (1998).

12.7 J. R. Bond, J. Centrella, A. S. Szalay and J. R. Wilson, "Dark matter and shocked pancakes", in *Formation and Evolution of Galaxies and Large*

Structures in the Universe, ed. by J. Audouze, and J. Tran Thanh Van (Reidel, Dordrecht, 1984) pp. 87–99.

12.8 J. R. Primack and G. R. Blumenthal, "What is the Dark Matter?", in *Formation and Evolution of Galaxies and Large Structures in the Universe*, ed. by J. Audouze and J. Tran Thanh Van (Reidel, Dordrecht, 1984) pp. 163–83.

12.9 H. Pagels and J. R. Primack, Phys. Rev. Lett. **48**, 223 (1982).

12.10 G. R. Blumenthal, H. Pagels and J. R. Primack: Nature **299**, 37 (1982).

12.11 J. R. Bond, A. S. Szalay and M. S. Turner, Phys. Rev. Lett. **48**, 1636 (1982).

12.12 K. A. Olive and M. S. Turner, Phys. Rev. D **25**, 213 (1982).

12.13 S. Colombi, S. Dodelson and L. M. Widrow, Astrophys. J. **458**, 1 (1996).

12.14 J. Sommer-Larsen and A. Dolgov, astro-ph/9912166.

12.15 C. J. Hogan, astro-ph/9912549; C. J. Hogan and J. Dalcanton, astro-ph/0002330.

12.16 G. R. Blumenthal, S.M. Faber, J. R. Primack and M. J. Rees, Nature **311**, 517 (1984).

12.17 N. A. Bahcall, J. P. Ostriker, S. Perlmutter and P. J. Steinhardt, Science **284**, 1481 (1999).

12.18 M. S. Turner, "Cosmological parameters", in *Particle Physics and the Universe (Cosmo-98)*, ed. by D. O. Caldwell, AIP Conf. Proc. 478 (American Institute of Physics, Woodbury, NY, 1999).

12.19 J. R. Primack, "Status of Cosmology", in *Cosmic Flows: Towards an Understanding of Large-Scale Structure*, ed. by S. Courteau, M. A. Strauss and J.A. Willick (ASP Conference Series), astro-ph/9912089 (1999).

12.20 S. S. Gershtein and Ya. B. Zel'dovich, JETP Lett. **4**, 174 (1966).

12.21 G. Marx and A. S. Szalay, in *Neutrino '72* (Technoinform, Budapest, 1971), vol. 1, p. 123.

12.22 R. Cowsik and J. McClelland, Phys. Rev. Lett. **29**, 669 (1972).

12.23 A. S. Szalay and G. Marx, Astron. Astrophys. **49**, 437 (1976).

12.24 V. S. Lyubimov, E. G. Novikov, V. Z. Nozik, E. F., Tretyakov and V. S. Kosik, Phys. Lett. B **94**, 266 (1980).

12.25 C. Caso et al., Eur. Phys. J. C **3**, 1 (1998).

12.26 C. Weinheimer et al., Phys. Lett. B **460**, 219 (1999).

12.27 V. M. Lobashev et al., Phys. Lett. B **460**, 227 (1999).

12.28 S. M. Bilenky et al., Phys. Lett. B **465**, 193 (1999); L. Baudis et al., Phys. Rev. Lett. **83**, 41 (1999).

12.29 P. Fisher, B. Kayser and S. MacFarland, hep-ph/9906244 (1999); R. G. H. Robertson, hep-ex/0001034 (2000).

12.30 http://www.hep.anl.gov/NDK/Hypertext/nuindustry.html.

12.31 S. Weinberg: *Gravitation and Cosmology* (Wiley, New York, 1972).

12.32 D. J. Fixsen et al., Astrophys. J. **473**, 576 (1996).

12.33 B. W. Lee and S. Weinberg, Phys. Rev. Lett. **39**, 165 (1977).

12.34 M. I. Vysotsky, A. D. Dolgov and Ya. B. Zeldovich, JETP Lett. **4**, 120 (1977).

12.35 P. Hut, Phys. Lett. B **69**, 85 (1977).

12.36 K. Sato and H. Kobayashi, Prog. Theor. Phys. **58**, 1775 (1977).

12.37 J. E. Gunn, B. W. Lee, I. Lerche, D. N. Schramm and G. Steigman, Astrophys. J. **223**, 1015 (1978).

12.38 H.-S. Kang and G. Steigman, Nucl. Phys. B **372**, 494 (1992).

12.39 S. D. Tremain and J.E. Gunn, Phys. Rev. Lett. **42**, 407 (1979).

12.40 S. M. Faber and D. N. C. Lin, Astrophys. J. **266**, L17 (1983).

12.41 C. Pryor and J. Kormendy, Astron. J. **100**, 127 (1990).

12.42 D. N. C. Lin and S. M. Faber, Astrophys. J. **266**, L21 (1983).

12.43 O. E. Gerhard and D. N. Spergel, Astrophys. J. **389**, L9 (1992).

12.44 D. N. Spergel, D. H. Weinberg and J. R. Gott, Phys. Rev. D **38**, 2014 (1988).

12.45 G. S. Bisnovatyi-Kogan and I. D. Novikov, Sov. Astron. **24**, 516 (1980).

12.46 J. R. Bond, G. Efstathiou and J. Silk, Phys. Rev. Lett. **45**, 1980 (1980).

12.47 J. R. Bond and G. Efstathiou, Astrophys. J. **285**, L45 (1984).

12.48 A. Melott, Mon. Not. Roy. Astr. Soc. **202**, 595 (1983).

12.49 J. Centrella and A. Melott, Nature **305**, 196 (1983).

12.50 A. A. Klypin and S. F. Shandarin, Mon. Not. Roy. Astr. Soc. **204**, 891 (1983).

12.51 C. Frenk, S. D. M. White and M. Davis, Astrophys. J. **271**, 417 (1983).

12.52 Ya. B. Zel'dovich, J. Einasto and S.F. Shandarin, Nature **300**, 407 (1982).

12.53 A. Dekel and S. J. Aarseth, Astrophys. J. **283**, 1 (1984).

12.54 N. Kaiser, Astrophys. J. **273**, L17 (1983).

12.55 P. R. Shapiro, C. Struck-Marcell and A. L. Melott, Astrophys. J. **275**, 413 (1983).

12.56 S. D. M. White, in *Inner Space/Outer Space*, ed. by E. W. Kolb et al. (University of Chicago Press, Chicago, 1986).

12.57 J. R. Bond, A. S. Szalay and S. D. M. White, Nature **301**, 584 (1083).

12.58 G. Smoot et al., Astrophys. J. **396**, L1 (1992).

12.59 U.-L. Pen, U. Seljak, N. Turok: Phys. Rev. Lett. **79**, 1611 (1997).

12.60 A. Albrecht, R. A. Battye and J. Robinson, Phys. Rev. D **59**, 023508 (1999).

12.61 W. Hu, D. Eisenstein and M. Tegmark, Phys. Rev. Lett.**80**, 5255 (1998).

12.62 D. Brodbeck, D. Hellinger, R. Nolthenius, J. R. Primack and A. A. Klypin, Astrophys. J. **495**, 1, with accompanying videotape (1998).

12.63 C.-P. Ma, "Neutrinos and Dark Matter", in *Neutrinos in Physics and Astrophysics: from 10^{-33} to 10^{+28} cm*, ed. by P. Langacker (World Scientific, Singapore, 2000), astro-ph/9904001.

12.64 J. Holtzman, Astrophys. J. Suppl. **71**, 1 (1989).

12.65 A.A. Klypin and J. Holtzman, astro-ph/9712217 (1997).

12.66 J. Holtzman and J.R. Primack, Astrophys. J. **405**, 428 (1993).

12.67 U. Seljak and M. Zaldarriaga, Astrophys. J. **469**, 437 (1996).

12.68 C.-P. Ma and E. Bertschinger, Astrophys. J. **455**, 7 (1995).

12.69 C.-P. Ma, Astrophys. J. **471**, 13 (1996).

12.70 D. Eisenstein and W. Hu, Astrophys. J. **511**, 5 (1998).

12.71 A. A. Klpin, J. Holtzman, J. R. Primack and E. Regős, Astrophys. J. **416**, 1 (1993).

12.72 R. K. Schaefer and Q. Shafi, Phys. Rev. D **49**, 4990 (1994).

12.73 R. Nolthenius, A. A. Klypin and J. R. Primack, Astrophys. J. **422**, L45 (1994).

12.74 S. Ghigna, S. Borgani, S. Bonometto, L. Guzzo, A. Klypin, J. R. Primack, R. Giovanelli and M. Haynes, Astrophys. J. **437**, L71 (1994).

12.75 G. Kauffmann and S. Charlot, Astrophys. J. **430**, L97 (1994).

12.76 J. Miralda-Escude and H.-J. Mo, Astrophys. J. **430**, L25 (1994).

12.77 C.-P. Ma and E. Bertschinger, Astrophys. J. **434**, L5 (1994).

12.78 A. A. Klypin, S. Borgani, J. Holtzman and J. R. Primack, Astrophys. J. **444**, 1 (1995).

12.79 A. A. Klypin, R. Noltheius and J. R. Primack, Astrophys. J. **474**, 533 (1997).

12.80 S. Ghigna, S. Borgani, M. Tucci, S. A. Bonometto, A. A. Klypin and J. R. Primack, Astrophys. J. **479**, 580 (1997).

12.81 R. Nolthenius, A. A. Klypin and J. R. Primack, Astrophys. J. **480**, 43 (1997).

12.82 C.-P. Ma et al., Astrophys. J. **484**, L1.

12.83 M. G. Haehnelt, M. Steinmetz and M. Rauch, Astrophys. J. **495**, 647 (1998).

12.84 J. X. Prochaska and A. M. Wolfe, Astrophys. J. **487**, 73 (1997).

12.85 D. Pogosyan and A. Starobinsky, astro-ph/9502019.

12.86 Y. Fukuda et al., Phys. Lett B **335**, 237 (1994).

12.87 Y. Fukuda et al., Phys. Rev. Lett. **81**, 1562 (1998).

12.88 M. Apollonio et al., Phys. Lett. B **420**, 397 (1998).

12.89 X. Shi, D. N. Schramm and B. D. Fields, Phys. Rev. D **48**, 2563 (1993); C. Y. Cardall and G. M. Fuller, Phys Rev D **54**, 1260 (1996); T. I Izotov and T. X. Thuan, Astrophys. J. **500**, 188 (1998).

12.90 C. Athanassopoulos et al., Phys. Rev. Lett. **77**, 3082 (1996); Phys. Rev. C **54**, 2685 (1996).

12.91 C. Athanassopoulos et al., Phys. Rev. Lett. **81**, 1774 (1998).

12.92 M. Fukugita, C. J. Hogan and P. J. E. Peebles, Astrophys. J. **503**, 518 (1998).

12.93 W. C. Haxton, Ann. Rev. Astro. Astroph. **33**, 459 (1995); V. Castellani et al., Phys. Rep. **281**, 309 (1997).

12.94 S. L. Glashow, P. J. Kernan and L. M. Krauss, Phys. Lett. B **445**, 412 (1999).

12.95 F. Vissani, hep-ph/9708483; H. Georgi and S. L. Glashow, hep-ph/9808293 (1998).

12.96 D. O. Caldwell and R. N. Mohapatra, Phys. Rev. D **48**, 3259 (1993); Phys. Rev. D **50**, 3477 (1994); J. T. Peltoniemi and J. W. F. Valle, Nucl. Phys. B **406**, 409 (1993); V. Barger, T. J. Weiler and K. Whisnant, Phys. Lett. B **427**, 97 (1998); S. C. Gibbons et al., Phys. Lett. B **430**, 296 (1998).

12.97 D. O. Caldwell, in 23rd Johns Hopkins Workshop, *Neutrinos in the Next Millennium*, hep-ph/9910349 (1999); D. O. Caldwell, G. M. Fuller and Y.-Z. Qian: astro-ph/9910175 (1999).

12.98 G. M. Fuller, J. R. Primack and Y.-Z. Qian, Phys. Rev. D **52**, 1288 (1995).

12.99 A. Zee, Phys. Lett. B **93**, 389 (1980); Phys. Lett. B **161**, 141 (1985).

12.100 A. Smirnov, hep-ph/9611465 (1996).

12.101 K. S. Babu, J. C. Pati and F. Wilczek, Nucl. Phys. B **566**, 33 (2000).

12.102 P.M. Garnavich et al., Astrophys. J. **493**, L53 (1998); Astrophys. J. **509**, 74 (1998); S. Perlmutter et al., Astrophys. J. **517**, 565 (1999).

12.103 R. S. Somerville, J. R. Primack and S. M. Faber, astro-ph/9806228, Mon. Not. R. Astron. Soc., in press.

12.104 S. Burles and D. Tytler, Astrophys. J. **499**, 699 (1998); Space Sci. Rev. **84**, 65 (1998).

12.105 A. E. Evrard, Mon. Not. R. Astron. Soc. **292**, 289 (1997).

12.106 J. E. Carlstrom, astro-ph/9905255 (1999).

12.107 V. R. Eke, S. Cole, C. S. Frenk and J. P. Henry, Mon. Not. R. Astron. Soc. **298**, 1145 (1998).

12.108 J. P. Henry, Astrophys. J., in press, astro-ph/0002365 (2000)

12.109 C. C. Smith, A. A. Klypin, M. A. K. Gross, J. R. Primack and J. Holtzman, Mon. Not. R. Astron. Soc. **297**, 910 (1998).

12.110 C. B. Netterfield et al., Astrophys. J. **474**, 47 (1997).

12.111 P. F. S. Scott et al., Astrophys. J. **461**, L1 (1996).

12.112 J. C. Baker, in *Proceedings of the Particle Physics and the Early Universe Conference*, www.mrao.com.ac.uk/ppeuc/proceedings.

12.113 A.D. Miller et al., Astrophys. J. **524**, L1 (1999).

12.114 C. Baugh and G. Efstathiou, Mon. Not. R. Astron. Soc. **265**, 145 (1993).

12.115 M. Girardi, S. Borgani, G. Giuricin, F. Mardirossian and M. Mezzetti, Astrophys. J. **506**, 45 (1998).

12.116 M. A. K. Gross, R. S. Somerville, J. R. Primack, S. Borgani and M. Girardi: in *Large Scale Structure: Tracks and Traces*, ed. by V. Müller, S. Gottlöber, J.P. Mücket and J. Wambsganss (World Scientific, Singapore 1998).

12.117 R. Valdarnini, T. Kahniashvili and B. Novosyadlyj, Astron. Astrophys. **336**, 11 (1998).

12.118 M. Fukugita, G.-C. Lin and N. Sugiyama, Phys. Rev. Lett. **84**, 1082 (2000).

12.119 S. A. Bonometto and E. Pierpaoli, New Astron. **3**, 391 (1998).

12.120 J. R. Primack and M. A. K. Gross, "Cold + hot dark matter after SuperKamiokande", in *The Birth of Galaxies*, ed. by B. Guiderdoni (Editions Frontieres, Gif-sur-Yvette, France, 2000), astro-ph/9810204.

12.121 J. R. Primack and M. A. K. Gross, astro-ph/0007187 (2000).

12.122 R. A. C. Croft, W. Hu and R. Davé, Phys. Rev. Lett. **83**, 1092 (1999).

12.123 A. A. Klypin, J. R. Primack and J. Holtzman, Astrophys. J. **466**, 1 (1996).

12.124 A. Jenkins et al., Astrophys. J. **499**, 20 (1998).

12.125 P. Colin, A.A . Klypin, A. Kravtsov and A. Khokhlov, Astrophys. J. **523**, 32 (1999).

12.126 R. Davé, D. Hellinger, J. R. Primack, R. Nolthenius and A. A. Klypin, Mon. Not. R. Astron. Soc. **284**, 607 (1997).

12.127 D. Eisenstein, M. Zaldarriaga, astro-ph/9912149 (1999).

12.128 E. Carretta, R. G. Gratton, G. Clementini and F. Fusi Pecci, astro-ph/9902086 (1999).

12.129 A. R. Cooray, Astron. Astrophy. **348**, 31 (1999).

13 High Energy Neutrino Astronomy: Towards Kilometer-Scale Detectors

Francis Halzen

13.1 Introduction

Of all high-energy particles, only neutrinos can directly convey astronomical information from the edge of the Universe – and from deep inside the most cataclysmic high-energy processes. Copiously produced in high-energy collisions, traveling at the velocity of light and not deflected by magnetic fields, neutrinos meet the basic requirements for astronomy. Their unique advantage arises from a fundamental property: they are affected only by the weakest of nature's forces (but for gravity) and are therefore essentially unabsorbed as they travel cosmological distances between their origin and us.

Many of the outstanding mysteries of astrophysics may be hidden from our sight at all wavelengths of the electromagnetic spectrum because of absorption by matter and radiation between us and the source. For example, the hot, dense regions that form the central engines of stars and galaxies are opaque to photons. In other cases, such as supernova remnants, gamma ray bursters and active galaxies, all of which may involve compact objects or black holes at their cores, the precise origin of the high-energy photons emerging from their surface regions is uncertain. Therefore, data obtained through a variety of observational windows – and especially through direct observations with neutrinos – may be of cardinal importance.

The sun is an intense source of electron neutrinos (ν_e), albeit of relatively low energy. Solar neutrino astronomy began with its first experiments in the mid-1960s; today there are five complementary neutrino detectors viewing the nuclear reactions in the core of the sun and, at the same time, studying the fundamental properties of neutrinos. The sun remained the only astronomical object studied with neutrinos until neutrinos from Supernova 1987A were observed. These are still the only two sources marking the astronomical neutrino spectrum.

Suggestions to use a large volume of water for high-energy neutrino astronomy were made as early as the 1960s [13.1]. In this case, a muon neutrino (ν_μ) interacts with a hydrogen or oxygen nucleus in the water and produces a muon travelling in nearly the same direction as the neutrino. The blue Cerenkov light emitted along the muon's trajectory of order 1 km long, would be detected by strings of photomultiplier tubes deployed deep below the surface. DUMAND, a pioneering project located off the coast of Hawaii, demonstrated that muons could be detected by this technique [13.2], but the planned detector was never realized. A detector composed of about ninety photomultiplier tubes (or optical modules) located deep in Lake Baikal was the first to demonstrate the detection of neutrino-induced muons in natural water [13.3]. The European collaborations ANTARES [13.4] and NESTOR [13.5] plan to deploy large-area detectors in the Mediterranean Sea within the next few years. The NEMO collaboration is conducting a site study for a future kilometer-scale detector in the Mediterranean [13.6].

The AMANDA collaboration, situated at the US Amundsen–Scott South Pole Station, has strikingly demonstrated the merits of natural ice as a Cerenkov detector medium [13.7, 13.8]. In 1996, AMANDA was able to observe atmospheric neutrino candidates using only 80 eight-inch photomultiplier tubes [13.7]. With 1997 data, with more optical modules in place, AMANDA extracted over 100 atmospheric neutrino events. AMANDA was the first neutrino telescope with an effective area in excess of 10 000 square meters. The collaboration has already augmented the 400 module detector with some 250 additional optical modules, producing an instrument dubbed AMANDA-II. AMANDA has met the key challenge of neutrino astronomy: it has developed a reliable, expandable and affordable technology for deploying a kilometer-scale neutrino detector, named IceCube.

IceCube and the detectors to be located in the Mediterranean Sea will be complementary in several respects. First and foremost, they will cover different portions of the sky. (From its location at the South Pole, IceCube observes the northern sky.) Because of the different properties of Antarctic ice and ocean water – ice scatters light more but absorbs it less – IceCube should have better energy resolution and a larger effective area for neutrino detection, while the deep-sea detectors should have somewhat higher pointing resolution. Finally, one of the key challenges in neutrino astronomy has been the ability to deploy and maintain optical modules far below the surface; here, too, ice and ocean are essentially different.

13.2   Scientific Goals

The many questions a high-energy neutrino telescope can address naturally begin with astronomy, both nearby in the Universe and far away. Fascinating issues in particle physics and other branches of science are accessible as well.

Neutrino telescopes will investigate the engines which power active galaxies, the nature of gamma ray bursts and the origin of the highest-energy cosmic rays. They will search for galactic supernovae, for the births of the supermassive black holes which power quasars and for the annihilation products of halo cold-dark-matter particles (WIMPS and supersymmetric particles). They can perform coincidence experiments with Earth- and space-based gamma ray observatories, cosmic-ray detectors, and even future gravitational-wave detectors such as LIGO.

With high-energy neutrino astrophysics, we are poised to open a new window into space and back in time to the highest-energy processes in the Universe. For guidance in estimating expected signals, we make use of the observed energy in high-energy cosmic-ray protons and nuclei as well as known sources of nonthermal, high-energy gamma radiation. Some fraction of cosmic rays will interact in their sources to produce pions. These interactions may be hadronic collisions with ambient gas or photoproduction with intense photon fields near the sources. In either case, the neutral pions decay to photons, while charged pions include neutrinos among their decay products, with spectra related to the observed gamma ray spectra. Estimates based on this relationship show that a kilometer-scale detector is needed to see the neutrino signals [13.9]. High-energy gamma rays may be produced by radiative processes from accelerated electrons, whereas neutrinos must be produced by hadronic processes. Therefore, high-energy neutrino astronomy has the potential to discriminate between hadronic and electronic models of interesting objects such as supernova remnants (SNRs), gamma ray burst sources (GRBs) and active galactic nuclei (AGNs).

The potential scientific payoff of neutrino astronomy arises from the great penetrating power of neutrinos, which allows them to emerge from dense inner regions of energetic sources. It necessarily follows that the expected interaction rates are small, even with a kilometer-scale detector. History has shown, however, that the opening of each new astronomical window has led to unexpected discoveries. Thus, for example, there could be hidden particle accelerators from which only the neutrinos escape. We include some examples of this type in the following discussion of possible neutrino sources.

13.2.1 Galactic Sources

The origin of cosmic rays is one of the oldest puzzles in science. The prevalent theory is that most cosmic rays, at least those with energies up to perhaps 100 to 1000 TeV, are accelerated in supernova blast waves. The argument is based largely on circumstantial evidence that the power available from supernova explosions is about right, and that strong shock waves naturally produce a spectrum consistent with what is observed (after accounting for effects of propagation). Confirmation of this theory could come by observing evidence of pion production at the correct level in the gas surrounding supernova remnants [13.10]. Photons with energies up to several GeV have been detected by the EGRET detector [13.11], but there are only upper limits on these objects in the TeV range and higher, which has raised uncertainties about the standard picture of the origin of galactic cosmic rays. TeV emission is now established [13.12] from the Crab Nebula, SN1006 and PSR B1706-44. At least for the first two, however, the TeV radiation is thought to originate from radiation by high-energy electrons (from the pulsar wind of the Crab and from blast-wave-accelerated electrons in the case of SN1006). Detection

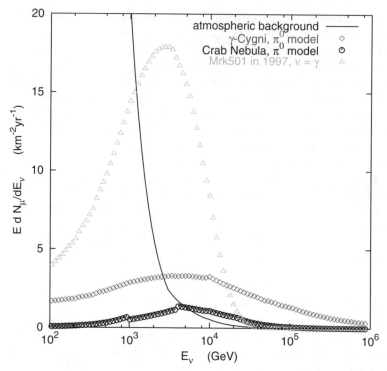

Fig. 13.1. Model of expected neutrino signal from a variety of sources compared with atmospheric background

of TeV neutrinos from these sources would confirm their role as accelerators of cosmic-ray ions as well.

We illustrate the challenge of detecting neutrino signals from cosmic-ray interactions in shells of SNRs by considering a possible extrapolation to high energy of a hadronic model for gamma ray emission from γ Cygni [13.13]. By assuming a relatively steep differential spectrum ($E^{-2.3}$), this model gives the maximum extrapolation to > 1000 TeV consistent with nonobservation of gamma rays with energy of several hundred GeV and higher. Figure 13.1 shows the expected signal of neutrino-induced muons from this source as a function of neutrino energy compared with the atmospheric background. The total signal above 50 GeV is about 20 events per year, but about half of this is lost in the atmospheric background within the assumed 1 degree acceptance cone. In addition, the high-energy tail of the signal will be cut off about one order of magnitude below the maximum energy of protons accelerated at the source. Thus the energy region ~ 10 TeV will be crucial for detecting this kind of weak source above background. The lower set of points in Fig. 13.1 illustrates the still more pessimistic situation for a hadronic model for production of the high-energy end (~ 10 TeV) of the photon spectrum of the Crab Nebula [13.14].

A more optimistic, though entirely hypothetical, possibility is an obscured source that might be discovered through its neutrino radiation. For example, a one-solar-mass black hole accreting at the Eddington limit releases $\sim 10^{38}$ erg/s. If 10 percent of this energy goes into neutrinos via accelerated protons with an E^{-2} differential spectrum interacting to produce secondaries, one would expect a neutrino-induced signal of order 50 events per year at 10 kpc.[1] A more massive black hole in the galactic center could conceivably produce a much bigger signal.

Cosmic rays also produce neutrinos as they propagate in the galaxy. EGRET has mapped the disk of the galaxy in diffuse GeV gamma radiation [13.17]. The region of the galactic center is observed to have a somewhat harder spectrum ($E^{-2.4}$) than the locally observed cosmic ray spectrum ($E^{-2.7}$), making it accessible for observation in neutrinos. For equal production of $\nu_\mu + \bar{\nu}_\mu$ and gamma rays, we expect roughly 100 neutrino-induced muons with $10 < E_\mu < 100$ TeV and an equal number in the atmospheric background in the angular region of 0.73 sr defined by the EGRET observation.

Although neutrino telescopes are intended primarily for TeV (and higher) energy neutrino astronomy, AMANDA and IceCube also have the potential to detect the burst of several-MeV neutrinos from a galactic supernova and possibly even an extragalactic burst of low-energy neutrinos associated with the birth of a massive black hole. By continuously monitoring the summed counting rate of its nearly 5000 optical modules, ice detectors will monitor the sky for such cataclysmic phenomena. The interactions of this host of low-energy neutrinos will be distributed uniformly throughout the detector, a signal rising above the background noise level. The very low noise rate for an optical module in ice (as opposed to the very high noise rates in ocean water) makes it possible to continue this "supernova watch", already begun by AMANDA [13.18].

13.2.2   Extragalactic Sources

Here again it is natural to use the energy density in cosmic rays for guidance as to the magnitude of expected neutrino signals. Somewhere in the Universe, Nature accelerates particles to astonishing energies of 10^{20} eV and even higher. Although there are plausible models for the origin of these particles in the halo of our own galaxy (e.g. [13.19, 13.20]), the predominant opinion is that cosmic rays with energies greater than $\sim 3 \times 10^{18}$ eV come from extragalactic cosmic accelerators.

---

[1] For a review, see [13.15]. Cross sections tabulated in [13.16] were used to obtain the estimates of rates.

Active Galactic Nuclei. One possible source is active galactic nuclei, among the brightest gamma ray sources in the Universe. AGNs emit as much energy as entire bright galaxies, but they are extremely compact; within time periods as short as hours, their luminosities are observed to flare by over an order of magnitude [13.12]. The standard model involves a massive, accreting black hole more than a million times as massive as our sun. AGNs emit at all wavelengths of the electromagnetic spectrum, from radio waves to TeV gamma rays, largely through the interactions of accelerated electrons with magnetic fields and ambient photons in the source. If protons are accelerated along with electrons to energies of PeV to EeV, they will produce high-energy photons by photoproduction of neutral pions. Near the central black hole, the ultraviolet thermal background provides the target photons; in the jets non-thermal photons may also act as targets. The relative merits of the electron and proton acceleration models are hotly debated, but with a kilometer-scale detector the issue can be settled experimentally. Proton acceleration turns AGNs into sources of high-energy cosmic rays and neutrinos, not just gamma rays. Weakly interacting neutrinos, unlike high-energy gamma rays and high-energy cosmic rays, can reach us from the heart of the accelerating site, even from the most distant and powerful AGNs.

Models for high-energy AGN emission have mostly focused on acceleration in the jets beaming perpendicular to the accretion disc. It is assumed that particles are accelerated by Fermi shocks in clumps of matter traveling along the jet with a bulk Lorentz factor Γ of about 10, possibly larger, relative to the observer. In order to accommodate bursts lasting a day or less in the observer's frame, the size of the clump in its rest frame must be less than $R' = \Gamma c \Delta t \approx 10^{-2}$ parsecs. The clumps may be more like sheets, thinner than the jet's size of roughly one parsec. The observed radiation at all wavelengths is produced by the interaction of the accelerated particles in the clump with the ambient radiation in the AGN. From the photon luminosity L_γ received over a time Δt, the energy density of photons in the rest frame of the clump can be inferred:

$$\rho_E = \frac{L_\gamma \Delta t}{\Gamma (4/3) \pi R'^3} \sim \frac{L_\gamma \Delta t}{\Gamma} \frac{1}{(\Gamma c \Delta t)^3} \sim \frac{L_\gamma}{\Gamma^4 \Delta t^2} \, . \tag{13.1}$$

With high luminosities L_γ emitted over short a Δt, the large photon density renders the blob opaque to photons of 10 TeV energy and above unless Γ is very large. A large Γ factor, typically larger than 10, is the agent that dilutes the blob until 10 TeV γs fall below the $\gamma\gamma \to e^+e^-$ threshold in the blob. Only transparent sources with large boost factors emit TeV photons. Examples are the nearby blazars Markarian 421 and 501. (A blazar is an AGN in which the jet illuminates the observer.) These will be relatively weak neutrino sources at best, because one expects at most one neutrino per photon. For example, if Mrk 501 emitted neutrinos at the rate $\phi_{\nu_\mu + \bar{\nu}_\mu} = \phi_\gamma(1997)$, where (1997) refers to the high state of this TeV gamma source during 1997 [13.21], one would expect about 30 events above 500 GeV in a detector with an effective area of 1 km^2, with only a small atmospheric background (see Fig. 13.1).

A source with the same morphology but with $\Gamma \simeq 1$ would be opaque to high-energy photons and protons. It would be a "hidden" source, with reduced or extinguished emission of high-energy particles, but undiminished neutrino production by protons on the high-density photon target. Nature presumably makes AGNs with a distribution of boost factors and intensities, so discovery of such a hidden source is an interesting possibility.

Another possibility for hidden sources involves neutrinos produced in the accretion disk near a black hole by interactions of accelerated protons [13.22]. In this model the degraded photon radiation is emitted in the X-ray band after electromagnetic cascading in the source. Stecker et al. [13.22] assumed that the accumulated radiation from all such AGNs is the source of the diffuse extragalactic X-radiation, thus providing the normalization for a predicted diffuse flux of high-energy neutrinos. The normalization of the neutrino-induced signal in this class of models depends significantly on details such as whether the accelerated protons which initiate the cascade in the accretion disk are confined in the acceleration region until they lose all their energy. Figure 13.2 shows a variety of such models [13.23], with the sensitivity of IceCube highlighted by a broad, nearly horizontal band showing approximately equal energy per decade of energy. The broad dark band with a steep spectrum denotes the atmospheric background. An existing measurement [13.24] rules out diffuse neutrino fluxes at a level greater than 5×10^{-6} cm^{-2} s^{-1} sr^{-1} GeV around 2–3 GeV. Thus the highest models, not shown in Fig. 13.2, are already eliminated. The Stecker et al. [13.22] prediction corresponds to several thousand upward-moving neutrino-induced muons per year in a kilometer-scale detector.

Waxman and Bahcall [13.25] have pointed out that sources such as AGNs which contribute to the observed ultra-high-energy cosmic rays are limited to an energy flux $< 5 \times 10^{-8}$ cm^{-2} s^{-1} sr^{-1} GeV around 10^8–10^9 GeV. Only "hidden source" models, in which both protons and neutrons have optical depths greater than unity, can exceed this bound. Whether some types of AGN satisfy the conditions to be hidden sources is at present under discussion [13.26–13.28].

Gamma Ray Bursts. Gamma ray bursts (GRBs) are another potential source of extragalactic cosmic rays with the right scale of energy to account for the observed cosmic rays above 3×10^{18} eV [13.29]. Mounting evidence suggests that GRB emission is produced by a relativistically expanding fireball, energized by a process involving neutron stars or black holes (for a recent review, see [13.30]). In the early stages the fireball, its radiation trapped by the very large optical depth, cannot emit photons efficiently. The fireball's kinetic energy is therefore dissipated until it becomes optically thin – a sce-

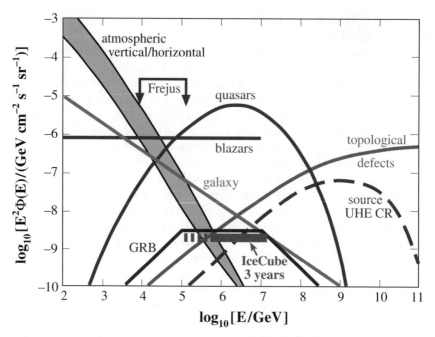

Fig. 13.2. Plot of diffuse neutrino flux versus energy for various models, with a *band* ("IceCube 3 years") indicating the sensitivity of a km² detector above 100 TeV after three years of operation. Note that the sensitivity to point sources and time varying sources, such as GRBs, is greater than shown. Realistic detector energy resolution has not been taken into account

nario that can explain the observed energy and time scales of GRBs, provided the bulk Lorentz factor of the expanding flow, Γ, is ≥ 300.

Protons accelerated in shocks in the expanding fireball interact with photons to produce charged pions, the parents of high-energy neutrinos. Assuming that particles accelerated in the GRB sources produce the observed cosmic rays above the "ankle" of the spectrum, one expects a signal at the level of the Waxman–Bahcall bound mentioned above, of order 50 upward-going muons per year in a kilometer-square detector [13.25, 13.31]. Although the expected rate is low, the neutrino signal will have a hard spectrum extending well beyond the atmospheric background and, even more important, some of the high-energy GRB neutrino events should coincide with observed GRB photon events within a narrow time window.

Because the opacity of the source depends strongly on the Lorentz factor of the outflow (compare (13.1)), the neutrino yield is not necessarily given by the rate of GRBs multiplied by the yield of a single GRB with average Lorentz factor $\Gamma \approx 300$. In addition, relatively nearby bursts and bursts with higher than average intrinsic luminosity would give bigger signals. Taking these factors into account, if fluctuations in Γ are responsible for observed burst-to-burst variations in luminosity, one might expect over 200 events per year, with two individual bursts each producing more than 10 up-going muon tracks in a km^2 detector [13.32].

In summary, neutrino observations will be a direct probe of the fireball model of GRBs, testing the plausible but unproven assumption that the highest-energy cosmic rays are produced in GRBs. If neutrinos from GRBs are observed, they will yield information on the distribution of bulk Lorentz factors in GRB events.

13.2.3   Particle Physics

Cold-Dark-Matter Search. Large neutrino telescopes occupy a unique place in the diverse landscape of dark-matter searches [13.33]. If cold-dark-matter particles such as neutralinos constitute the nonluminous halo of our galaxy, neutrino telescopes will detect them if their mass is sufficiently high. For neutralino masses above 0.5 TeV, the sensitivity of neutrino telescopes reaches a maximum because of the high energy of the neutrinos produced by neutralino annihilation, increasing roughly linearly with mass up to 10 TeV (higher masses are inconsistent with standard cosmology). Direct searches, whose sensitivity decreases with mass, are unlikely ever to cover this mass range; nor will the Large Hadron Collider, which will operate below the threshold for producing neutralinos of 0.5 TeV mass and higher. The rate of neutrino events from dark-matter annihilation is calculable; failure to observe such events can be translated into constraints on models, supersymmetric or otherwise. For example, AMANDA's 1997 data indicate that the instrument's potential for discovering WIMP masses exceeding 200 GeV already matches that of existing detectors which have operated for five to ten years [13.18].

Topological Defects, Monopoles, Charm. Neutrino telescopes will also search for ultra-high-energy neutrino signatures from topological defects predicted by grand unified theories, and for magnetic monopoles. Using the 1997 data, the AMANDA group established a record limit on relativistic monopoles, one order of magnitude below the Parker bound [13.18]. Neutrino telescopes observe muons and neutrinos with energies exceeding those produced at accelerators or observed in other underground experiments, such as Gran Sasso, where hints of a prompt charm component in the muon spectrum may have been identified in both experiments. A neutrino telescope has the potential to search for prompt cosmic-ray muons, including those produced by the decay of heavy quarks produced in high-energy cosmic-ray interactions in the atmosphere.

Neutrino Physics. As instruments of particle physics, kilometer-scale neutrino detectors have the potential to discover the ν_τ; they also offer the possibility of studying neutrino oscillations over baselines much larger than an astronomical unit, up to cosmological distances. Observing neutrino oscillations (which implies a nonzero neutrino mass) provides a window into physics beyond the Standard Model, a model which is known to be incomplete. Because cosmic sources of neutrinos produce beams of ν_e and ν_μ, neutrino telescopes can perform the ultimate long-baseline experiment: ν_τ appearance studies from cosmological sources such as a GRB can probe the ν_τ mass with a sensitivity of $\Delta m^2 > 10^{-17}$ eV^2 [13.31]. PeV ν_τ can be identified by the unique signature of a "double-bang event", consisting of a pair of separated showers associated with the charged-current production of a tau lepton (τ), followed by its decay [13.34].

Above 1 PeV, ν_e and ν_μ are absorbed by charged-current interactions in the earth before reaching a detector at the opposite surface. In contrast, the earth never becomes opaque to ν_τ, since the τ produced in a charged-current ν_τ interaction decays back into a ν_τ with smaller energy and interaction probability [13.35]. The process repeats until the ν_τ can penetrate the earth. The appearance of a ν_τ component in a pure $\nu_{e,\mu}$ beam would be signaled by a flat angular dependence at the highest neutrino energies. Since point sources are always at constant zenith angle as viewed from the South Pole, making use of this second signature would require a diffuse flux that extends from well below to well above 100 TeV. The signature is more powerful for the Mediterranean detectors because neutrinos from a celestial source travel through the earth over a distance which varies with time of day.

Neutrino telescopes may shed light on the question of whether Super-Kamiokande has indeed observed ν_μ to μ_τ oscillations. In the 10 GeV energy region, ν_τ charged-current interactions signaling the appearance of ν_τ in the atmospheric neutrino flux, can be identified by the additional energy released by the prompt τ decay [13.36]. Such measurements probably require modifications of the planned detectors in order to achieve good energy measurement at low threshold; see the later discussion of ANTARES, however.

13.2.4   Other Science

IceCube, because of its size and structure, will be a three-dimensional kilometer-scale air shower detector that will be useful in studying the primary cosmic-ray spectrum from below the knee to approaching the ankle ($\sim 10^{15}$ to $> 10^{18}$ eV). The development of highly efficient hot-water drilling has also encouraged the possibility of deploying large arrays of radio detectors for the detection of ultra-high-energy neutrinos. Hot-water drilling also lends itself to the deployment of a seismic array in Antarctica, at the earth's axis; the subcontinent itself is seismically quiet, and seismic sources (mostly earthquakes) are distributed fairly evenly in all directions. Efficient drilling technology can even facilitate the search for life in the lakes below the Antarctic ice sheet.

Similarly will the infrastructure of "water-based" telescopes support novel studies of oceanography and the environment.

Finally, if neutrino telescopes are successful in identifying distant flaring sources of photons and energetic neutrinos, it will be possible to make some fundamental physics observations. First, the relative timing of photons and neutrinos over cosmological distances will allow unrivaled tests of special relativity, to an accuracy of a part in 10^{16}. In addition, the fact that photons and neutrinos of the same energy should suffer the same time delay in traveling through the gravitational field of our galaxy will lead to tests of the weak equivalence principle to one part in 10^6 [13.25].

13.2.5 Summary

The exciting science that we can anticipate with neutrino telescopes makes the case for commissioning kilometer-scale neutrino observatories. Yet the science these instruments may do which we cannot anticipate may well fuel the quest for astronomical knowledge in the 21st century. Unexpected discovery has followed the inauguration of most new astronomical instruments. Large reflecting telescopes on mountaintops led to the discovery of distant galaxies and an expanding Universe; radio telescopes found the cosmic microwave background; X-ray and gamma ray satellites have uncovered a bestiary of awesome cosmological objects. Even the first modest solar-neutrino observatory revealed a paradox about the nature of fundamental interactions which is yet to be fully explained. No one can predict all that a high-energy neutrino observatory will find in the sky – except that it is very likely to amaze us.

13.3 Large Natural Cerenkov Detectors

The first generation of neutrino telescopes, launched by the bold decision of the DUMAND collaboration to construct such an instrument, is designed to reach a large telescope area and detection volume for a neutrino threshold of order 10 GeV. This relatively low threshold permits calibration of the novel instrument on the known flux of atmospheric neutrinos. The architecture is optimized for reconstructing the Cerenkov light front radiated by an up-going, neutrino-induced muon. Only up-going muons made by neutrinos reaching us through the earth can be successfully detected. The earth is used as a filter to screen the fatal background of cosmic-ray muons. This makes neutrino detection possible over the lower hemisphere of the detector. Up-going muons must be identified against a background of down-going, cosmic ray muons which are more than 10^5 times more frequent for a depth of 1–2 km. The method is sketched in Fig. 13.3.

The optical requirements on the detector medium are severe. A large absorption length is required because it determines the spacing of the optical sensors and, to a significant extent, the cost of the detector. A long scattering length is needed to preserve the geometry of the Cerenkov pattern. Nature has been kind and offered ice and water as adequate natural Cerenkov media.

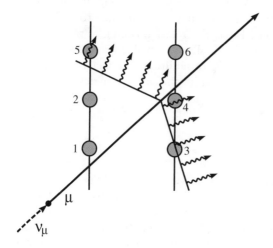

Fig. 13.3. The arrival times of the Cerenkov photons in six optical sensors determine the direction of the muon track

Table 13.1. Optical properties of South Pole ice at 1700 m, Lake Baikal water at 1 km and the range of results from measurements in ocean water below 4 km ($\lambda = 385$ nm [a])

	AMANDA (1700 m)	Baikal	Ocean
Attenuation	~ 30 m [b]	~ 8 m	25–30 m [c]
Absorption	95 ± 5 m	8 m	–
Scattering length	24 ± 2 m	150–300 m	–

[a] Peak PMT efficiency.

[b] Same for bluer wavelengths.

[c] Smaller for bluer wavelengths.

Their optical properties are, in fact, complementary. Water and ice have similar attenuation lengths, with the roles of scattering and absorption reversed; see Table 13.1. Optics seems, at present, to drive the evolution of ice and water detectors in predictable directions: towards very large telescope area in ice, exploiting the long absorption length, and towards lower threshold and good muon track reconstruction in water, exploiting the long scattering length.

13.3.1   Baikal, ANTARES, Nestor and NEMO: Northern Water

Whereas the science is compelling, the real challenge is to develop a reliable, expandable and affordable detector technology. With the termination of the pioneering DUMAND experiment, the efforts in water are, at present, spearheaded by the Baikal experiment [13.3]. The Baikal Neutrino Telescope is deployed in Lake Baikal, Siberia, 3.6 km from shore at a depth of 1.1 km. An umbrella-like frame holds eight strings, each instrumented with 24 pairs of 37 cm diameter QUASAR photomultiplier tubes (PMTs). Two PMTs in a pair are switched in coincidence in order to suppress background from natural radioactivity and bioluminescence. Operating with 144 optical modules since April 1997, the NT-200 detector has been completed in April 1998 with 192 optical modules (OMs). The Baikal detector is well understood, and the first atmospheric neutrinos have been identified.

The Baikal site is competitive with deep oceans, although the smaller absorption length of Cerenkov light in lake water requires a somewhat denser spacing of the OMs. This does, however, result in a lower threshold, which is a definite advantage, for instance for oscillation measurements and WIMP searches. The Baikal group has shown that the shallow depth of 1 km does not represent a serious drawback. By far the most significant advantage is a site with a seasonal ice cover, which allows reliable and inexpensive deployment and repair of detector elements.

With data taken with 96 OMs only, the group has shown that atmospheric muons can be reconstructed with sufficient accuracy to identify atmospheric neutrinos; see Fig. 13.4. The neutrino events are isolated from the cosmic-ray muon background by imposing a restriction on the chi-square of the Cerenkov fit, and by requiring consistency between the reconstructed trajectory and the spatial locations of the OMs reporting signals. In order to guarantee a minimum lever arm for track fitting, only events with a projection of the most distant channels on the track larger than 35 m are considered. This does, of course, result in a higher threshold.

In the following years, NT-200 will be operated as a neutrino telescope with an effective area between 10^3 and 5×10^3 m^2, depending on energy. Presumably too small to detect neutrinos from extraterrestrial sources, NT-200 will serve as the prototype for a larger telescope. For instance, with 2000 OMs, a threshold of 10–20 GeV and an effective area of 5×10^4–10^5 m^2, an expanded Baikal telescope would fill the gap between present underground detectors and planned high-threshold detectors of cubic-kilometer size. Its key advantage would be low threshold.

The Baikal experiment represents a proof of concept for deep-ocean projects. These have the advantage of larger depth and optically superior water. Their challenge is to find reliable and affordable solutions to a variety of technological challenges for deploying a deep underwater detector. Several groups are confronting the problem; both NESTOR and ANTARES are developing detector concepts, rather different from one another, in the Mediterranean.

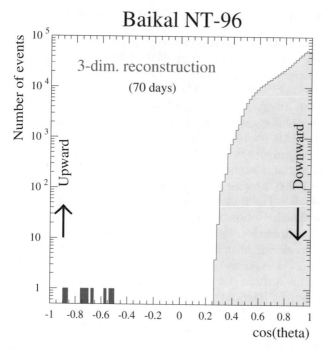

Fig. 13.4. Angular distribution of muon tracks in the Lake Baikal experiment after the cuts described in the text

The NESTOR collaboration [13.5], as part of a series of ongoing technology tests, is testing the umbrella structure which will hold the OMs. The collaboration has already deployed two aluminum "floors", 34 m in diameter, to a depth of 2600 m. Mechanical robustness was demonstrated by towing the structure, submerged below 2000 m, from shore to the site and back. This test should soon be repeated with fully instrumented floors. The actual detector will consist of a tower of 12 six-legged floors vertically separated by 30 m. Each floor contains 14 OMs, with four times the photocathode area of the commercial 8 inch photomultipliers used by AMANDA and ANTARES.

The detector concept is patterned on the Baikal design. The symmetric up/down orientation of the OMs will result in uniform angular acceptance, and the relatively close spacings in a low threshold. NESTOR does have the advantage of a superb site off the coast of southern Greece, possibly the best in the Mediterranean. The detector can be deployed below 3.5 km relatively close to shore. With the attenuation length peaking at 55 m near 470 nm, the site is optically superior to that of all other deep-water sites investigated for neutrino astronomy.

The ANTARES collaboration [13.4] is investigating the suitability of a 2400 m deep Mediterranean site off Toulon, France. The site is a trade-off between acceptable optical properties of the water and easy access to ocean technology. The detector concept indeed requires remotely operated vehicles for making underwater connections. The first results on water quality are very encouraging, with an attenuation length of 40 m at 467 nm and a scattering length exceeding 100 m. Random noise, exceeding 50 kHz per OM, is eliminated by requiring coincidences between neighboring OMs, as is done in the Lake Baikal design. Unlike the case in other water experiments, all photomultipliers will be pointed sideways or down in order to avoid the effects of biofouling. The problem is significant at the Toulon site, but affects only the upper pole region of the OM. The relatively weak intensity and long duration of bioluminescence results in an acceptable dead time of the detector. The collaboration has demonstrated its capability to deploy and retrieve a string.

With the study of atmospheric neutrino oscillations as a top priority, the collaboration had planned to deploy in 2001–2003 ten strings, instrumented over 400 m with 100 OMs. After a study of the underwater currents, the collaboration decided that it could space the strings by 100 m, and possibly by 60 m. The large photocathode density of the array will allow the study of oscillations in the range $255 < L/E < 2550$ km GeV^{-1} with neutrinos in the energy range $5 < E_\nu < 50$ GeV. More recent plans call for a different deployment of the 1000 OMs, on 15 strings separated by 80 m. Each string is instrumented with triplets of PMTs spaced by 8–16 m.

A new R&D initiative, based in Catania, Sicily, has been mapping Mediterranean sites, studying mechanical structures and low-power electronics. One must hope that with a successful pioneering neutrino detector of 10^{-3} km^3 in Lake Baikal and a forthcoming 10^{-2} km^3 detector near Toulon, the Mediterranean effort will converge on a 10^{-1} km^3 detector at the NESTOR site [13.37]. For neutrino astronomy to become a viable science, several of these or other projects will have to succeed besides AMANDA. Astronomy, whether in the optical or in any other waveband, thrives on a diversity of complementary instruments, not on "a single best instrument". When, for instance, the Soviet government tried out the latter method by creating a national large-mirror project, it virtually annihilated the field.

13.3.2 AMANDA: Southern Ice

Construction of the first-generation AMANDA detector was completed in the austral summer of 1996–97. It consists of 300 optical modules deployed at a depth of 1500–2000 m; see Fig. 13.5. Here the optical module consists of an 8 inch photomultiplier tube and nothing else. It is connected to the surface by a cable, which transmits the high voltage as well as the anode current of a triggered photomultiplier. The instrumented volume and the effective telescope area of this instrument match those of the ultimate DUMAND Octagon detector, which, unfortunately, could not be completed.

As predicted from transparency measurements performed with strings near 1 km depth [13.38], it was found that the ice was bubble-free below

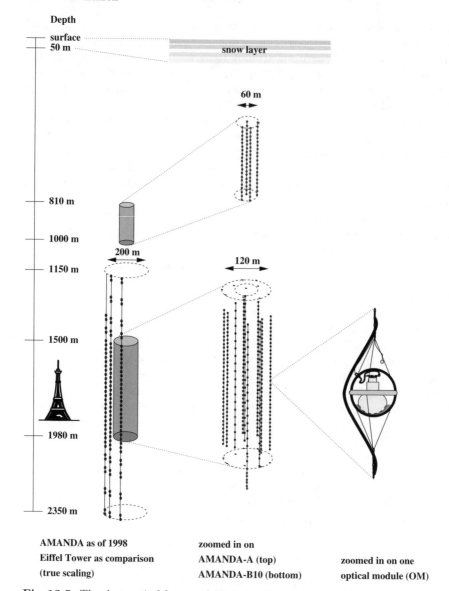

Depth

surface

50 m

snow layer

60 m

810 m

1000 m

200 m

1150 m

120 m

1500 m

1980 m

2350 m

AMANDA as of 1998
Eiffel Tower as comparison
(true scaling)

zoomed in on
AMANDA-A (top)
AMANDA-B10 (bottom)

zoomed in on one
optical module (OM)

Fig. 13.5. The Antarctic Muon and Neutrino Detector Array (AMANDA)

1400 m. Calibration of the detector (optical properties of the ice, geometry of the detector, cable time delays) was completed in the austral summer of 1997–98. The AMANDA group found the following.

- The absorption length is 100 m or more, depending on depth [13.38]. Because of the unexpectedly large absorption length, OM spacings should be similar to or larger than those of proposed water detectors.
- The scattering length varies between 15 m and 40 m with color and depth and is on average ∼ 25 m. These values may include the combined effects of deep ice and the refrozen ice disturbed by the hot-water drilling.
- Operated at a rate of 70 Hz, 20 OMs report in an average trigger. In a typical neutrino analysis one requires that more than five out of the 20 photons are "not scattered". Such a "direct" photon is required to arrive within 25 ns of the time predicted by the Cerenkov fit. This allows for a small amount of scattering and includes the dispersion of the anode signals over the 2 km cable. In a full reconstruction, additional information is extracted from scattered photons by minimizing a likelihood function which matches their measured and expected time delays.

The most striking demonstration of the quality of natural ice as a Cerenkov detector medium is the observation of atmospheric neutrino candidates with the partially deployed AMANDA detector which consisted of only eighty 8 inch photomultiplier tubes [13.7]. The up-going muons are separated from the down-going cosmic ray background once a sufficient number of direct photons and a minimum track length guarantee adequate reconstruction of the Cerenkov cone. For details, see [13.7]. The analysis methods were verified by reconstructing cosmic ray muon tracks registered in coincidence with a surface air shower array.

After completion of the AMANDA detector with 300 OMs, a similar analysis led to a first calibration of the instrument using the atmospheric neutrino beam. The separation of signal and background is shown in Fig. 13.6 after requiring, sequentially, five direct photons, a minimum 100 m track length, and six direct photons per event. The details are somewhat more complicated; see [13.8]. A neutrino event is shown in Fig. 13.7. If the long muon track is required, the events are gold-plated, but the threshold is high, roughly $E_\nu \geq 50$ GeV.

This type of analysis now allows AMANDA to harvest roughly one high-energy atmospheric neutrino per day, adequate for calibration of the detector. It is impressive that three analyses with two independent sets of software tools are able to extract largely overlapping neutrino samples from the data.

While water detectors exploit the large scattering length to achieve sub-degree angular resolution, ice detectors can more readily achieve large telescope area because of the long absorption length of blue light. In early 1998 three 2 400 m deep strings, with optical modules spread over the lowest kilometer, were deployed. They form part of an intermediate detector, AMANDA II, which was completed in 1999–2000 with the addition of six

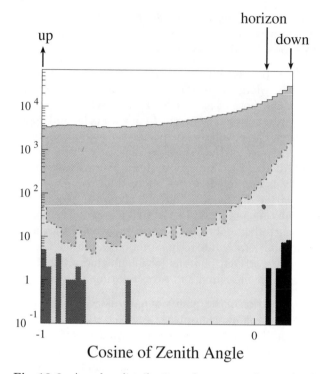

Fig. 13.6. Angular distribution of muon tracks in AMANDA at three levels of quality cuts. These are, roughly, $N_{\text{direct}} \geq 5$, muon track longer than 100 m, and $N_{\text{direct}} \geq 6$. Details [13.8]

more strings. Construction of IceCube will be staged over approximately five deployments in 2001–02 to 2005–06 of 16 strings per year for a total of 4800 optical modules. A straw-man design for IceCube calls for strings with 60 PMTs in 80 holes spaced laterally by 125 m.

Extrapolating from the performance of AMANDA, the achievement of degree resolution of long muon tracks in IceCube has been demonstrated by Monte Carlo simulations. Doing nothing more sophisticated than adjusting the direct-hit and track length cuts used in AMANDA, simulation establishes the straw-man IceCube design as a kilometer-scale detector over most of the solid angle; see Fig. 13.8. An angular resolution of 2.5° is achieved at the trigger level.

IceCube will offer great advantages over AMANDA and AMANDA II beyond its larger size: it will have a much higher efficiency of reconstruction of tracks, map showers from electron and tau neutrinos (events where both the production and the decay of a τ produced by a ν_τ can be identified) and, most importantly, measure neutrino energy. Initial simulation of a strawman IceCube design indicate that the direction of contained showers

Event # 1197960

This is a display of the event after cleaning, i.e., with only the hits used in the reconstruction shown.

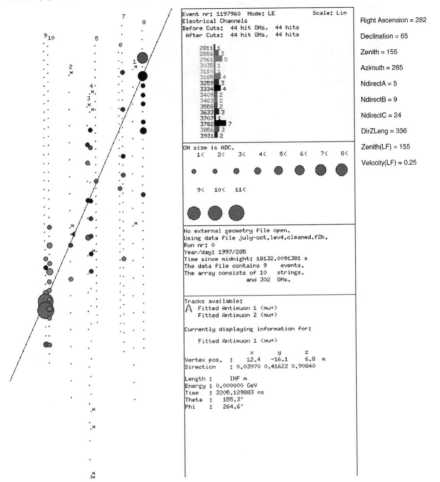

Fig. 13.7. Neutrino candidate in AMANDA. The shading and size of the *circles* represents the time and amplitude, respectively, of triggered photomultipliers. The reconstructed muon track moves upward over more than 300 m

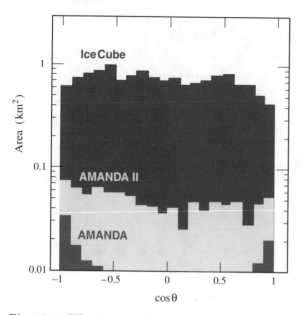

Fig. 13.8. Effective area of the straw-man IceCube detector described in the text, as a function of zenith angle. We required six or more photons delayed by no more than 75 nanoseconds and a muon track length exceeding 150 m. We assumed the expected doubling of the collection area on the basis of the use of larger PMTs with a wavelength-shifting coating

can be reconstructed to better than 10° in both θ and ϕ above 10 TeV. Energy reconstruction is expected to be superior in ice, in part because of the scattering. Energy resolution is critical because, once one establishes that the energy exceeds 100 TeV, there is, in practice, no background in a kilometer-scale detector. Simulations predict a response in energy of about 25%, which is linear. This has to be contrasted with the logarithmic energy resolution of first-generation detectors.

In the next decade, high-energy neutrino telescopes will move us beyond the sun:

> Never concerned that the answer may prove disappointing, with pleasure and confidence we turn over each new stone to find unimagined strangeness leading on to more wonderful questions and mysteries – certainly a grand adventure!
> (Richard Feynman).

This quotation is originally from a public address, "The Value of Science", given by Richard Feynman at the Autumn 1955 meeting of the National Academy of Sciences. The address was subsequently published in the book "What Do You Care What Other People Think? Further Adventures of a

Curious Character", ed. by R. Leighton (W.W. Norton & Co., New York, 1988, pp. 240–248.

Acknowledgments

Large sections of this chapter have been inspired by the unpublished proposal *IceCube: a Kilometer-Scale Neutrino Observatory*. This work was supported in part by the University of Wisconsin Research Committee with funds granted by the Wisconsin Alumni Research Foundation, and in part by the US Department of Energy under grant no. DE-FG02-95ER40896.

References

13.1 K. Greisen, Ann. Rev. Nucl. Science, **10**, 63 (1960); F. Reines, Ann. Rev. Nucl. Science, **10**, 1 (1960); M. A. Markov and I. M. Zheleznykh, Nucl. Phys. **27**, 385 (1961); M. A. Markov, in *Proceedings of the Tenth 1960 Annual International Conference on High Energy Physics at Rochester*, ed. by E. C. G. Sudarshan, J. H. Tinlot and A. C. Melissinos (University of Rochester, Rochester, NY, 1960).

13.2 J. Babson et al. (DUMAND collaboration), Phys. Rev. D **42**, 3613 (1990).

13.3 I. A. Belolaptikov et al., Astropart. Phys. **7**, 263 (1997); V. A. Balkanov et al., Nucl. Phys. Proc. Suppl. A **75**, 409 (1999).

13.4 E. Aslanides et al., astro-ph/9907432, 1999.

13.5 L. Trascatti, in *Proceedings of the 5th International Workshop on Topics in Astroparticle and Underground Physics (TAUP 97)*, Gran Sasso, Italy, 1997, ed. by A. Bottino, A. di Credico and P. Monacelli, Nucl. Phys. B Proc. Suppl. **70**, 442 (1998).

13.6 Talk given by G. Riccobene (NEMO collaboration), proceedings to be published by American Institute of Physics, ed. by M. Diwan. Transparancies available at http://superk.physics.sunysb.ed/NNN99/talk_slides.

13.7 AMANDA collaboration, Astropart. Phys. **13**, 1 (2000).

13.8 A. Karle (AMANDA collaboration), "Observation of atmospheric neutrinos with the AMANDA experiment", to be published in *Proceedings of the 17th International Workshop on Weak Interactions and Neutrinos*, Cape Town, 1999.

13.9 F. Halzen, "The case for a kilometer-scale neutrino detector", in *Nuclear and Particle Astrophysics and Cosmology, Proceedings of Snowmass 94*, ed. by R. Kolb and R. Peccei (World Scientific Press, Singapore, 1995); "The case for a kilometer-scale neutrino detector: 1996", in *Proceedings of the Sixth International Symposium on Neutrino Telescopes*, ed. by M. Baldo-Ceolin, Venice (University of Padua, Padua, Italy, 1996).

13.10 L. O'C. Drury, F. A. Aharonian and H. J. Völk, Astron. Astrophys. **287**, 959 (1994).

13.11 J. A. Esposito, S. D. Hunter, G. Kanbach and P. Sreekumar, Astrophys. J. **461**, 820 (1996).

13.12 R. A. Ong, Phys. Rep. **305**, 93, (1998); C. M. Hoffman, C. Sinnis, P. Fleury and M. Punch, Rev. Mod. Phys. **71**, 897 (1999), T. C. Weekes, "High energy astrophysics", to appear in *Proceedings of DPF'99*, UCLA, 1999; E. Lorenz, talk at TAUP99, Paris, 1999.

13.13 T. K. Gaisser, R. J. Protheroe and T. Stanev, Astrophys. J. **492**, 219 (1998).

13.14 W. Bednarek and R. J. Protheroe, Phys. Rev. Lett. **79**, 2616 (1997).

13.15 T. K. Gaisser, F. Halzen and T. Stanev, Phys. Rep. **258**, 173 (1995).

13.16 R. Gandhi, C. Quigg, M. H. Reno and I. Sarcevic, Astropart. Phys. **5**, 81 (1996).

13.17 S. D. Hunter et al., Astrophys. J. **481**, 205 (1997).

13.18 AMANDA collaboration, in *Proceedings of the 26th International Cosmic Ray Conference*, Salt Lake City, 1999.

13.19 A. V. Olinto, R. I. Epstein and P. Blasi, in *Proceedings of the 26th International Cosmic Ray Conference*, Salt Lake City, 1999.

13.20 V. S. Berezinsky, P. Blasi and A. Vilenkin, Phys. Rev. D **58**, 103515 (1998).

13.21 F. Aharonian et al., astro-ph/9903386.

13.22 F. Stecker, C. Done, M. Salamon and P. Sommers, Phys. Rev. Lett. **66**, 2697 (1991); erratum, Phys. Rev. Lett. **69**, 2738 (1992); F. W. Stecker and M. H. Salamon, Space Sci. Rev. **75**, 341 (1996).

13.23 G. C. Hill, Astropart. Phys. **6**, 215 (1997).

13.24 W. Rhode et al. (Frejus collaboration), Astropart. Phys. **4**, 217 (1996).

13.25 E. Waxman and J. Bahcall, Phys. Rev. D **63**, 023003 (2001).

13.26 K. Mannheim, R. J. Protheroe and J. P. Rachen, astro-ph/9812398, Phys. Rev. D **63**, 023003 (2001).

13.27 J. Bahcall and E. Waxman, hep-ph/9902383.

13.28 J. P. Rachen, R. J. Protheroe and K. Mannheim, astro-ph/9908031.

13.29 E. Waxman, Phys. Rev. Lett. **75**, 386 (1995); M. Milgrom and V. Usov, Astrophy. J. **449**, L37 (1995); M. Vietri, Astrophys. J. **453**, 883 (1995).

13.30 T. Piran, astro-ph/9810256, Phys. Rep. **314**, 575 (1999).

13.31 E. Waxman and J. N. Bahcall, Phys. Rev. Lett. **78**, 2292 (1997); M. Vietri, Phys. Rev. Lett. **80**, 3690 (1998); M. Boettcher and C. D. Dermer, astro-ph/9801027.

13.32 F. Halzen and D. Hooper, astro-ph/9908138, Astrophys. J. Lett. **527**, 93 (1999).

13.33 L. Bergstrom, J. Edsjoe and P. Gondolo, astro-ph/9906033.

13.34 J. G. Learned and S. Pakvasa, Astropart. Phys. **3**, 267 (1995).

13.35 F. Halzen and D. Saltzberg, Phys. Rev. Lett. **81**, 4305 (1998).

13.36 T. Stanev, astro-ph/9907018, Phys. Rev. Lett. **83**, 5427 (1999).

13.37 M. Spiro, presentation to AASC Committee of the National Academy of Sciences, Atlanta (1999).

13.38 AMANDA collaboration, Science **267**, 1147 (1995).

Index

Addendum on the SNO Result

Because of the potential importance of the results of the SNO experiment, publication of this book was delayed, but postponements of the SNO announcement persuaded us to proceed. Although the rest of the book has already been printed, we have included this brief addendum on the effect of the recently revealed SNO results.

The Sudbury Neutrino Observatory (SNO) detected solar neutrinos from the decay of ^{8}B via the charged current (CC) reaction on deuterium ($\nu_e + d \to p + p + e^-$) and by the elastic scattering (ES) of electrons ($\nu_x + e^- \to \nu_x + e^-$). The CC reaction is sensitive to ν_e's only, whereas the ES reaction also has small sensitivity to ν_μ's and ν_τ's, so that any solar ν_e conversion to ν_μ or ν_τ will give a larger flux in the ES case than in that of CC. SNO [1] measures a CC flux $\phi_{SNO}^{CC}(\nu_e) = [1.75 \pm 0.07(\text{stat.})^{+0.12}_{-0.11}(\text{sys.}) \pm 0.05(\text{theo.})] \times 10^6$ cm^{-2}s^{-1}. Their ES flux, $\phi_{SNO}^{ES}(\nu_x) = 2.39 \pm 0.34(\text{stat.})^{+0.16}_{-0.14}(\text{sys.})] \times 10^6$ cm^{-2}s^{-1}, is consistent with but far less accurate than that obtained by Super-Kamiokande [2], which is $\phi_{SK}^{ES}(\nu_x) = [2.32 \pm 0.03(\text{stat.})^{+0.08}_{-0.07}(\text{sys.})] \times 10^6$ cm^{-2}s^{-1}. The difference between this Super-Kamiokande result, $\phi_{SK}^{ES}(\nu_x)$, and the SNO CC rate, $\phi_{SNO}^{CC}(\nu_e)$, is $(0.57 \pm 0.17) \times 10^6$ cm^{-2}s^{-1} or 3.3 standard deviations, giving a probability that the SNO measurement is not a downward fluctuation from Super-Kamiokande's of 99.96%.

This ES/CC difference is independent of solar neutrino models, but confidence in the result is increased by agreement with such models. The flux of non-electron active neutrinos, $\phi(\nu_{\mu\tau})$, can be inferred from $\phi_{SK}^{ES}(\nu_x)$, $\phi_{SNO}^{CC}(\nu_e)$, and the relative neutrino scattering cross sections for SK ($\sigma_{\nu_\mu,\nu_\tau}/\sigma_{\nu_e} \approx 0.171$), giving $\phi(\nu_{\mu\tau}) = 3.69 \pm 1.13) \times 10^6$ cm^{-2}s^{-1}. Using this and $\phi_{SNO}^{CC}(\nu_e)$, the total flux of active ^{8}B neutrinos is then $\phi(\nu_x) = (5.44 \pm 0.99) \times 10^6$ cm^{-2}s^{-1}, in good agreement with the calculated [3] value of 5.05×10^6 cm^{-2}s^{-1}, further indicating dominance of an active-active solar transition.

Despite this agreement there is so much uncertainty in the ^{8}B flux that the solar neutrino oscillation, which clearly demonstrates that this deficit of solar neutrinos is a particle physics and not a solar physics effect, may not all be ν_e's turning into the active neutrinos ν_μ and ν_τ. The SNO data are inconsistent with maximal mixing to sterile neutrinos at the 3.1 standard deviation level, but there could be a relatively large branch of oscillations to sterile neutrinos [4].

Since the SNO result is so new at the time of writing, its effects on neutrino mass-mixing scenarios is very much under discussion, and the interpretation offered below may change as more work is done. The $\phi_{\mathrm{SK}}^{\mathrm{ES}}(\nu_x)$, $\phi_{\mathrm{SNO}}^{\mathrm{CC}}(\nu_e)$ difference makes clear that the solar neutrino deficit is largely due to $\nu_e \to \nu_\mu$ and/or ν_τ, and hence the simplest neutrino scheme need involve only the three known active neutrinos, with $\nu_e \to \nu_\mu$ for the solar process and $\nu_\mu \to \nu_\tau$ for the atmospheric one. That requires the LSND experiment (see Chap. 7) to be wrong. Should the MiniBooNE experiment confirm LSND, however, at least one light sterile neutrino would be required, and that can certainly be accommodated as a subdominant process in either the solar or atmospheric neutrino transitions. To include LSND, the most likely scheme (usually designated 2+2) required $\nu_e \to \nu_s$ (the sterile neutrino) for the solar case, $\nu_\mu \to \nu_\tau$ for the atmospheric one, and that these two pairs of neutrinos be split by the $0.3 \geq \Delta m^2 \geq 10$ eV^2 mass-squared difference required by LSND. The SNO results are incompatible with that simple version of the 2+2 scenario. However, if one allows mixing among all four of the constituents, there appears to be enough freedom for 2+2 to survive [5], and if three sterile neutrinos – a natural outcome of some theoretical approaches – are added, instead of one, the survival of 2+2 is even more likely.

The alternative to 2+2 for including the LSND result is the so-called 3+1 scheme, in which the three active neutrinos are close in mass as in the simplest neutrino scenario, while the sterile state is more massive by the amount required by LSND. Of course, more sterile neutrinos could be added to this structure also. The 3+1 scenario has been disfavored on two grounds. First, since the $\nu_\mu \to \nu_e$ of LSND has to be accomplished via an indirect transition, $\nu_\mu \to \nu_s$ followed by $\nu_s \to \nu_e$, a product of mixing angles is involved. Both the $\nu_\mu \to \nu_s$ and $\nu_s \to \nu_e$ mixings must be small enough so that the ν_s is not brought into equilibrium in the early universe, and yet their product must be large enough to match that required by LSND. While a lepton number asymmetry in that epoch would reduce the constraint, also the new LSND analysis has produced a shift to somewhat smaller mixing angles, easing this restriction. That shift also helps the second problem with 3+1, but an analysis [6] of the limitations on the mixing angle product (or transition amplitude for $\nu_\mu \to \nu_s \to \nu_e$) from other experiments at the 95% CL has no overlap with the 99% CL region allowed by the latest LSND analysis. According to [6], there are small overlaps of a 99% CL bound with the LSND 99% CL region at $\Delta m^2 \sim 6$, 1.7, and 0.9 eV^2, which still gives a small probability that this solution describes nature. However, as discussed in Chap. 7, the 6 eV^2 region is within the 90% CL if the LSND more accurate decay-at-rest $\bar{\nu}_\mu \to \bar{\nu}_e$ analysis is used solely, and the poorly determined decay-in-flight $\nu_\mu \to \nu_e$ data are left out, as the writer strongly believes should have been done. In that case the 3+1 scheme becomes much more probable.

Thus, the SNO results not only do not rule out sterile neutrino contributions to the solar neutrino transitions, but also they do not contradict LSND, although including the latter now requires a more complex and perhaps less likely scenerio. More information on neutrino mass-mixing schemes is presented in Chaps. 9 and 10, and little of that material is altered by the SNO developments. Generally, except for mention of $\nu_e \rightarrow \nu_s$ as a possible solution to the solar problem, as occurs in those chapters, as well as in 4, 7, and 11, there is not much that is affected by this late-breaking news, although knowledge of the result earlier would have changed some emphases.

References

1. Q.R. Ahmad et al.; nucl-ex/0106015,
   http://www.sno.phy.queensu.ca/sno/first_results/.
2. S. Fukuda et al., Phys. Rev. Lett. **86**, 5651 (2001).
3. J.N. Bahcall, M.H. Pinsonneault and S. Basu, astro-ph/0010346v2.
4. V. Barger, D. Marfatia and K. Whisnant, hep-ph/0106207; J.N. Bahcall, M.C. Gonzalez-Garcia and C. Peña-Garay, hep-ph/0106258v2.
5. M.C. Gonzalez-Garcia, M. Maltoni, and C. Peña-Garay, hep-ph/0105269. About 20% admixture of sterile to the active transition seems to work.
6. W. Grimus and T. Schwetz, Eur. Phys. J. C **20**, 1 (2001).

July 11, 2001   *David O. Caldwell*

Printing (Computer to Film): Saladruck, Berlin
Binding: Stürtz AG, Würzburg